Cheeney
1989. January.

The Stability of Slopes

To my son, Nicholas, and my wife, Caryl

The Stability of Slopes

E N BROMHEAD
Reader in Geotechnical Engineering
Faculty of Engineering
Kingston Polytechnic
Kingston upon Thames

SURREY UNIVERSITY PRESS

USA: Chapman and Hall, New York

Published by Surrey University Press
A member of the Blackie Group
Bishopbriggs, Glasgow G64 2NZ
Furnival House, 14–18 High Holborn, London WCIV 6BX

Published in the USA by
Chapman and Hall
in association with Methuen, Inc.
29 West 35th Street, New York, NY 10001

British Library Cataloguing in Publication Data
Bromhead, E. N.
    The stability of slopes: a study of slope
    1. Slopes (Soil mechanics). 2. Soil
    stabilization
    624.1′51363   TA710

    ISBN 0-903384-55-8

Library of Congress Cataloging-in-Publication Data

Bromhead, E. N.
    The stability of slopes.

    Bibliography: p.
    Includes index.
    1. Slopes (Soil mechanics)   2. Soil stabilization.
    I. Title.
    TA710.B744   1985   624.1′513   85-27970
    ISBN 0-412-01061-5

Photosetting by Thomson Press (India) Limited, New Delhi.
Printed in Great Britain by Bell & Bain (Glasgow) Ltd.

Contents

4 Measurement of shear strength   84

5 Principles of stability analysis   105

6 Techniques used in stability analysis   137

Preface

This book examines our present understanding of the behaviour of natural and man-made slopes. Information which was previously scattered throughout numerous technical journals is here put into a logical and modern framework. The book describes various forms of slope instability and discusses those aspects of soil and rock behaviour leading to different types of failure.

The stability of a slope is accurately assessed by careful consideration of the geology, material strengths, porewater pressures and utilization of relevant analytical models. Where appropriate, computer techniques are used to facilitate analysis. A selection of case histories sets the theory in context, underpinning the major themes of the book.

The book is directed at the practising geotechnical engineer or the engineer geologist and postgraduate commencing research. Geomorphologists seeking to quantify the mechanics of 'slope-forming processes' will also find much of interest.

I owe a great deal to my research colleagues, and indeed to many academics and practising engineers with whom I have worked on slope-stability problems. It is not possible to name them all, but certainly without the help of John Hutchinson, Michael Kennard, Richard Pugh, Len Threadgold, Martin Chandler, Nick Lambert and Neil Dixon, this work would not successfully have reached completion.

ENB

1 An introduction to slope instability

1.1 Slope instability and landslides

Landslides, slips, slumps, mudflows, rockfalls—these are just some of the terms which are used to describe movements of soils and rocks under the influence of gravity. These movements can be merely inconvenient, but can from time to time become seriously damaging or even disastrous in their proportions and effects. In the United Kingdom, we normally think of hazards arising from the earth's surface processes in terms of flooding and short-term climatic effects, but in other parts of the world slope instability, too, is recognized as an ever-present danger. Even in Britain, landslides and other gravity-stimulated mass movements are an important and costly problem, and they are a continual source of concern for geotechnical engineers and engineering geologists.

In view of the vast range of different ways in which these movements can occur, some sort of descriptive scheme for these must exist, in order that the reports of one observer be clearly understood by others. Such a consensus unfortunately does not exist. Indeed, many systems of classification for the different types of slope instability have been proposed. These include the notable schemes by Sharpe (1938), Varnes (1958), and Hutchinson (1967) to which Skempton and Hutchinson (1969) give a comprehensive list of illustrative case records. A major source of difficulty with these schemes is the limited range of words in English which can be combined to describe different types of mass movement: this gives a superficial degree of similarity to the various classification systems. Mass movements come in such an enormous range of sizes, shapes and types that even if there were only one classification system, it would be difficult to decide in many cases just what class a particular mass movement lay in. A descriptive term in one scheme may represent something completely different in another. It is neither the time nor the place here to introduce another scheme—and yet the available systems lack a universal applicability. Perhaps the systematic classifiers of mass movements will excuse the following compendium of their efforts.

We will subdivide the whole of mass movement into three major classes: *slides*, *falls* and *flows*. The major differences between these three are in the way in which movement takes place. In a *slide*, the moving material remains largely in contact with the parent or underlying rocks during the movement, which takes place along a discrete boundary shear surface. The term *flow* is used when the material becomes disaggregated and can move without the concentration of displacement at the boundary shear. Although a flow can remain in contact with the surface of the ground it travels over, this is by no means always the case. Lastly, *falls* normally take place from steep faces in soil or rock, and involve immediate separation of the falling material from the parent material, with movement involving only infrequent or intermittent contact thereafter, until the debris comes finally to rest.

Small mass movements tend to be of one type alone: either a slide, a fall or a flow. Larger movements may often change from one to another as they progress. For instance, a large rockfall may develop into a flow and finish up as a slide as the debris is brought to rest.

The reader may well be forgiven for thinking that this section dwells mainly on natural slopes, and forgets failures in man-made earthworks. This is not a deliberate omission, but it reflects the fact that most of the forms of instability that occur in man-made slopes are also experienced in natural slopes, and the wider range of scale present in the latter, the far longer time for which they have existed, and the great variety of naturally occurring materials (many of which would be spurned by even the most inexperienced engineer) make them more readily available as examples. Indeed, the causes of landslides in natural slopes, and their mechanics, are frequently far better understood than failures in earthworks, since the need to apportion blame when a man-made earthwork fails to perform as intended may lead to hurried stabilization and investigation works with the result that the technical facts of the case become obscured. (There may well also be deliberate obfuscation.)

1.2 Classification for mass movement: falls

A fall of material, soil or rock, is characteristic of extremely steep slopes. The material which moves can break away from the parent rock by an initial sliding movement: some shear surfaces may develop in response to gravity stresses and, in moving, the material is projected out from the face of the slope. Alternatively, due to undermining at a low level in the slope, an overhang may form. Causes for the undermining may include wave action, river or stream erosion, erosion of an underlying bed by seepage, weathering or careless excavation; they therefore include both internal and external agencies. Then, either because the rock is jointed, or because it has insufficient strength *en masse*, there comes a point at which the undermining causes a fall to occur. Progressive weakening, perhaps by weathering of a susceptible unit in a cliff, can also allow joint-bounded blocks to rotate until they pass through a position of equilibrium and overtopple. As the blocks rotate, they throw more stress on the outside edge and this must accelerate such a process.

The effect of water pressures in a joint-bounded rock mass should not be underestimated. Where the natural outlets become blocked, by ice formation as a typical example, extremely large thrusts can be developed. Ice wedging itself, if the water *in* the joints freezes, can also generate significant forces. These may be sufficient to rupture unjointed rock. Failure may follow immediately, or wait until the thaw. Ice formation is not absolutely necessary to the mechanism: unfrozen water exerts high thrusts if the joints are full. An example of the combined effects of high joint-water thrusts and scour can be seen in the American Falls at Niagara (Figure 1.1) where massive joint-bounded blocks of Lockport Dolomite lie piled against the slope face. These have been dislodged partly by the undermining of the sandstones and shales beneath the dolomite, but a consideration of the forces involved shows that water thrusts have an important role.

Even in arid regions, a form of wedging can occur. In response to daily, or annual, temperature changes, joints open and close. When they are open, small pieces of debris can fall down the joint, preventing proper closure. The effect of this would be progressive. Seismic shocks, too, can dislodge debris from steep slopes. Examples of this are given later in the book.

When a fall occurs, the material involved will break up: if not while in motion, then on impact. The resulting debris may form a scree or talus tidily stacked against the cliff face, or if more mobile materials are involved, then lobes and tongues of debris with much larger run-out can be formed. The

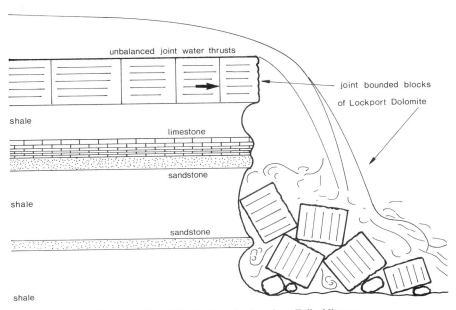

Figure 1.1 Rockfalls affecting the American Falls, Niagara.

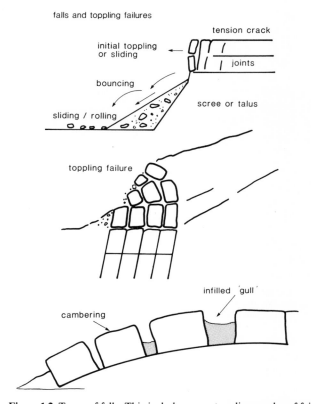

Figure 1.2 Types of falls. This includes some toppling modes of failure.

mobility of a given material is dependent on its dynamic strength, the behaviour of the air or water in its pores or joints, and the energy input. Some falls involve only a few blocks, which may be sufficiently strong to remain intact. They may move by rolling, sliding or bouncing, controlled by slope angle and block shape. Debris from earlier movements is more likely to be reactivated and to take place in subsequent events, and it may be necessary to subdivide falls into *primary* (involving 'new' material), and *secondary* types (reactivated debris).

A diagrammatic section, showing the elements of a typical slope subject to failure by a series of small falls, is shown in Figure 1.2. The onset of a fall is often accelerated by toppling as changes occur in the support conditions for different rock blocks. Accordingly, some of the types of toppling failure, not yet developed as falls, are shown in this composite figure.

Usually, the fall itself is not witnessed, only its results. A good example of this is the cliff of Bridport Sand at West Bay, Dorset, shown in Figure 1.3.

Figure 1.3 Rockfall in sandstone cliffs, West Bay, Dorset. The weathering process picks out hard bands in the sandstones, and the freshness of the scar, even if the debris were to be removed by the sea, would show that a fall had taken place.

Although well protected from marine attack by a good shingle beach (the western end of the Chesil Beach), and rarely acted on directly by waves, isolated storm conditions can give rise to rapid erosion at the base of the cliff. Softer layers in the cliff are also etched out differentially by weathering, giving the cliff its rugged appearance, and making the fresh joint-bounded scar of the rockfall all the more conspicuous by contrast. The piled-up debris of the fall, so different from the bare foot of adjacent lengths of cliff, is evidence for the recent nature of this fall.

On occasion, however, rockfalls or other mass movements are observed. In a small proportion of cases the observer can record the event. Figure 1.4 shows an extraordinary sequence of photographs taken of a rockfall from Gore Cliff on the Isle of Wight by a holidaymaker who had been alerted by the premonitory opening of clifftop cracks and noises emanating from the cliff. Debris from this fall loaded the rear of an old landslide complex, and set off a series of slides which continued for some years after.

1.3 Classification for mass movement: slides

Slides are often translational in nature, i.e. they involve linear motion, especially if fairly shallow. A rock block may slide down intersecting joint planes which daylight in the face of a cutting, or a block may move down

Figure 1.4 Rockfall at Gore Cliff, Isle of Wight. This sequence of three photographs was taken by a holidaymaker in September 1928.

a steeply inclined joint or bedding plane. Where such a block is joint-bounded on its sides, a lateral thrust from water filling some of the joints can push a block even along a low-angle surface. This is identical to the thrusts mentioned in connection with falls, and indeed, some movements may occur which are transitional between these two classes, starting as a slide, and developing into a fall.

In weaker rocks and soils, shearing can take place through the rock mass as well as along joints and other discontinuities in the rock. This shearing may tend to follow along curved shear surfaces, and all or part of the slide may rotate. In soft soils, the sliding surface may be very nearly the arc of a circle in cross-section, but the presence of different lithologies in a stratified deposit invariably causes a slide to adopt a flat-soled shape. Terms such as *toe, foot, heel, crown, head*, etc. are often applied to the parts of slides (Figure 1.5) and the exposed slip surface is often termed the *back scarp* or *rear scarp*. In the example of Figure 1.14, for example, the cliff forms the rear scarp of a substantial slide complex.

Often, the presence of 'corners' in a slip surface, resulting from the interaction between a bedding-controlled flat sole and the curved rising rear part of a slip surface causes great internal distortion in the sliding mass, which may be reflected by breaks in the ground surface, or *counterscarps*. In especially severe cases, a *graben* feature may form. Several of these features will be seen in the illustrations, notably the graben of the Miramar landslide at Herne Bay (Figures 2.8, 2.9) and in several of the sections in subsequent parts of the book.

Where the slides are *retrogressive*, i.e. eat back into the slope, slides are *multiple* in form. The frontal elements of such slides are often much more rotational in character and far more active than the overall movement. Sometimes a big slide will precipitate consequent movements in the disturbed material in its toe, or by overriding a lower slope, cause a downslope *progressive* slide. A fall of material from the temporarily oversteepened rear scarp of a rotational landslide would cause movements to occur in the slide debris downslope, and would be a source of such a downslope progression.

Disturbed soils are very susceptible to the infiltration of rainwater. Furthermore, backtilted elements or grabens in a slide may cause ponds to form by intercepting surface water flows. The discharge from these is a focus for secondary slide-type movements. Where any of these things happens, slide activity is increased locally in the form of *mudslides*. Mudslides are slides of debris at a high water content. They exhibit a high mobility, but are not flows in the sense of this classification because of the existence of discrete boundary-shear surfaces. They are often loosely described as mudflows, and occasionally, mudruns, particularly in the early literature predating the finding of the boundary-shear surfaces. Mudslides can also be caused by the careless discharge of water on to a slope. Domestic drainage, or the collection of even modest amounts of precipitation over extensive paved

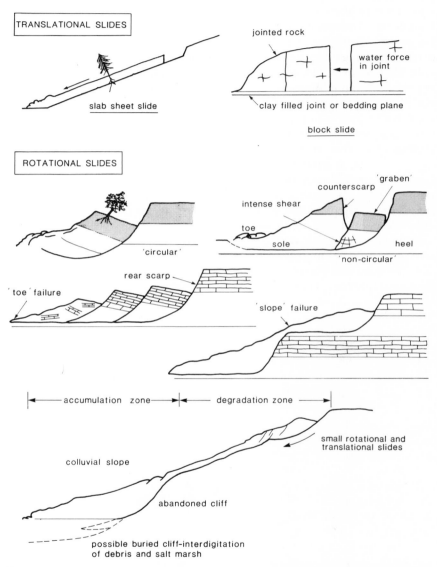

Figure 1.5 Types of slides. Flat-soled non-circular slides are probably the commonest in sedimentary rocks with low dip, because of the effect of bedding on the mode of formation of the slip surface.

areas or roofs, when discharged on to a slope, can be the focus for mudslide generation.

Compound landslides occur in response to complex structures in the slope. Double and triple forms of these 'multistoreyed' landslides are commonplace, but particularly where the mantle of debris is thick, the geomorphic expression of the underlying structure may be muted, and the subsurface details recovered only after extensive investigation works. For example in the Coal Measures sequence in South Wales with a repetitive sequence sandstone–seatearth–coal–sandstone, elements of a multistoreyed landslide can form on each seatearth. A long history of landsliding with negligible removal of material from the toe has led to the individual elements of the landslide complex coalescing so that at a cursory inspection they cannot be separated. In contrast, in the alternating clay–sand sequences to be found on the coast in South Hampshire (including parts of the Isle of Wight, Figure 1.6), marine erosion at the toe of the slope removes some transported debris so that the individual stratigraphically controlled benches can clearly be distinguished. They are termed 'undercliffs' locally.

Ordinarily, however, one would take the occurrence of a set of clearly distinguishable, stratigraphically controlled landslide benches as evidence that the more competent beds in the sequence were those that dominated the geomorphology. The corollary of this is widely used in geological mapping.

Figure 1.6 Interbedded Cretaceous clays and sands, forming a series of landslide benches in each of the argillaceous horizons. Blackgang, Isle of Wight.

Figure 1.7 Rotational slide in weathered London Clay affecting railway cutting. Many of the railway cuts on the approaches to London are scarred by these small slides, which pose a continual maintenance problem to the railway engineering staff.

Figure 1.8 Rotational slide in weathered overconsolidated clay fill affecting a highway. Clay fills are susceptible to long-term failure as construction-induced suctions are lost. These slides are usually only obvious to the motorist at interchanges and overbridges, and the extent of the problem is not always apparent.

Skempton (1977) reviewed the failure of a large set of cuttings of various ages in weathered London Clay, a stiff fissured overconsolidated clay of Eocene age. Typical of these are the rotational slides adjacent to a railway tunnel portal shown in Figure 1.7, and in the highway fill shown in Figure 1.8. Such slides are troublesome, and often costly to rectify, although their occurrence in small numbers is more an indication that the correct balance between first cost and maintenance has been properly achieved in the design of the earthworks, than a serious cause for concern.

A deeper-seated rotational landslide in largely unweathered London Clay in a coastal cliff is shown in Figures 1.9 and 1.10. The first of these shows an aerial view, taken shortly after failure in November 1971, and the second, some years later, is a terrestrial view showing the degradation of the rear scarp in the intervening period, and the development of a landslide pond trapped behind the tilted mass of the main slide. Such ponds are a character-

Figure 1.9 Aerial photograph of deep-seated rotational slide in unweathered London Clay on the north coast of the Isle of Sheppey, Kent (1971). (Courtesy Cambridge Air Photographs Collection.)

Figure 1.10 The slide of Figure 1.9 after some degradation of its rear scarp and the formation of a landslide pond. In the latter stages of the erosion and landsliding cycle, the characteristic morphology of these slides is lost.

istic, if often ephemeral, feature of these landslides. In the aerial view the damage caused to buildings, and the truncation of a road, can be seen. In the past century, a number of other buildings, seaward of the present coastline, have been lost, including a church, together with about 100 metres of the road.

Damage to property involved can be severe. In the Warden Point landslide described above, a wartime concrete bunker used as a shed, and a small house, were lost. Another house was left so close to the rear scarp that it had to be demolished for safety reasons, followed by another some years later as the rear scarp retrogressed. Although of reinforced concrete construction, the bunker was not entirely located on the slide mass, and its landward extremity dragged on the rear scarp and caused the severe cracking shown in Figure 1.11. The pronounced tilt of the bunker reflects the rotation of the slide.

Figure 1.12 shows an early Victorian house destroyed by the slide of 1978 in a reactivation of movements of part of the complex shown in Figure 1.6. During the last century, landsliding in this vicinity has obliterated several kilometres of roads, and a number of houses. This particular house was founded on the seaward part of an old rotational slide in the uppermost of the benches. The bedding-controlled slip surface, almost horizontal at this location, was some 5m below foundation level. A movement of 15m laterally

Figure 1.11 Reinforced concrete wartime bunker affected by rotational slide. The left-hand edge of this bunker scraped along the slip surface as the slide took place.

Figure 1.12 House at Blackgang, Isle of Wight, damaged by landslide of 1978. The occupants escaped safely.

Figure 1.13 Step in road caused by landslide movements, Blackgang, Isle of Wight. Although obviously in a minor road, such steps pose an obstacle to persons escaping from a slide, and can be more of a hazard than the slide itself.

Figure 1.14 Folkestone Warren, Kent–train caught in landslide of December, 1915. The train was halted by soldiers and railwaymen because of falls of chalk down the line. While it was stationary, the main slide took place. [Contemporary photograph.]

Figure 1.15 Mudslide, North Kent coast. These tend to alternate with the deep-seated rotational landslides, occurring where natural or artificial drainage is discharged on the slope face.

Figure 1.16 Shallow, sheet-type translational landslide.

resulting from a reactivation of the slide caused the damage shown. Large lengths of the road, formerly (until breached irreparably on another section by landsliding) the main coast road, collapsed onto lower benches. At the margin of the slide, the road (Figure 1.13) was broken by the formation of both a vertical step, and a lateral offset. This sort of damage, occurring early in the development of a slide, can impede rescue attempts, although the effect in respect of rescuing property is more serious than in respect of the rescue of persons.

Railways can also be affected by dislocation in line and level (Figure 1.14).

Shallow forms of slides illustrated in Figures 1.15, 1.16 include mudslides and sheet slides. The first of these illustrations shows an elongate or lobate mudslide in clay cliffs on the Kent coast, and the second a sheet slide. Such slides may well be the reactivation of sheet mudslides first formed under quite different climatic or environmental conditions to those acting today.

1.4 Classification for mass movement: flows

In this classification, a flow is a mass movement which involves a much greater internal deformation than a slide. The important characteristics of a flow could be obtained through movements taking place on a large number of discrete shear surfaces, or by the water content of the moving mass being so high that it behaved as a fluid, the latter being the case in clay soils at water contents above the liquid limit. Some soils will absorb water readily when disturbed, fractured and cracked by an initial *landslide*, and this leads to the sliding movements breaking up into earth- or mudflows. Genuine flows of clay soils require the moisture content to be appreciably above the liquid limit, or there is a tendency for basal shear surfaces to form and the movement is properly termed a slide. Flows can also take place in fine-grained non-cohesive soils. For this, high moisture contents are not required, and flow-like movements can even take place in loose, dry, silts and sands. They are a combination of sliding and individual particle movements.

Debris from falls and slides which have high energy can also behave as a flow, especially if air or water is entrained in the initial movement, since this trapped fluid may develop high pore pressures which buoy up the individual debris particles. Alternatively, the numerous interparticle impacts may produce an effect analogous to intermolecular motion in a real fluid, and thus enable even dry, non-cohesive, debris to flow. A third explanation for flows of very high mobility is the collapse of an initially very loose and metastable soil grain structure, with a resulting compression of the pore fluid and the generation of high pore-fluid pressures. With entrapped air or water, or internally generated pore-fluid pressures, the debris has low strength; but when the fluid escapes and the pore-fluid pressures dissipate, the regain in

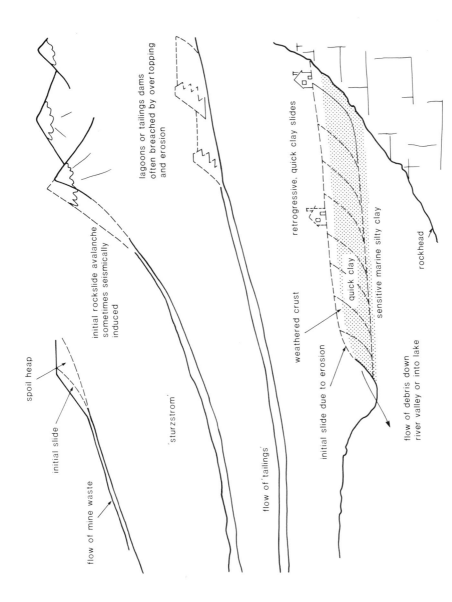

Figure 1.17 Types of flows. Usually lobate (tongue-like) in plan.

strength can be very rapid. The effect may be accounted for by an interparticle impact model—when the number of impacts per unit time per unit volume of the debris drops below a certain threshold, there is a net decrease in the number of particles in motion, and the chain reaction halts rapidly.

Large-scale mass movements have been observed from satellite photography on the surfaces of the moon and the nearer planets. In some cases, these are slides, but in more than one case the debris has a run-out from the toe of the slope comparable to flow-side avalanche movements on earth (Murray *et al.*, 1978). This points to a mechanism for the 'fluidization' of the debris which does not necessarily involve pre-existing pore fluids in the soil. The interparticle impact model may be the best explanation of this high degree of debris mobility.

Regardless of which theory proves correct, there is no doubt that many mass movements do have large run-outs, and can occur at high velocity. Equally certain is that in some of these, high pore-fluid pressures play an active part. At the time of writing, the scale of many of these regrettably rules out their treatment through engineering works, and the only practical engineering solutions lie in the identification and avoidance of hazardous locations.

Other flows may result from the disturbance of very sensitive soils, for example under-consolidated recent sediments, such as are found in rapidly accreting deltas. Some of these recent sediments can have an initially stable structure, which becomes metastable through changes in the chemistry of the soil minerals or the pore fluid. The quick clays of Scandinavia and eastern Canada are classic examples in this category, and result from a decrease in the salinity of the pore fluid. Other, flocculated, recent marine sediments can deflocculate in a strongly alkaline environment. Examples are known of such soils collapsing when saturated with effluent from chemical works.

Flow-type mass movements, termed flow slides by Bishop (1973), occur in loose non-cohesive debris such as solid mine and quarry waste and tailings.

1.5 Cost and frequency of occurrence: failures of natural slopes and of man-made slopes

The cost to the community of landslides and other related mass movements is high, but unquantifiable on a national scale. Information collected in the USA suggests that the annual costs run into billions of dollars, and accounts of this by Schuster (1978) estimate total losses due to slope movements in California alone at about $330M per annum. In addition to economic costs, there is loss of life—perhaps 25 or so per annum on average in the United States, although individual disasters can greatly exceed this average. An example of this is the collapse of a series of coal-mine waste-slurry impounding dams at Buffalo Creek, West Virginia, which killed 125 people and caused extensive damage to private property and public utilities.

Elsewhere in the world, data comes from records of single events, rather than from a systematic collection of data. The Japanese are particularly badly affected by landslides: theirs is a densely populated series of islands, susceptible to landslides and shaken by earthquakes. The death toll there can be dramatic—some 171 in 1971 and 239 the following year. The *Guinness Book of Records* details a landslide in Kansu Province, China, with a death toll of 200 000, but this may not be a genuine record.

By far and away the most devastating mass movement disasters are those caused by flowslides. In mountainous areas, flowslides of debris can sweep down from high mountainsides in the form of *sturzstroms* (Hsu, 1978). Examples of these are the two events of 1962 and 1970 in which sturzstroms from the slopes of the Andean mountain Huascaran killed more than 4000 in the first disaster, and some 18 000 more in the second avalanche some eight years later. Similar movements in the Alps led Heim (see Hsu, 1978) to a study of such phenomena, providing a source of important historic data on their occurrence.

Second to the sturzstrom in its incredible violence and destructive power comes the collapse of a large dam, releasing debris as well as water to scour downstream. The Buffalo Creek collapse has already been referred to: this was the progressive collapse of relatively poorly built embankments, and the attention which collapses such as this have focused on spoil disposal in the whole mining and quarrying industry, and the imposition of legislation, have made it a little less likely that such events will occur so commonly in the future. Nevertheless, even professionally designed and constructed water-retaining embankments and structures can and do suffer from stability problems with the subsequent release of impounded water.

The least controllable 'dam breachings' occur when the dam is not man-made at all, but is the consequence of a landslide forming a natural dam. One example in the Andes temporarily dammed the Mantaro river, impounding water to a depth of 170 m over a length of 31 km of the river. The death of 450 in the initial landslide was followed by extensive devastation downstream when the landslide dam was overtopped, and rapidly scoured away. Similar landslide dams form in all the mountainous regions of the world, and pose major hazards to habitations and engineering works downstream, not least of which are man-made embankment (or other) dams for river regulation, irrigation, water supply or hydroelectric generation. If the breaching of a landslide dam in a valley upstream releases a flood, the artificial dam may not have adequate spillway capacity, and may be overtopped and breached in turn. Such a landslide dam may not even be in the same country as the reservoir.

Holmes (1978) records the earthquake-stimulated landslide from Nanga Parbat (8117 m) in the schist–migmatite–granite complex at the northwestern end of the Himalayas in Kashmir, of December 1840. A landslide dam impounded water to a depth of over 300 m, forming a lake 65 km long which, when breached, swept away a Sikh army camped at Attock. The destructive effects of the wave from the rapidly emptying lake spread destruction for hundreds of kilometres. He also records the prehistoric landslide which blocked the Upper Rhine and its tributaries at Flims. Debris from this slide extended 14 km from the mountain, covering an area of 50 km^2, and impounding a lake within which more than 820 m thickness of sediments built up.

As well as the breaching of a dam, slides into a reservoir can cause damaging waves. Landslide-generated waves can also occur in lakes and coastal inlets (lochs and fjords). On an exposed coastline the effect is unlikely to be significant, as the wave energy is dissipated into the open sea. Most spectacular among the landslide-generated waves was the Vaiont slide of 1962 in Northern Italy: this virtually filled the reservoir, causing a wave which overtopped the dam by some 100 m (it ran up the opposite slope to the slide by about 260 m). Downstream, this wave destroyed a small town and caused more than 2000 fatalities.

Quick-clay landslides in the sensitive marine deposits of Scandinavia and Eastern Canada can also take place with astonishing rapidity. At Rissa, near Trondheim, a quick-clay landslide starting from a small slip on a lakeside, enlarged to several hundred metres square in the course of about 45 minutes, and then retrogressed about a kilometre in the course of the next five minutes. Fortunately, the death toll was small (1) because the slide took place during the hours of daylight. Landslide-generated waves from this damaged a village at the other end of Lake Botnen, several kilometres away. Other quick-clay landslides have been more costly in life and property. The landslide at St Jean-Vianney, Quebec, destroyed 40 houses and killed 31.

Whereas the most major disasters are almost uncontrollable, it is striking that so many smaller landslides are, if not caused by, then provoked by, the activities of man. There are untold instances of relatively minor slope-grading works stimulating quite major movements, and nearly as many cases where rainfall from large paved areas has been discharged without due regard to the consequences and has caused instability.

The United Kingdom data have not been collected in a systematic fashion, and the inflation of the last two decades renders research back into the records of lesser value than quoting costs direct. Some figures are available to give some idea of the scale of problems encountered. Where these have been rendered into 1985 money values a measure of rounding has been introduced, and the reader will no doubt forgive the resulting imprecision.

Landslides are common along the coastline of Britain. On the North Kent coast, where coastal landslides occur in the 30- to 40-m high clay cliffs, the current costs of regrading and stabilization together with a reinforced concrete seawall at the toe can exceed £1.5 M per kilometre. At Folkestone, a recent investigation into the stability of the Warren cost in excess of £100 K for the drilling alone. At Ventnor, on the Isle of Wight (Chandler and Hutchinson, 1984) gradual movements of the massive landslide complex have caused damage amounting to at least £1.5 M over the last 20 years.

Slides on inland slopes are equally expensive. A small landslide perhaps 50 m square in South London, triggered by the excavation at the toe of an eight-degree slope in London Clay containing pre-existing shear surfaces cost £20 K to investigate, and some £150 K to construct remedial measures including piling through the slide, surface drainage, a piled retaining wall and underpinning to structures upslope threatened by undermining. The likely total cost, if the properties downslope whose construction excavations triggered the slide need to be demolished, may double.

The failure of earthworks can be equally costly. Carsington Dam in Derbyshire suffered a slide in its upstream face just before completion in July 1984. The main embankment, which cost approximately £15 M, was almost completely destroyed in the slide, detailed investigation of which has cost of the order of £500 000 to date, including the costs of extra muckshifting.

Areas subject to landslide hazard are widespread. The damage from slope movements often amounts to between 25 and 50% of all the costs arising from geological and other natural causes, and they are therefore a major concern to the geotechnical profession. The geotechnical engineer and engineering geologist need not feel that they are neglecting the rest of their professional duties if they concentrate on the stability of slopes: the potential costs of failure, and savings if it is avoided, are justification enough.

Gravity-stimulated mass movements have the potential to change history. What might have come of Hannibal's assault on Rome across the Alps if parts of his army had not been lost to avalanches? Holmes (1978) cites an account of the foundation failure of a volcanic cone built up on a basement of late Tertiary Clays. The deep sweeping slip surfaces lowered the summit of the Merapi volcanic cone (now 3911 m) by about 400 m, releasing additional volcanic activity which in combination with the forming of a large dam by the toe heaves destroyed a large and flourishing Hindu culture in central Java. The evidence is largely geological, but does include some brief inscriptions dating back to the immediate aftermath of the calamity in 1006.

1.6 Some disasters and their impact on knowledge

Attention was dramatically focused on the problem of the stability of tips and heaps of spoil and other wastes resulting from mining and quarrying in 1966, when a slide originating in a colliery tip in South Wales caused the loss of 144 lives. Most of these were young children, trapped when their school was partly demolished by the debris. Other mass movements, also arising from inadequate treatment of mine waste and fly ash, have been responsible for severe loss of life in the USA and in Belgium respectively.

The principles which apply to the stability of such waste tips are exactly the same as those applying to natural slopes or to other man-made earthworks, and the scant attention given to their stability reflected a general air of indifference that was widespread at the time. That this should have been so is still surprising, not only with the benefit of hindsight. Mining and quarrying operations can and do produce enormous volumes of waste materials. Bishop (1973), writing at the height of a major motorway construction programme in the UK, compared the production of waste from mining and industrial sources taken together with domestic waste production with the magnitude of earthmoving in heavy civil engineering works being undertaken at that time. It is striking that waste is produced equivalent to nearly 60% of the then totals for dam and highway construction.

Half of the waste is produced in the UK as a result of coal mining, amounting to about 60 M tonnes per year in Bishop's survey. Of that, about 5 M tonnes per year was produced in the form of slurry or tailings. Even though coal production has consistently declined in recent years, the amount of spoil has increased. This increase is usually attributed to mechanization of the mining process.

Other mining and quarrying operations produce their share of solid and liquid wastes. More recently, Penman (1985) has estimated, on the basis of world production of non-ferrous metals and the average quality of ores, that world production of tailings exceeds 5000 M tons per annum. This quantity certainly exceeds the tonnage of all other materials used by man.

At first sight, spoil heaps would seem to be identical to other engineering fills. They do, however, have one important difference: a lack of adequate compaction. Therefore, even intrinsically good materials can have undesirable engineering properties when handled improperly. Much mine and quarry waste is in aqueous suspension, and has to be stored behind *tailings dams* in a *lagoon*. Penman notes that of the world's dams, that of greatest volume is the New Cornelia Tailings Dam near Ajo in Arizona. Furthermore, in an (incomplete) register of mine and industrial tailings dams, 8 are higher than 150 m, 22 are higher than 100 m and 115 are more than 50 m high. Six of these impoundments have a surface area greater than 100 km^2, and a storage volume greater than 50 km^3.

Other calamities, such as the failure of rail and roadway embankments crossing old landslip areas, have been expensive mistakes from which the engineering profession has learnt much, but which do not count as disasters. On the other hand, some disasters should have been foreseeable, and should have been avoided: they occurred nonetheless because of a lack of vigilance. From such there is no input of new knowledge, merely a timely reminder to guard against careless or foolish design, and to think deeply about the technical aspects of all earthworks design.

1.7 History of the understanding of slope stability

The understanding of slope instability has not come about in an ordered manner. Ideas have been advanced from time to time; some in being accepted have passed into the body of knowledge; others, because they represent concepts too advanced for the times have been rejected, only to be resurrected at a later date; or because of their patent absurdity have been consigned to the rubbish bin. Skempton (1979) reviews some of the published landmarks in early soil-mechanics development, many of which are valid today, and without which none of the concepts described below would have been possible.

Empirical advances were made by the French military engineers building the massive earth fortifications of the seventeenth to nineteenth centuries, work which culminated in the earth-pressure theories of Culman, Coulomb and others which are still taught to undergraduates today. However, even a very large number of isolated fortresses cannot bring their constructors into contact with so very many different types of ground as can an extended linear construction such as a canal, road or railway. It is then perhaps not surprising that the principles of modern lithostratigraphy were first enunciated by the canal engineer William Smith, nor that the first serious work on the stability of slopes came from the French canal engineer Alexander Collin (1808–1890). Collin (1846, tr. Schriever, 1956) recognized the curved shape of sliding surfaces in both cut and fill slopes in clays. He even had some of these dug out to measure them, and attempted some stability analyses.

Dam and canal slopes pose several extra problems in stability, notably the condition known as *rapid drawdown* (section 7.9) where the supporting effect of the water pressures is removed as the level of water is lowered. Collin described this phenomenon in qualitative detail. He certainly also understood that sliding surfaces in clays definitely exist, a concept that was still being argued about in civil engineering circles a hundred years later, although he could add little to the debate as to whether they existed in the ground all the time, or were created by the development of a slide.

At the end of the nineteenth century, British interest turned more to earth pressures than to the stability of slopes, despite early work of considerable importance, Gregory (1844), for instance, showing clearly the shape of the sliding surface in a failure of a cutting at New Cross (Figure 1.18). The canal network was already in decline, the railways were nearly all finished, or were being built to established empirical designs. Even the dams under construction, needed to supply water to industry and the urban complexes made possible by industrialization, tended to be situated in the upland areas of Britain where the geology and construction technology of the time favoured masonry or concrete structures—although, to be fair, earth dams were constructed in fairly large numbers, but to standardized designs in which there was little innovation.

The landslides at Folkestone Warren affecting the railway (described in

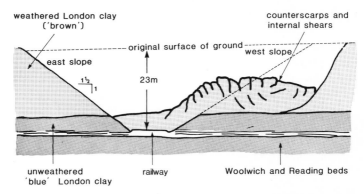

Figure 1.18 Section through a slope failure in a cutting at New Cross. After Gregory (1844). This was one of the first detailed accounts of an earthworks failure, and showed quite clearly the shape of the slip surface and how this reflected the geology.

detail later in the book) had some impact on the study of slopes, but their great size, rendering investigation and the construction of remedial measures extremely expensive, together with uncertainty as to their age and origin, meant that they were not a suitable vehicle for the pursuit of a wider under-standing of landslide mechanics at that time. The same could be said of the large landslides affecting the Panama Canal under construction during the early years of this century.

It was a much smaller event in the scheme of things that brought soil mechanics back to the neglected lines of approach pioneered by Collin. This was the failure of a quay wall and jetty in Gothenburg Harbour (Figure 1.19). Timber piles, severed by the slide, were extracted, and these showed that the surface on which the slide had taken place was approximately a circular arc in section. Now it happens that analyses of stability to an adequate engineering degree of accuracy can be most easily made if the slip surface is of this shape, and the *Swedish slip circle* became an integral part of the geotechnical engineer's repertory. In some ways, it was a retrograde step, because it was seized upon with fervour as a universal solution to slope stability problems; even today, one sees analyses based on a slip circle applied to natural slopes containing landslides which have slip surfaces not remotely resembling a circular arc in section.

Modern treatment of slope stability really then awaited one more develop-ment. This was the enunciation of the principle of effective stress by Terzaghi, and its application to the shear-strength behaviour of soils and rocks by his followers, which included an active school in England.

In the late 1930s and early 1940s, Terzaghi came to England to advise on several slopes problems. These included the further stability problems experienced by the railways at Folkestone Warren, to which was added the treatment of a large landslide in the tunnel portal near Sevenoaks, and the failure in the reservoir embankment then under construction at Chingford.

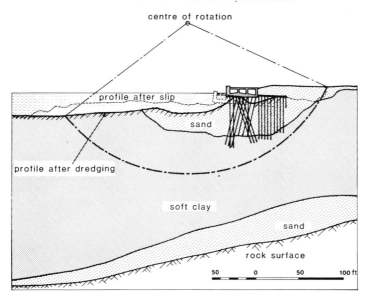

Figure 1.19 Quay wall failure at Gothenburg Harbour, 1916. After Petterson and Fellenius. Modern slope stability analysis techniques have evolved from the study of this particular slide.

From the Chingford work emerged three things which were to alter the course of soil mechanics development in the United Kingdom. Firstly, there was the involvement of the Building Research Station, which has continued to this day to be a centre of excellence in geotechnics, and a fertile breeding-ground for leading figures in the profession; there was the impact of Terzaghi's ideas on the young Skempton, which led to his founding the internationally important school of soil mechanics in the Department of Civil Engineering at Imperial College; and last, but by no means least, there was the impression on the contractors for that project, John Mowlem, which led to the founding of Soil Mechanics Ltd. This firm was the forerunner of many geotechnical contracting companies, underwritten by the financial and manpower resources of major civil engineering contractors, that were able to act as a repository of geotechnical expertise often not retained by consulting firms with a much lighter geotechnical workload. Such geotechnical contractors were able to afford well-equipped, high-grade laboratories, and to utilize the new laboratory tools being designed in the universities. It is a matter of some regret, therefore, to note that many of the leading geotechnical contractors in this country have suffered from the financial pressures imposed on their parent companies by the economic situation of the last decade.

At Imperial College, Skempton was able to attract the brightest talent, and to investigate many cases of slope instability. Often, these were natural slopes as at Jackfield and Monmouth, or at Sevenoaks, in some cases stimulated by the activities of man, but in others, taking place in response to natural processes. (Coincidentally, the greatest stability problems in the Sevenoaks slides took place immediately above the railway tunnel, just up-slope of the tunnel portal and cutting whose instability was investigated by Terzaghi.) The programme of new road construction undertaken in the 1950s and 1960s gave rise to a number of these stability problems, and cases which supplemented the many case records of long-term instability in railway-cutting slopes. The finding of slip surfaces in these slides was the key to the development of the theory of residual strength expounded so eloquently in Skempton's Rankine Lecture of 1964, and a wider appreciation of the effects of the Quaternary on slope stability in the northern part of Europe than before.

The finding of slip surfaces of tectonic origin in the Siwaliks of Northern India and Pakistan during dam construction confirmed the general applica-bility of the residual strength ideas. And, of course, it gives the answer to the question of whether slip surfaces are present in the ground, or are caused by a slide, which so taxed early thinkers. The answer, as so often in geological sciences, is 'maybe'. At least we now know *why*.

Meanwhile, the rest of the world had not been idle. At the Geotechnical Institutes of Norway and Sweden, work on quick clays, and the extremely violent mass movements to which these are subject, had taken place. Many leading geotechnical scientists had researched there with them, and the description of quick-clay landslides later in this book owes much to their work. (One of them, Hutchinson, later moved to Imperial College, and continued the work of Skempton on natural slopes. References to these two, and to other members of that team, abound in this book.) Also, in the alpine mountainous areas of the world, landslides and avalanches of immense size are an ever-present danger. Awareness of these, and the work of others outside the United Kingdom, was stimulated by the Aberfan disaster of 1966, in which the collapse of a tip of colliery discard became a flowslide, analogous in its mechanics to the devastating mountain rockslides and avalanches, bringing home more clearly to the British geotechnical community that even in this relatively geologically stable island we are not immune to such violent mass movements.

Also emanating from Imperial College was early work on the application of electronic computers to the solution of slope stability problems. In 1958, Little and Price described the first computer solution of slope stability problems using Bishop's theory. At the time, this was the most complicated program ever to have run on a British-built computer. Nowadays, such software is commonplace. In 1958, it cost about a shilling (5p) to analyse one slip circle on the computer as against about £1 by hand. Nowadays, with inflation, the manual analysis costs have risen significantly (even though the engineer of today has calculation aids undreamed of 30 years ago) and the cost of analysing a single slip surface is negligible, once the costs of data acquisition and input have been met.

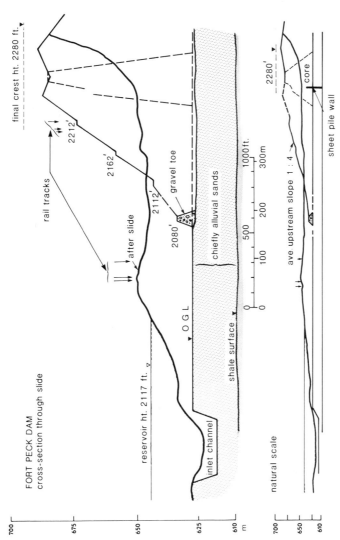

Figure 1.20 Flow slide at the Fort Peck Dam. After Bishop (1973). Loose fine sands and silts are particularly vulnerable to this type of failure, particularly when saturated and shaken or sheared. Fort Peck Dam was constructed originally by hydraulic filling, and was therefore similar to many lagoons or tailings dams in the mining and quarrying industry.

1.8 Previous work

All elementary textbooks on soil mechanics, and many in engineering geology, address themselves briefly to the problem of slope stability, but there are very few books that concern themselves solely with this theme. Collin's book, translated ably by Schriever (1956), sets the historical perspective, as indeed does the collection of Skempton's papers published in 1984 by the British Geotechnical Society to commemorate his 70th birthday: many of his writings are specifically directed to slope problems.

The first work specifically commissioned to help engineers was the Highway Research Board Special Publication No. 29, *Landslides in Engineering Practice*, edited by E. B. Eckel in 1958. Like Zaruba and Mencl's *Landslides and their Control*, it stresses the importance of the recognition of the impact of geological structures and lithologies on slope stability of natural slopes or those affected by engineering and building works. It does, however, give more prominence to the use of aerial photographs in the detection of unstable slopes to be avoided in the planning stage of road construction than do the Czech authors, no doubt reflecting the differing economic and historical perspectives of the countries in which they work. Zaruba and Mencl, however, have the advantage of later work on which to draw, and so are able to emphasize properly the influence of the Quaternary on slope development and landsliding.

Later work includes the completely revised version of the HRB book, now as a Transportation Research Board Special Publication (No. 176) which draws on later work, and has each of its chapters written by a different distinguished North American author. Hoek and Bray's *Rock Slope Engineering* recounts the results of a long-term project on rock-slope stability carried out in the Royal School of Mines (Imperial College). This extensive and detailed work concentrates on rock slopes and provides much useful data. With rock-slope stability so exhaustively covered in this latter work, it would be redundant for me to attempt to cover it, to inevitably a lesser depth and breadth, in these pages.

Finally, there is Voight's two-volume *Rockslides and Avalanches*. More a collection of papers than a text, these volumes are a source of extremely valuable data on large-scale mass movements of extreme violence and rapidity. Fortunately, in Britain, such large-scale events are unlikely to be encountered, but the cases described act as a reminder that slope instability elsewhere in the world can be an affair of unexpectedly grave danger to life and property.

Appendix D lists conference proceedings and symposia specifically devoted to landslides and slope stability generally, in the hope that this will draw the information contained within them more readily to the reader's attention.

2 Natural slopes

2.1 Slopes showing a response to present-day conditions

A natural, as distinct from man-made, slope may have reached its present shape by one of several routes. It may, for instance, be the result of the long-term action of a set of processes which are still active at the present. This is a 'steady-state' model of the development of that slope. Alternatively, a set of processes might act for a relatively short duration, forming that slope, and subsequently much less active processes may be at work. These then more or less subtly modify its morphology, so that whereas the main subsurface structures are the result of the formational processes, they may not be evident from the modified surface features.

In the first case, the erosive agents will be readily identifiable, and their individual effects should be more or less readily quantifiable, but in the second case it might call for a great deal of detective work to discover the original causes for the formation of the slope. Some examples may well explain this better.

As an example of the steady state, consider the development of the coastal landslides in the London Clay of the north Kent coast, as can be seen typically on the coast of the Isle of Sheppey. All the stages in the following process can be seen in one section of the cliff: 'space' therefore replaces 'time' in a walk-over survey. The details of this cyclical slope development process were first described by Hutchinson (1970), and the account below generally follows his scheme, but with some observations of my own.

Along the Sheppey coast, the London Clay is about 140 m thick and the top 40 to 50 m is preserved in the cliffs with about 100 m below sea level. In places, the cliffs are capped with later Tertiary sands. The London Clay is a stiff, fissured, silty clay of Eocene age which, although a blue-grey colour when unweathered, is a brown colour in the uppermost 10 m or so as a result of weathering of the abundant pyritized plant and animal remains. This weathering also attacks the calcareous septarian nodules, or claystones, which occur in quite well-defined horizons, leading to the formation of selenite crystals in the clay. The foreshore has a strongly developed, gently inclined, wave-cut platform. This is sometimes formed in the intact clay, but more often comprises the planed-off remnants of former slides.

Marine attack on the cliffs forms a steep sea cliff in the lowermost $\frac{1}{3}$ of the slope in the intact grey clay, above which is usually a flatter slope in old and exhausted mudslide remnants (of which more later). Continued erosion of the toe of the slope causes large and deep-seated rotational slides to occur. These eat back 20 m or so into the top of the cliff, and take a length of cliff up to 200 m in extent. Movements are quite rapid, taking probably a little over an hour for a 10 m displacement. Generally, the movements do not show a clear 'toe', but are accompanied by much heaving in the already sheared and disturbed foreshore materials. This is shown in diagrammatic form, along with the other stages, in Figure 2.1.

The slides leave a steep rear scarp which rapidly degrades to a flatter angle. Quite clear 'degradation' and 'accumulation' zones in this flatter slope can be distinguished initially, and the pile-up of debris on the rear of the main slide constitutes a loading which, together with the generation of undrained porewater pressures (section 7.6) and the continuance of marine erosion at the toe, acts to destabilize the main slide and cause further movements. In due course, the accumulated debris may form a considerable part of the slope, retained, as it were, by the ridge of much less disturbed material formed as the main slide mass subsided and rotated. The accumulation zone tends to keep the same slope angle on average throughout its formation, but the degradation zone gradually flattens until it reaches an angle too shallow to sustain the shallow slips and mudslides which transport soil to the accumulation zone.

While the rear scarp is so active, so too are the lower slopes. This tends to be partly as a result of rainfall (which has a larger catchment area following the slope rotation), but also as a result of the 'overspill' from ponds which tend to build up behind the backtilted ridge of the main slide. Such overspill is usually at each end of the slide mass, and the outflow of water encourages the formation of shallow and highly active mudslides in these locations. These secondary forms of instability are termed *lateral mudslides* in the following. Lateral mudslides push forward several metres each winter, and protect the toe of the slide from direct marine attack in their immediate vicinity. Inevitably, this leads to a concentration of marine attack in the centre of the length of coast affected by the slip, with a breaching of the original slide 'wall' and the formation for a short while of a *central mudslide* as the debris in the accumulation zone spills out.

In the later stages of erosion, all traces of the slipped mass above the wave-cut platform are removed, together with all of the disturbed material arising from the degradation of the rear scarp. At this time, the cliff reveals the bilinear profile of the start of a fresh cycle of sliding and erosion. In the more exposed parts of the cliff the cycle time is between 30 and 50 years in duration.

Some points arising from this are worthy of note. Firstly, in what is overall a steady retreat of the cliff there are cycles of behaviour. Most prominent of these is the major cycle of big landslide followed by slower erosion, but of almost equal importance is the annual climate-driven cycle. In the summer,

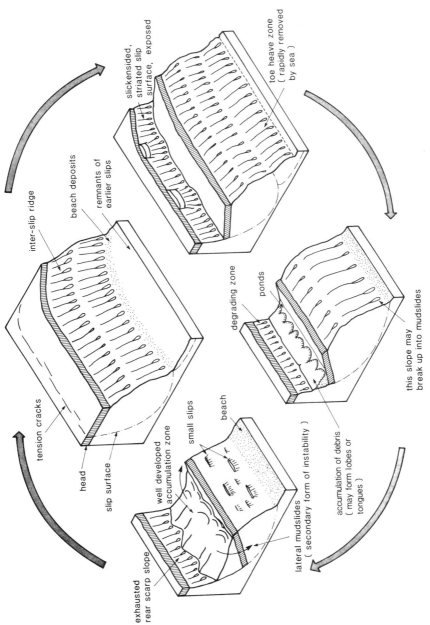

Figure 2.1 Cyclic erosion and landsliding in an actively eroding clay cliff. Typical of relatively uniform clay slopes eroding under strong marine or fluvial attack.

the clay in the mudslides is baked and hard: it can be walked on easily. In contrast, during the winter, mudslide deposits several metres thick can trap the unwary. Secondly, arising from the fact that slides are not synchronous along the whole coast there are two further results, namely that all of the stages in the process can be observed just by looking at different areas, and also, once adjacent sections of the cliff have got 'out of step' they tend to remain so, and then the cliff is divided into regular sections in which the process described above tends to repeat as the cliffs retreat, but in a way which is largely independent of its neighbours.

Soil slopes subject to active toe erosion tend to fall into two main classes: those where the slope retreat occurs with relatively shallow processes, and those where the erosion triggers the above deep-seated modes of failure. In contrast to the longer time cycles of deep-seated landsliding, the cycle of events in the former case tends to follow an annual, seasonal sequence. Hutchinson (1973) attributes the change-over from one type of behaviour to the other to a change in the rate of toe erosion. Shallow slide movements, he argues, take place in response to moderate rates of erosion; and the cyclical, deep-seated landslides take place in areas of intense erosion. The

Figure 2.2 Regular embayments occupied by mudslides at Herne Bay, Kent. The extreme regularity of these may reflect some input from land drainage.

threshold occurs at a rate of 1 m per year retreat in stiff clay cliffs from 35 to 45 m in height.

There are, however, alternative views that suggest that the transition from one type of behaviour to the other owes more to localized lithological factors, or in many cases to local hydrological conditions, than to the rate of toe erosion. Additionally, where slopes are really subject to deep-seated mass movements, but are inspected during only one phase of a cycle of long duration, they may appear to exhibit solely shallow mass movements.

Those slopes which do show shallow mudslide type of mass movement are described in north Kent and south Hampshire by Hutchinson and Bhandari (1971), and are common elsewhere. Individual mudslides tend to occupy full cliff-height embayments, sometimes described as corries by analogy with the glacial landform. A very active set of these is shown in an aerial photograph (Figure 2.2). There are often two or more levels where mudslide activity takes place: the size of each is a delicate balance between debris supply via small slides and falls from the rear and sides of the corrie, and the mass transport capacity of the mudslide. Movement takes place along a basal shear, although some incorporation of bed material appears to take place: the mudslide erodes its own channel.

The mudslides move with some rapidity. In Antrim for example (Hutchinson *et al.*, 1974) the onset of movement in the winter months has been described as 'potentially dangerous surges', and elsewhere similar rapid movements have been encountered. Such rapid thrusting involves overriding at the toe, with the formation of multiple shear zones (Figures 2.2, 3.6). Under present-day climatic conditions, the onset of winter, elevated groundwater tables, and the increased incidence of small falls from the rear scarp of the mudslide providing some loading *and* generation of porewater pressure at

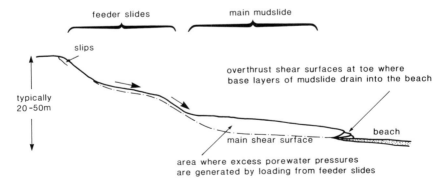

Figure 2.3 Section through a typical mudslide in overconsolidated clay. The intense shearing at the (relatively) well-drained toe is as characteristic as the high undrained porewater pressures at the head of the slide. The latter are essential to the mechanics of movement of these mudslides on low-angled slope.

the head of the mudslide all contribute to the renewed mobility of the slide, and can keep it in motion on relatively low overall slopes.

2.2 Examples of slopes preserving evidence of former conditions

The classic examples of slopes preserving evidence of former conditions fall into several categories. First of all, there are slopes which owe their present morphology to a period of degradation under periglacial conditions. Such slopes are widespread in the United Kingdom, and the presence of shear surfaces at shallow depth running nearly parallel to the ground surface is a major hazard to engineering works. Some examples of such slopes are reviewed below. Sometimes the slip surfaces are relatively deep seated, and give rise to even more serious problems.

Another type of slope which contains the evidence of former conditions is one which has been formed by toe erosion, but where the eroding influence—marine, river, glacier, lake, etc.—has ceased to be active. Such a slope would then degrade naturally by landsliding at a gradually decreasing rate until an ultimately stable slope is achieved. The process is termed 'free degradation', and the resulting degraded slope is called an 'abandoned cliff'. The degradation of an abandoned cliff may quite simply be the response to present-day conditions, or if abandonment took place under different conditions than those acting at the present, for example under periglacial conditions at the close of the last glaciation, then the slopes may owe their degraded form to periglacial, rather than modern, environments. The problems of abandoned slopes which have degraded freely under present-day conditions are rather different from those degraded under, say, periglacial conditions, because the former will have only a very small margin of stability and will therefore be destabilized with quite limited engineering works. They are therefore considered separately in a later section.

In every case, it will be found that the understanding of those processes which formed the slope, in however rudimentary a fashion, is an essential prerequisite for a successful engineering investigation and treatment of problems which arise as a result of the instability of that slope.

2.3 The impact of Pleistocene conditions on slope development

In many areas of the United Kingdom, there is evidence of past cold climate. This can be obvious, as in the glaciated terrain of the upland areas; or it may be less evident, becoming exposed only when detailed mapping is undertaken, or when it is exposed by construction works.

Large-scale structures in many of the slopes in the English Midlands show the influence of past climate in a graphic way. These structures include *cambering* and *valley-bulging* (Figure 2.4).

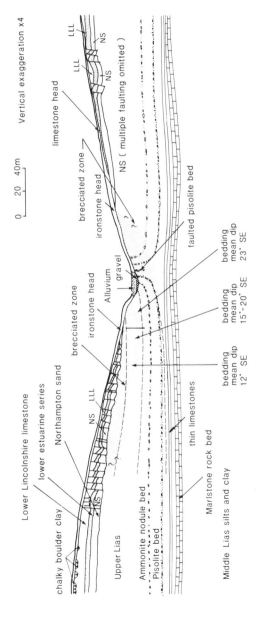

Figure 2.4 Cambering and valley-bulging: section at Empingham dam. (After Horswill and Horton, 1976.)

Cambering and valley-bulging can be considered to be two aspects of the same phenomenon. They occur where valleys have been cut into predominantly argillaceous formations which are overlain by more competent strata. First described in detail by Hollingsworth *et al.* (1944), these large-scale structures are extremely widespread where the appropriate geological structures exist. They can have a major impact on engineering in their vicinity both in valley bottoms, where the disturbed soils affect foundation design, and higher up the valley side where slope stability is certain to be a problem.

Valley-bulging is an upward displacement (anticlinal folding) of deformable strata in the valley floor. Higginbottom and Fookes (1970) refer to valley-bulging as 'superficial', qualifying this as 'not directly related to the deep tectonic structure of the area', whereas the valley-bulged strata may be extremely deep-seated in terms of engineering structures such as dams in the valley. A modern account of dam construction in a valley subject to valley-bulging in its floor and cambering in its upper slopes is given by Horswill and Horton (1976). An interesting appendix to this paper by Vaughan considers various explanations for the mechanics of the formation of these features. It discounts the additional weight of ice on hill slopes as a causative mechanism.

Valley-bulged features had, however, been recognized much earlier in cut-off trench excavations for Pennine dams in the early years of this century (Lapworth, 1911), and in many dams since. Where more competent strata are interbedded with the argillaceous rocks of the valley floor, these become fractured by the valley-bulging, and can require extensive grouting to seal flow-paths under dams.

Cambering is the downslope movement of the caprock on the slope together with superficial slides of the mantle of mudslide debris derived from the clay strata forming the hillside. It is best seen where blocks of the caprock have become detached from the parent strata, and have moved downslope, progressively grading into mudslide debris. Fracturing parallel to the contours of the valley, caused by the cambering, forms gulls (tension cracks). In many cases these gulls are infilled with later material, of Pleistocene or Recent age. This may show that they are features of considerable antiquity, and therefore not necessarily a cause for serious concern over stability. Small-scale gulls can occur distant from the slopes with which they are associated, in some cases several hundred metres away.

In slopes which are not cambered, other forms of mass movement are present. These include dip and fault structures, deep rotational landslides and solifluction (the term used here, in agreement with modern practice, to describe mudslides which have taken place under periglacial conditions in a partly thawed active layer above a permafrost horizon, and which are preserved in relict form by modern climatic conditions). Many of the larger landslides followed the rapid downcutting and enlargement of valleys by streams and rivers fed by meltwaters as the ice sheets retreated. They are

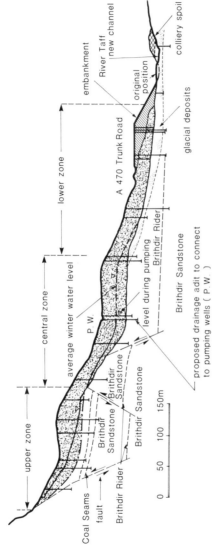

Figure 2.5 Cross-section through the Taren Landslide. (After Morgan, 1985.) This natural landslide complex, about 2 km distant from the site of the disastrous Aberfan tip complex slide, was at one stage thought by the local inhabitants to be an old quarry! In common with many other Welsh landslides caused by rapid downcutting of rivers or in valley sides oversteepened by ice and subsequently left unsupported, present-day movements were small.

present in huge numbers in the valleys of south Wales (Knox, 1939; Forster and Northmore, 1985), in the Bath area (Hawkins, 1973) and in the Pennines, the Cotswolds and the Weald.

The town of Bath, built on extensively landslipped ground, has been affected by numerous movements. Indeed, its development has been influenced to a remarkable extent by landsliding (Kellaway and Taylor, 1968). They describe the phases of landslipping starting with valley-bulging and cambering followed by two periods of rotational landsliding: an early deep-seated period, and a later, shallower series of slides. Further shallow slipping, during the Georgian development of the town in the late eighteenth century, prevented the completion of areas to their original design (Hawkins, 1978). Other areas were stabilized by the construction of drainage adits.

Sheets and lobes of solifluction debris are common on the slopes and toes of natural hillsides. Their stability under present-day climatic conditions may be marginal, and they are liable to reactivation with the interference from engineering works, or following undercutting by streams.

Apart from the large-scale structures in the slopes described above, many shallow features also exist. Glacial and periglacial conditions give rise to a wide variety of geomorphological features, and a number of these can be recognized in relict form in the United Kingdom. However, the feature of most engineering significance in terms of slope behaviour is the widespread occurrence of the remnants of periglacial solifluction. Chandler (1972) discusses the mechanics of solifluction, a term coined early in this century to describe shallow downslope movements of waterlogged soils. In modern usage by the engineering community, the term is applied in the context of *periglacial* solifluction, to describe a process which has given rise to mantles of sheared and landslipped debris on low-angled clay slopes. The fabric of the sheared debris is so similar to that in present-day mudslides that it is certain that *solifluction* was a process akin to modern mudsliding. However, since mudslides as presently examined are found to contain artesian (or in excess of merely hydrostatic) porewater pressures, the mechanics of the formation of the periglacial equivalent must by inference also demand the generation of such high porewater pressures.

In Chandler's study to relict solifluction deposits in Britain, he classifies them into three groups:

sheet movements of principally clayey materials (Figure 2.7)

sheet movements of granular materials, overlying clays

individual sub-horizontal 'successive emergent' shears aligned *en echelon* in section relative to the ground surface.

Of these, the first class seem not to require groundwater levels significantly higher than present ground level to become unstable on their present slopes, although they do to have a high mobility. The generation of these excess

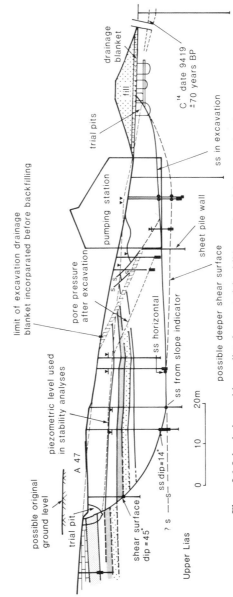

Figure 2.6 Inland slope with a relict deep-seated landslide reactivated by excavation for a pumping work. (After Chandler, 1979.) Low-angled slopes are normally hazardous by virtue of shallow, fossil mudslide-type slides. This slide was deep-seated, and had no surface expression at all before being re-activated by the pumping station excavation.

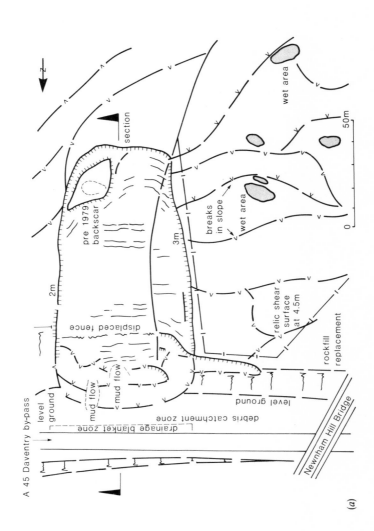

(a)

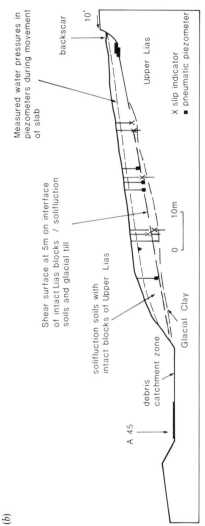

Figure 2.7 Inland slope in Lias clay with a solifluction sheet reactivated by cutting at the toe. (After Bicysko, 1980.) (*b*) Plan of slide shown on facing page.

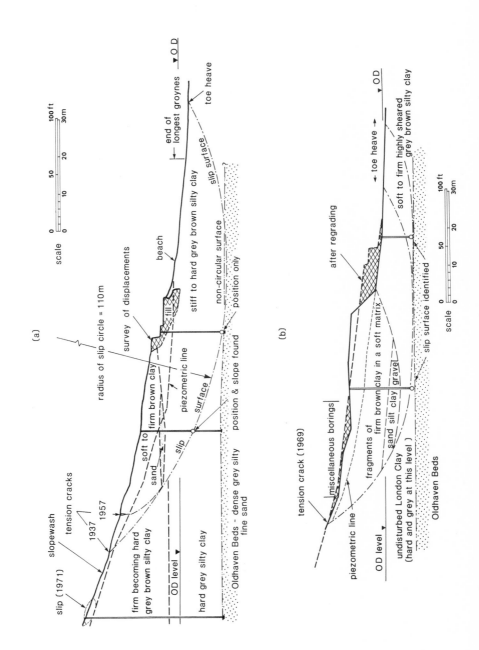

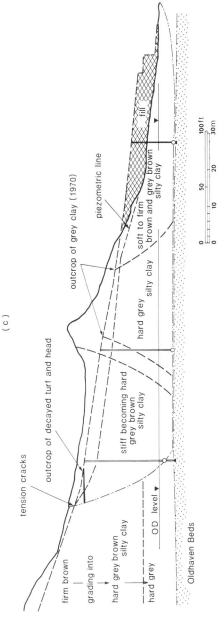

(c)

Figure 2.8 Slides at Herne Bay showing influence of structure on slide type (Bromhead, 1978). The underlying Oldhaven Beds (weak sandstone) dip to the east, and slide surfaces change from the graben form of the Miramar slide, to the very nearly circular section of the Beacon Hill slide where the contact is well below sea level.

porewater pressures is likely to be a result of the repeated freezing and thawing that the surface layers were subject to during their formation. Examples of periglacially soliflucted Lias Clay slopes destabilized by the excavation of highway cuts is given by Biczysko (1981). In contrast, the second class (*vide* the Sevenoaks examples, Chapter 10, and other examples in the same vicinity listed by Weeks, 1969) demands rather higher excess porewater pressures for their development. It is suggested that this may be the result of the formation of a frozen surface to the high-permeability debris, so that the unfrozen pore water below could pond up beneath. This would seem to be supported by piezometric measurements and other observations made by Chandler in active solifluction features in Spitzbergen.

His third class of periglacially initiated shearing is most probably the remnants of slightly deeper-seated slide movements initiated by the same high groundwater pressures, but arrested before they could fully develop and provide debris to be incorporated in a solifluction sheet.

Case records are detailed by Brunsden and Jones (1972), Chandler (1970, 1979) and Chandler *et al.* (1974).

2.4 Types of failure and the importance of geological structure

The internal structure of a slope has an important bearing on its stability, or on the mode of failure which might occur. In many sedimentary sequences, clays or clay shales are interbedded between sandstones or limestones. A geologist might describe the clays as *incompetent* and the other rock as *competent*, a loose way of indicating that slope failures are more likely to be seated in the clay strata.

During an investigation into the stability of a coastal slope at Herne Bay in Kent (Bromhead, 1978) a series of deep-seated landslides in an Eocene clay (the London Clay) were investigated. Although the London Clay is a fairly thick stratum, the Herne Bay slopes were formed in the lowest 50 m of the clay, with a weak sandstone underlying the clay. The sandstone outcropped over part of the site. The contact dipped across the site and the landslides occurred at sections where the contact was respectively well below, slightly below, and at, sea level. Other, and much smaller, landslides occurred as *slope* types of failure where the sands rose well above sea level. The shape of the sliding surface of each landslide reflected the geological structure, varying from an almost perfect 'slip circle' where the contact with the sands was deep below sea level, to a flat-soled graben-type landslide where the contact was close to sea level.

A graben is usually developed at ground surface above a 'corner' in the slip surface. Although the graben surface itself may rotate and tilt back upslope slightly, its dominant component of movement will usually be settlement, and this demands the formation of counterscarps where the graben

Figure 2.9 Miramar landslide, 1953 (Bromhead, 1978). The counterscarp and graben may be clearly seen in this photograph.

moves relative to the downslope parts of the landslide which slide in a different direction, often sub-horizontally. There may be an analogous feature at the toe of the slope where there is a change in direction of the slip surface from its flat sole to the rising, passive resistance, section; this is normally much less in evidence than the counterscarp at the top of the slope. Figure 2.9 shows the graben-type Miramar landslide, illustrated in section after considerable degradation in Figure 2.8.

Another graben-type landslide, this time of the slope-failure mode, rather than the toe- or base-failure mode of the Miramar slide, is illustrated in Figure 2.10. This slide occurs in the Corallian series of rocks at Red Cliff, east of Weymouth in Dorest. The graben appears to be disproportionately wide in this section, although matters are complicated by an east–west component of movement as well as a movement normal to the roughly east–west trend of the coastline. The slip surface may penetrate deeper than that shown, and go down to a bedding plane in the Nothe Clay, although surface mapping revealed outcropping slip surfaces only at the location shown. However, since movement would be along the bedding in this variable sequence of rocks, covered with a mantle of debris, the failure to record such a slip surface is perhaps not surprising. Alternatively, the graben may here reflect the interaction of faulting with this sequence of rocks. In the case of this landslide, the tipping of debris to infill the graben has certainly accelerated the movements.

Another well-known coastal graben-type landslide in Dorset is at the western end of the stretch of coastal cliffs running from Lyme Regis to

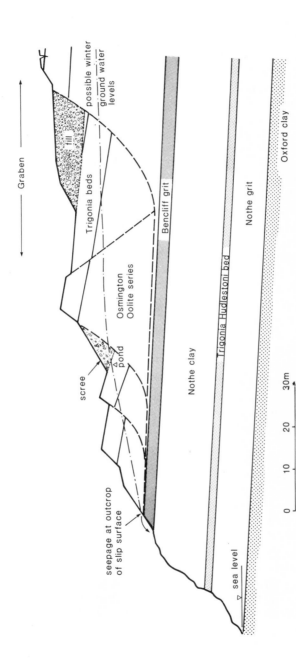

Figure 2.10 Section through graben landslide in Corallian rocks at Red Cliff, Dorset. The western end of this slide affects the grounds of a holiday complex. Tipping of waste material on the graben is a serious additional destabilizing factor.

Exmouth. This slide, which occurred in 1799 (Arber, 1941; Pitts, 1979) has now become covered in vegetation, but this cannot obscure the large ridge feature known locally as 'Goat Island', and the deep chasm of the graben. A section through this slide would be similar to, but on a much larger scale than, the Red Cliff slide.

Elsewhere, local geological structure similarly controls the types of landslide that develop. It will be seen in Chapter 10, where the Folkestone Warren and Isle of Wight landslides are considered, that these too are non-circular-shaped landslides with a flat sole. At Folkestone, this occurs near the base of the Gault immediately above the sands of the Folkestone Beds, but on the Isle of Wight, it occurs at a minor lithological change in the Gault rather than at the major change at its base. An example can be seen close to the Red Cliff landslide discussed above. About 300 metres to the west, nearer Weymouth, in Jordan Cliff, the Oxford Clay is involved in rotational landslides. As at Red Cliff, the subsurface details are conjectural, but the occurrence of subdued counterscarps seaward of the two back-tilted rotational slip segments, and again in the toe area, are pointers to the presence of a flat sole to the slip surface at about the level indicated. Also at Red Cliff, the frontal elements of the graben slide are involved in a multiple rotational form of slip, but on a relatively minor scale.

Figure 2.11 Multiple rotational landslide on the Isle of Portland. The last retrogressive slip elements may be seen just beneath the rear scarp. Lower slopes are obscured by the effects of quarrying.

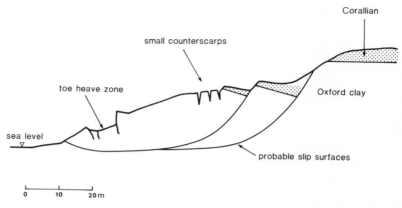

Figure 2.12 Multiple rotational landslide in Oxford Clay at Jordan Cliff, Dorset. Subsurface details are conjectural, but in accordance with the general morphology.

Some landslides in interbedded sands and clays adopt a *multistorey* form. This takes the form of landslides seated in each of the clay strata in turn. Where these are widely spaced, the individual slides can be discerned. In the nomenclature used in southern England, related to coastal slopes, each of these forms an individual 'undercliff'. An example of this, described in detail by Barton (1973, 1977) is shown in section in Figure 2.13. Where the clay beds are thicker, or the mantle of debris deeper (possibly resulting from slower erosion allowing a build-up of colluvium) the slides merge, and it is only with difficulty that the individual elements can be separated. Not only is the geomorphology likely to be more indistinct in a case like this, but stability analysis is also complicated. This is because the individual slide elements cannot be treated independently, but the interaction between them must be assessed too.

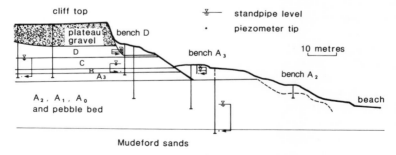

Figure 2.13 Stratigraphically controlled 'multistorey' landslides at Barton-on-Sea (Barton, 1973). Some horizons on which sliding has taken place are quite major lithological boundaries, others are much more subtle.

Figure 2.14 Black Venn mudslide complex. The mudslides plunge over a large number of stratigraphically-controlled benches in the cliffs.

Each element in one of these multistorey slide complexes could be a straight-forward rotational slide, but since the overall form is almost inevitably the result of bedding control, the individual component slides tend also to be in response to that bedding, either a graben slide, or a flat-soled, multiple-rotational, non-circular slide.

An example of one such multistorey landslide exists in the Taff valley in South Wales. This consists of two bedding-controlled elements, with a third element in the form of an accumulation of landslide debris which has overridden the glacial infill to the Taff valley.

In rock slopes, the discontinuities control the mode of failure. These dis-continuities can include the bedding, but faulting, tectonically-induced shearing and jointing in the rock mass each have a role to play in the development of failures. A mode of failure characteristic of rock slopes occurs where two intersecting discontinuities daylight in the slope face and the resulting discontinuity bounded wedge moves along the V-shaped notch formed by those discontinuities. Where they are open joints, such sliding is assisted by joint-water pressures. Commonly, only the V-shaped notch is seen, as the material will have slipped out and, unlike an earth or soil slide, will have parted company with the rock face in so doing, necessitating the debris being cleared away.

Another common type of failure occurs in steep rock faces where the

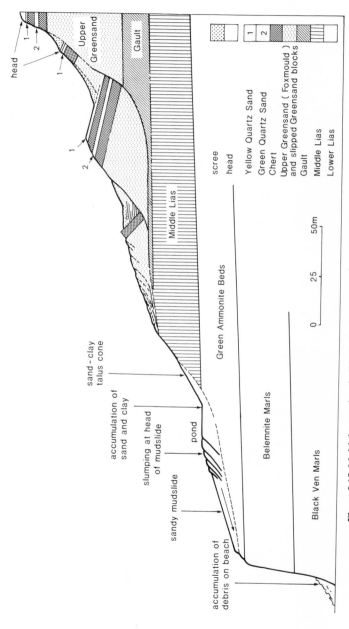

Figure 2.15 Multiple rotational slide in Cretaceous rocks at Stonebarrow Hill, Charmouth, Dorset (slope failure). (After Brunsden and Jones, 1976.)

discontinuities are near vertical. Rockfalls and toppling failures then occur by collapse of individually joint-bounded blocks.

Patterns or failure involving more than two discontinuities can occur, but failures involving two joints, or at most, two joint sets, are the commonest by far.

2.5 Free degradation, colluvial and scree slopes: the abandoned cliff

Important evidence for the processes of slope formation starting from initially steeply cut slopes (coastal cliffs, river cliffs and so on) come from a series of investigations undertaken in the last 20 years or so. The most important of these was the investigation by Hutchinson and Gostelow (1976) of the abandoned cliff at Hadleigh, which is part of a line of such slopes extending west from Southend-on-Sea. This cliff line is formed in the London Clay, rising generally to a height of 40 m or more.

Initial interest centred on the retrogression of the crest of the cliff in recent times, affecting the walls of the ruined medieval Hadleigh Castle. A slip as recently as 1965 caused the collapse of one of the two remaining towers, and throughout this century good records of retreat have been compiled. The situation is complicated because the castle was built on a spur for good defensive reasons, and the slopes have retrogressed on both sides.

The present morphology shows a trilinear profile. At the bottom of the

Figure 2.16 General view of the abandoned slopes at Hadleigh from the air.

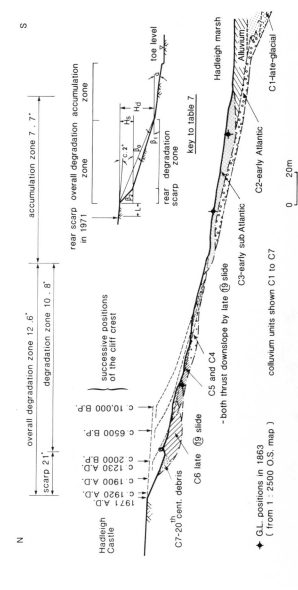

Figure 2.17 Section through the abandoned cliff at Hadleigh, Essex. (After Hutchinson and Gostelow, 1976.) The different phases of mudsliding which formed this slope were identified in this investigation: in the main they were contemporary with the growth of marsh at the toe of the slope. Some retrogression of the slope crest is still in progress.

slope the accumulation zone is inclined at about 8° to the horizontal; this is followed by a degradation zone of about 11° (the difference showing that a final equilibrium state has not yet been achieved) and an upper, scarp slope of about 20°. A subsurface investigation, by means of trial pits and continuously sampled boreholes, showed that the cliff, with its toe at a depth of − 19 m OD and now buried by marsh, had been formed by erosion in the Middle and Late Devensian. By the end of this period, erosion ceased, and the cliff has degraded in an episodic manner. The available evidence, from the Hadleigh investigation and elsewhere, was that the various phases of movement had occurred in response to climatic changes.

Part of the evidence came from the interdigitation of tongues of mudslide material with the marsh at its toe. Four main phases of degradation, each followed by a period of relative stability, were discovered. These were identified, and dated as follows:

late glacial, periglacial mudsliding associated with the toe level of − 19 m OD, and complete by about 10 000 years BP

Figure 2.18 Landslipped, and partly collapsed, medieval tower of Hadleigh Castle.

early Atlantic, temperate mudsliding associated with a somewhat higher toe level of -9 m OD, the rise being caused by the accumulation of the salt marsh at the toe of the slope (^{14}C date on organic marsh clay, but occurring 7000–6500 years BP)

early Sub-Atlantic temperate mudsliding. The level of the toe of the slope was by then at its present level of about $+3$ m OD (dated from a ^{14}C date on wood fragments buried by the toe of the slide to 2100 $-$ 2000 years BP)

late nineteenth century, moderately deep-seated sliding at the crest of the slope. This was caused, at least in part, by disturbance to the slope from human activities at its toe.

The historical investigations for this study revealed that there had been stability problems with the curtain walls of the castle soon after construction in the thirteenth century, and that this was the result of construction on landslipped ground. Reconstruction of former profiles showed a retreat of the cliff crest by at least 50 m, and with the morphological and historical records showing that retreat of the slope is still in progress, the slope evolution to an angle of ultimate stability was shown to take significantly longer than the 10 000 to 15 000 years since abandonment.

This sort of picture is likely to be built up in any detailed investigation of the landsliding in an ancient slope. At the toe of the slope there will be an accumulation zone, within which is preserved the evidence of the accumulation of successive phases of landsliding, and in the upper slopes will be the more or less active rear scarp and a flatter transport zone which is nevertheless steeper than the accumulation zone in the lower slopes. The transport zone, which may in fact be contributing to the overall slope degradation process by basal erosion, can really be considered to be a part of the generic 'degradation' zone.

At Hadleigh, the investigators were fortunate in that not only were there several phases of landsliding which could be recognized from soil colour, texture and the overriding of dateable material, but the growth of marsh at the toe of the slope had separated several of the phases with alluvial materials, and this was an additional aid in separating those phases. In other colluvial accumulations, it is possible that none of these factors would be present, and it might well prove impossible to interpret the record, such as it is, preserved in the accumulation zone. Very little, if anything, is preserved as a record in the degradation zone except for the very last slide event.

Ultimately, however, the degradation process will be virtually complete. At this time, the gradients of both the degradation and accumulation zones will be the same, and slide movements will become smaller and less frequent, occurring only when the groundwater table is at its highest, or if the equilibrium of the slope is tampered with in any way. Small cuts and fills can do this, so can changes in the vegetation pattern. Clearance of woodland can be a cause of reactivation of movement. Evidence of the small movements

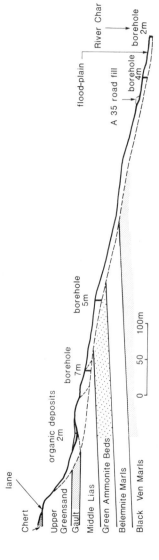

Figure 2.19 Soliflucted slopes in the Char valley. (After Brunsden and Jones, 1973.) Broadly similar to the Hadleigh slopes, there is an upper zone of rotational slips with sheet slides in the lower valley slopes.

c

Figure 2.20 Great Southwell Landslip: the tilt of these blocks is downslope.

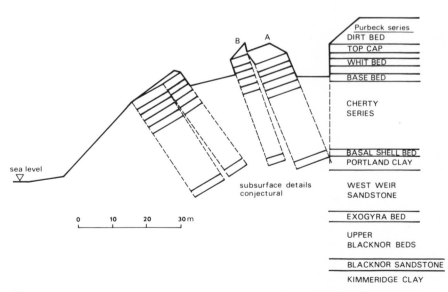

Figure 2.21 Section through the toppling failure at the Great Southwell Landslip, Isle of Portland. The subsurface details of this are largely conjectural.

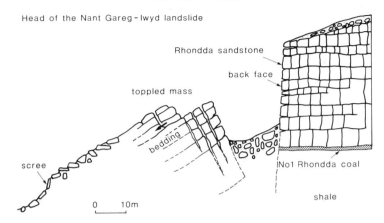

Figure 2.22 Section through a toppling failure. (After De Freitas and Watters, 1973.)

which do occur is preserved in the form of waves in the slope profile. These can easily be ploughed out, so their absence is no real indicator of stability, especially as they may well reappear if the land is returned to grass for a number of years.

The ultimate angle reached by slopes degrading under temperate conditions before the slope movements cease is a simple function of the residual strength of the soil, or its strength under repeatedly sheared conditions. The evaluation of such strength is covered later, in Chapter 4, and its application to slope stability calculations, including this very issue, in Chapter 5. Note, however, that the slopes reached by free, sub-aerial degradation under temperate conditions may be quite considerably steeper than those resulting from periglacial solifluction; a cautious approach to the stability of even fairly low-angled slopes is advisable.

2.6 Toppling failures and their occurrence

Toppling failures may be localized features, or may occur on a massive scale. De Freitas and Watters (1973) give a number of examples of this type of failure from the British Isles, and the subject has received considerable attention from research workers who have investigated the mechanics of such movements, including Goodman and Bray (1976) and Evans (1981). The classification of toppling failures involves either *primary* or *secondary* toppling. In the former case, downslope movements occur with the opening of near-vertical joints and rotation of individual rock blocks of slight shear displacements on surfaces close to the horizontal. Secondary toppling mechanisms include the thrust at the toe of a slide, or drag along its basal shear; they also include 'bearing capacity' failures in an underlying incom-

petent formation, or differential weathering and alteration of such an underlying bed so that its stiffness is reduced close to an exposed face. Joint-
separated blocks of the overlying rock are then free to rotate.

Two examples will illustrate the toppling process. These are the Nant
Gareg-Iwyd landslide, illustrated by De Freitas and Watters, and the similar
Great Southwell landslide on the Isle of Portland (Figures 2.20–2.22). In
both of these examples the subsurface details shown are something of a guess
because the detailed investigation of such failures has not yet been carried
out on a sufficiently large scale. However, it is possible to say with some
confidence that Evan's model to account for the development of these failures
is likely to be fairly close to the truth. Other causes for the differential
undermining of the blocks such as spring sapping of a non-cohesive bed
(seepage erosion) are also likely where the geological succession is favourable.

2.7 Rockslides, avalanches and other rapid mass movements

The subject of rockslides and avalanches in natural slopes is covered exhaustively by Voight and his co-authors (Voight, 1978). The unifying feature of the
case records presented therein is the occurrence of large rockslides, incorporating many millions of cubic metres of debris, which then degenerated
into a flow type of movement. As a result of this change into a flow, the
debris has been able to travel a considerable distance from its source. The
destructive potential of such mass movements is enormous, but since the
majority of cases come from the world's mountainous regions where population densities tend to be low, loss of life has been small in relation to the
size and violence of the slides.

In each case the presence of tectonically-induced discontinuities in the
rock mass appears to have played a role in the initiation of the movement.
Other effects—seismicity in several cases—have then led to collapse, in a
similar fashion to the failure of smaller slopes involved in engineering works.
The same types of movement—rotational slides, toppling and so forth—are
experienced, and they are therefore in principle open to the same methods
of investigation, analysis and treatment as related later in this book. The
mechanics of flow, too, have received the attention of several workers, and
indeed, are touched upon briefly elsewhere in following sections. They will
not therefore be examined further here. Suffice it to say that there is still
debate about the precise mechanisms involved, and the impossibility of
making field measurements (say of pore-fluid pressure) in such a mass movement will in all probability prevent a consensus view being arrived at in the
foreseeable future.

Two well-known examples of these large mass movements are the
Blackhawk landslide and the Sherman Glacier rockslide (Shreve, 1968). The
former of these is a flowslide which had its origins in a rockfall from the summit

Figure 2.23 Toe and lateral heaves from shallow landslides. This example is one of a series of shallow slides in a reactivated solifluction sheet of Wadhurst Clay near Robertsbridge, Sussex.

Figure 2.24 Sherman Glacier rockslide. (After Bishop, 1973.)

of the Blackhawk Mountain in southern California. Involving some 400 M m^3 of marble breccia, the remains occupy a lobe over 8 km long by more than 3 km wide. Over the last 5 km, the average slope is about 2°. The debris is between 10 m and 30 m in thickness, and shows marked lateral ridges where the debris stabilized most rapidly due to a loss of pore fluid during the slide. A minimum velocity of in excess of 110 km/h is indicated from topographic features overtopped by the debris in its passage.

The Sherman Glacier rockslide was triggered by the Alaska earthquake of 1964. Although one-tenth of the volume of the Blackhawk slide, at some 40 M m^3, it nevertheless had a run-out of 4 km. The debris carpet left after the flowslide was 3–6 m in thickness.

Other mass movements involving large masses of material, relatively high velocities, and large run-out are submarine landslides in unconsolidated recent sediments. There is a considerable literature on these, but a useful introduction by Bjerrum (1971) covers some examples from the Norwegian Fjords. These slides or flowslides of loose, saturated sand sediments are often started by the placing of fill for quays, land reclamation, or in one instance by the dumping of dredgings. In this example, the flowslide which developed took with it the anchor of a dredger, which was towed 400 m across the fjord at about 12 knots when its other moorings parted. The anchor buried in the submarine flowslide debris could not be winched back on board when the sand ceased to be mobile.

A particularly well-documented case is Voight *et al.*'s account (1983) of the Mt St Helens rockslide/avalanche and associated vulcanism.

2.8 Identifying and locating unstable slopes

An essential first step in the detailed investigation of landslides, whether this is undertaken purely for academic interest, or because of their possible impact on a proposed construction project, is to make a visit to the site to look at the surface expression. Failing this (or possible in addition), photographs may be examined.

The field survey, and an interpretation of such photographic information as may exist, can then be summarized on one or more maps or plans. Topographic information, measured by orthodox surveying techniques, can be recorded on a conventional contoured plan, or drawn direct on to a large-scale air photograph.

Vertical photographs, taken with high-precision cameras mounted in an aircraft, provide the best images for use in photogrammetric mapping, but are difficult for the unskilled to interpret. In particular, the lighting, and the resulting shadows that are cast, can cause detail to be obscured or revealed to an extraordinary degree. To see the exaggerated vertical relief effect one needs adequate stereoscopic vision in the first instance, and an acquired 'knack', so it is probably better to take rather low-angled oblique air photographs, the viewpoint of which will be much more familiar to the average engineer, and which will be much more comprehensible.

This view is in agreement with the findings of Clayton and Matthews (1984), who have even had success in taking such photographs from a small remotely-piloted drone. However, the conventional wisdom, e.g. Dumbleton and West (1970–72) is to prefer stereoscopic vertical air photographs, and they detail apparatus to assist in use in the field. The writer is invited to form his own conclusions as to which is the preferable technique by examining the many oblique air photographs in this book.

Terrestrial photographs can also be used for measurement as well as for qualitative record purposes. These measurements can vary from the simple to detailed analysis of stereoscopic pairs of photographs. Since one is more likely to be able to return to the same position on the ground than to regain the exact position of an aeroplane, it is possible to use other photographic techniques. For instance, two photographs taken in the same direction from the same viewpoint (but at different times) when viewed stereoscopically will show up deformations with an illusion of added relief: this can be measured with a parallax bar and related to the movements which have taken place (Butterfield et al., 1970). Kalaugher (1984) has developed the technique in the field, and using relatively simple and inexpensive apparatus makes it possible for an observer to relocate himself at an unmarked viewpoint, and to examine a slope for movement.

2.9 Dating mass movement

It is often necessary to put dates to old landslides, if for no other reason than to discover their present activity, or whether they were the result of conditions now past and unlikely to recur. For research reasons, academic interest, or even as part of a wider-reaching study, the dating of landslides may be an end in itself.

Archaeological and antiquarian records may provide some dating information. Hutchinson and Gostelow (1976) use old maps, air photographs and historical records (including a nineteenth-century etching) to trace the recession of the cliff edge of the slopes near Hadleigh Castle in Essex. Elsewhere, records of recession of cliff lines can be obtained from early editions of OS maps (e.g. Hutchinson et al., 1981).

There may be sufficient coverage of air photographs to take the place of maps, although this is unlikely except in particular situations. The case of the Stag Hill landslide, on which part of the University of Surrey was subsequently built, is a case in point. The cathedral standing on the crest of Stag Hill has been the focus for quite a number of air photographs, in which the landslide shows more or less clearly (Clayton and Matthews, 1984).

The technique of radiocarbon or ^{14}C dating is used to put approximate dates on organic material buried by landslides. Basically, the ratio of radioactive carbon ^{14}C to its more stable isotope ^{13}C in the atmosphere is kept relatively constant by cosmic radiation, but when incorporated in organic matter begins to decay. Since one half of the ^{14}C disappears in the half-life of 2574 years, and half of the rest in the same period thereafter and so on, the ratio of the two constituent isotopes must reflect the time that has elapsed since the organism ceased to take in atmospheric carbon (i.e. died). Therefore, by measuring the ratio of ^{13}C to ^{14}C, it is possible to estimate the period since that organic matter ceased to take in atmospheric carbon. Woody matter is the best, but organic soils or shells and animal remains such as bone or horn have been used.

Carbon-14 dating is useful where organic matter can be found *in situ* under landslip debris in a position of growth, because it may otherwise have been emplaced out of context by the slip. For example, some tree trunks of yew, radiocarbon dated at approximately 4500 years BP, were found in a thick horizon of landslide debris in a sea cliff at St Catherine's Point, Isle of Wight. Evidence which indicates that these have been located approximately in their position of growth before being overwhelmed by landsliding comes in several forms. Firstly, they are a native species, as would be expected from their age. Secondly, they were found in a series of palaeosols and tufas indicative of a period when landsliding was comparatively inactive, so that streams, ponds and vegetative cover could become established. Some plants and animals can ingest 'old' carbon from the rocks on which they live, or from 'hard' water, and it is also widely recognized that the burning of large quantities of fossil fuels in the recent past has altered the natural balance of the two carbon isotopes in the atmosphere. Care must be taken not to confuse dateable carbon with 'fusain' or fossil charcoal of great antiquity which will yield 'infinite' dates.

Evidence of a much more general sort about the date of landslide events comes from the study of pollen, or of faunal assemblages. There are some established events in post-glacial history which can be correlated to changes in the flora or fauna (the latter often consequent upon the former), and thus the preservation of a record can give evidence pre- or postdating a particular landslide. Pond sediments are particularly useful in preserving pollen, and a landslide pond will yield a pollen record postdating the slide. On the other hand, pond sediments overridden by the toe of an encroaching landslide give evidence which predates the slide.

Pollen, where preserved, is found in abundance. It may be used via statistical analysis of the quantities of each different species present as an indicator of the probable climatic environment at the time of deposition: the analysis depends on the presence of sufficient numbers of pollen grains to render the statistical treatment valid. Other floral remains are not usually found in sufficient quantity to permit more than the identification of a single species. Similarly, macrofaunal remains, bone, horn, etc. are found only in isolation, but smaller species with shells, such as molluscs, or with chitinous wing cases (beetles) may be present in quantity, and may be treated in a statistical fashion in the manner of pollen.

All of the dating techniques demand specialist knowledge and techniques not usually within the competence of the practising geotechnical engineer or engineering geologist. However, the correct interpretation of the results of such dating lies completely within his purview.

Other data which provide evidence on the date of landslide movements, particularly those where the total movement is small, come from the dislocation of linear artificial features. These features can include fence lines and walls, avenues of trees, road and railway routes, pipelines and electricity transmission lines, or tunnels. The latter are obviously only affected by relatively deep-seated ground strains, but the others will be as responsive to shallow movements as to deep ones. Clearly, the movement must postdate the building of that particular feature, although that is not necessarily evidence that some movements were not taking place before construction commenced.

The use of man-made artefacts to date movement presupposes that the slope has been developed. Vegetation can be used as a guide to the dating of the last movement: trees for instance often have a bend close to their roots (Figure 2.25) caused by shallow, downslope movements early in their life,

Figure 2.25 Trees disturbed by shallow movements early in their life. These show the characteristic bend in their trunks close to the ground surface.

followed by a period of quiescence during which they have resumed vertical growth. A core from the tree including all of its annual rings will provide an age within which the date of the last movement could clearly be bracketed. A similar effect, but with the initial bend indicating a tilt upslope, is indicative of a more rotational kind of movement.

In steep slopes, bent trees may not indicate a period of quiescence at all, but rather that once the tree has reached a certain size, its root network has stabilized the soil in its general vicinity. Most trees in mountainous areas show the effect to a greater or lesser degree, and one could only say with confidence that movements had ceased if trees less than a certain age were all upright, while all those older were bent. (One of the few human activities which is aided by slope instability is the carving of alpenhorns. These are made from 'bent' trees, and so the industry is dependent on a continuance of slope instability for its raw materials!)

Other evidence for *landslide susceptibility*, if not for the actual presence of landslides, comes from the *types* of vegetation present. Hydrophilic (water-loving) plants such as the horsetail (*Equisetum*) or the teazel (*Dipsacea sylvestris*) are both commonly found on waterlogged and landslipped clay slopes in Britain. Other plants, particularly those suited to rapid colonization of bare ground following landslipping, will also flourish. All of the varieties of reeds and rushes, brambles and some willows are useful indicators of high water tables and the associated problems when found on clay slopes.

3 Fundamental properties of soil and rocks

3.1 Stress–strain properties of soils and rocks

When stress is applied to a specimen of soil or rock some deformation is caused, and elementary theories of engineering mechanics postulate some simple relationship between the applied stress and the resulting deformation. In a linearly elastic theory, for instance, the strains in the material are directly proportional to the applied stresses, and are reversible: that is, removal of a stress causes the material to revert to its original, unstressed, shape. More complex deformation models have been proposed for geological materials in which stress and strain are related by non-linear relationships, and where the unloading characteristics are different from those for first loading.

When the shear stress on a specimen reaches a limiting value, which is termed the shear strength, larger, irreversible deformations take place.

Consider a soil placed or compacted into a simple shear box apparatus. By applying loads to a platen it is possible to influence the normal stress on the soil in the plane of separation of the box halves, and by imposing relative displacements on the two halves of the box, shear stresses can be applied to this plane. The soil in the box may be thought of as representing an element of soil in a developing landslide.

If the normal stress is held constant, and the shear load is measured as the box halves are moved relative to each other, then the stress–displacement relationship can be explored. Two basic patterns will emerge similar to those of Figure 3.1: those of *brittle* and *non-brittle* form.

In non-brittle soils, the load transmitted will rise to a point and then remain constant, or nearly so, regardless of further displacement. In contrast, the deformation required to achieve the maximum, or peak, load-carrying capacity of a brittle soil *can* be passed. Continuing displacement in this way causes the load carrying capacity to decline to an *ultimate* or *residual* value. This is usually accompanied by the formation of shearing surfaces in the soil on which strains are concentrated. Thus only part of the soil mass will develop these surfaces of weakness. For a brittle soil, therefore, it is possible to talk of both peak and residual strengths, each acting over part of the slope section. The selective development of these shear surfaces due to a non-uniform straining of the soil is what we would consider to be part of a *progressive failure* (Bishop, 1967).

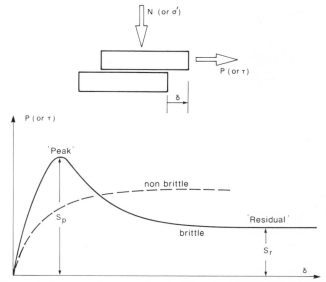

Figure 3.1 Stress–strain curves, demonstrating the difference between 'brittle' and 'non-brittle' failure modes. The former can only be captured with the use of stiff, strain-controlled measuring apparatus.

Naturally, the soil strength reflects both the normal stress applied and the pressure in the pore fluid, usually increasing with the normal stress, and decreasing with increasing fluid pressure. The combined effect of these (at least in saturated soils: in partly saturated soils the effects of fluid pressure in both air and water phases must be considered separately) is obtained if the numerical difference between normal stress and pore-fluid pressure is taken: this is termed the *effective* stress. A plot of shear strength against

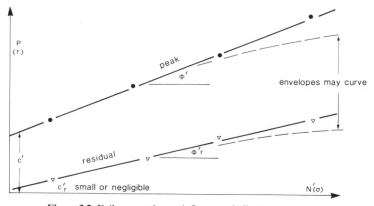

Figure 3.2 Failure envelopes: influence of effective stress.

normal stress may well form an arbitrary curve, but for a limited range of effective stresses it may be adequately represented by a single straight-line segment of intercept c and slope angle tan ϕ (Figure 3.2). These two parameters may be defined in terms of the gross normal stresses, or total stresses, or in terms of *effective* stresses. In the latter case, they are usually suffixed with an prime (') to denote the fact. Other suffices, usually as subscripts, denote peak (p), residual (r) or other conditions.

3.2 Influence of effective stress on shear strength

In the above, the soil strength has been considered as a simple soil parameter, which is fixed for a given soil at the outset. Nothing could be farther from the truth. The soil strength is the sum total of a number of different effects, and to understand it to any degree, it is necessary for the individual constituent elements of behaviour to be understood.

It helps in the understanding of the shear strength of soils to remember that they are composed of particles, and that the phenomenon we think of as 'strength' is the sum total of many particle-to-particle interactions. At the simplest level, we could imagine friction between the particles. Then, the magnitude of the friction forces must reflect the normal forces between particles, and some overall measure of the intensity of these forces *on average* throughout the soil is required. But such a measure already exists: it is the direct or normal stress on the soil. Returning to the individual soil particles for the moment, what then is the effect of pressure in the fluid(s) filling the pores of the soil? Do these pressures affect the particle-to-particle contact forces?

If a section is taken through the soil, and the forces carried across this section by the soil grain contact forces is considered, it will be evident that the introduction of fluid pressures acting in the pores does have the effect of reducing the grain-to-grain contact loads. Hence the frictional resistance to movement must depend in some way on the numeric difference between the applied *total* stress, and the pore-fluid pressure. This difference, or the part of the normal stress which is *effective* in generating shear resistance, is termed the *effective stress*. Giving direct stress the symbol σ, and pore-fluid pressure u, then the effective stress σ' is given by

$$\sigma' = \sigma - u \tag{3.1}$$

If it is then found that shearing resistance is directly proportional to this *effective stress*, then we might write

$$\tau_f = (\sigma - u)\mu \tag{3.2}$$

or to be slightly more general, and to include for some attraction between the particles (perhaps at the molecular level) which gives shearing resistance even at zero effective stress, we could write

$$\tau_f = c' + (\sigma - u)\tan\phi' \qquad (3.3)$$

In the above, the opportunity has been taken to replace a *coefficient of friction* with the angle whose tangent gives that coefficient. This is more in line with conventions in geotechnics, and gives a parameter which is more readily appreciated than the numeric value of a coefficient of friction.

Indeed, we will see subsequently that the application of direct stress can itself affect the pore-fluid pressures. If the addition of $\Delta\sigma$ causes a pore-fluid-pressure change of Δu, then the above should be written

$$\tau_f = c' + (\sigma + \Delta\sigma - u - \Delta u)\tan\phi' \qquad (3.4)$$

The approach outlined above is adequate to explain in simplistic terms why the shear strength of soil is dependent upon effective stress. Unfortunately, the very simplicity of the argument can be its undoing. For instance, consider a single, spherical soil particle to be a 'free body'. From such elementary considerations as Newton's laws of motion, it will be seen that there can be no net couple on this particle: but that is precisely what is implied by the explanation outlined above!

What is confused in the above is the appearance of 'friction' *en masse* with friction at the particle level. If the soil were composed of spherical particles we would need a much better explanation than friction. What we have in fact is a set of forces acting on the particles which have resultants directed to react against the combination of the applied normal and shear loads—and

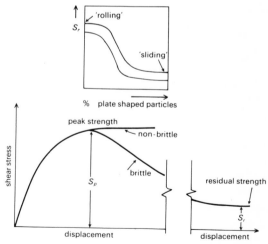

Figure 3.3 Influence of particle shape on residual strength. Important work by Lupini *et al.* has showed that a certain percentage of platy particles must be present for an aligned particle shear surface to form, the strength of which is controlled by interparticle friction rather than by particle mechanics.

interparticle friction is not *necessarily* an important element of this. These resultants, however, are as likely to be proportional to *effective stresses* in the soil mass as were the frictions in the original and simplistic account. For this reason, the use of 'angle of friction' is abandoned, and 'angle of shearing resistance' used instead.

It can be shown that the angle of shearing resistance in a granular soil is to all intents and purposes independent of the interparticle angle of friction in granular soil of rounded particles (Skinner, 1969). It is somewhat simpler to show that the angle of shearing resistance depends on such factors as density of packing (whether this is due to a more efficient arrangement of a singular particle size, or to the presence of a range of particle sizes in the soil), and to the shape and size distribution of the particles. The peak angle of shearing resistance is therefore a parameter which is likely to reflect the depositional origin of the soil, and its subsequent stress history (in that this can alter the state of packing of the individual grains), than on their fundamental interparticle friction characteristics.

3.3 Stress history effects: overconsolidation

Many soils have been subject to high stresses at some stage in their depositional history. Later, these high stresses may be alleviated or entirely removed through the erosion of overlying later sediments. This process is termed *overconsolidation*, and is an early stage in the lithification of a sediment. The terms 'soil' and 'rock' are used here in quite a different manner from that in which they are used in the pure geological sciences. For an engineer, geological materials are soil until they become so hard that they must be excavated with special tools, wedges, picks, pneumatic and hydraulic drills, or blasted with explosives. The overconsolidated (often Tertiary or Cretaceous in age in Britain) clays, silty clays and clay-shales therefore count as soils, not rocks, in this classification. And yet they are clearly different from the soft sediments of recent age. The latter would be classified as *normally consolidated* if they had never experienced higher stresses than those to which they are presently subject.

One major difference between the two, apart from consistency, or undrained strength, is the presence of fissures in the overconsolidated soils. The presence of these fissures, which are after all only one category in a wide variety of different discontinuity surfaces in the soil mass, are related to the overconsolidation process in some way, or rather to the stress relief element of it. Their genesis is by no means understood, but is no doubt due in some way to differential strains during the initial consolidation, and the later swelling as stress is removed. Fissures have a bearing on the shear strength of the soil, and this is discussed, together with aspects of the impact on shear strength of other discontinuities, in section 3.9 below.

A second major difference is the presence of higher lateral stresses in the ground which are left 'locked in' as vertical stresses are released. These then give greater stress relief effects if excavations are opened up in the ground.

Also, the overconsolidated soils inevitably have lower porosities than their normally consolidated counterparts. This leads to their having a stiffer behaviour in respect of deformations under applied load, and more dilatant behaviour under shear because of the denser packing of the particles.

Overconsolidation is normally caused by the removal of later sediments but can be caused in other ways. Tills, as an example, can be overconsolidated by the weight of overlying ice. Often, two or more tills may be found at a site, separated sometimes by laminated clays and silts. The lower till is invariably much more heavily overconsolidated than the upper. It may be mistaken for rockhead in some probings and borings, only to pose an unpleasant surprise when exposed in excavations. Occasionally, too, *drilling* in the lower till may be easy, and the two not discriminated. When bulk excavation is undertaken, the differences show up all too clearly. One should not, therefore, express too much amazement at finding fissures in tills.

Other sources of overconsolidation effects include desiccation. Salt-marsh deposits, and river alluvium on the flood plain, can be overconsolidated by the strong suctions of normal evaporation, and these may be supplemented by influence of plant roots. A rapidly accreting sediment may then have several strongly overconsolidated horizons in an otherwise normally consolidated profile.

3.4 Effect of drainage during shear

The porewater-pressure changes which take place in a soil under the application of the shearing stress have an influence on the effective stresses, and hence on the strength. Hence, if during a test efforts are made to eliminate any *changes* in the porewater pressure, all stress increments will be effective stress increments. This can be arranged in some tests by shearing at a low rate of strain, or by introducing additional drainage measures such as side filter drains in the triaxial test to shorten the distance that escaping water needs to travel. Such a test is termed a *drained* test. Note that the term 'drained' does not imply zero water content, nor does it imply zero water pressure, in this context (although in special cases, either or both of these could be true). What the term 'drained' means is that all porewater-pressure changes set up in the soil as a result of applied loading during the test are allowed to come into equilibrium with some external pressure source, and that no significant test-generated pore-fluid pressures exist during the shearing stage.

A drained test will yield effective stress-based shear strength parameters, and the applicability of these to field cases where the rate of loading is slow is obvious. Indeed, at normal rates of loading all granular soils, gravels and sands will be fully drained, and so will many silty soils. It is only when the soils are loaded extremely rapidly, as in an earthquake, or when they contain an appreciable clay fraction (appreciable in its effect: the percentage in weight or volume need only be quite modest) that *undrained* porewater pressures are important.

During a drained test, or in the field during the application of a drained load increment, the moisture content of the soil will change. It is easy to see that the application of direct stresses will usually cause an overall decrease in volume and the expulsion of water from the soil. Shearing stresses, on the other hand, can cause either a contraction (with further expulsion of water) or dilation (drawing water into the soil). To the layman it might seem paradoxical that soil might increase in water content during a drained test, but this is so.

At the other extreme, a test might be envisioned in which no moisture content change was permitted during shear. This would be termed an *undrained test*, and would be applicable to field conditions where a load was applied very rapidly, or to a soil of such low permeability that no water could escape or enter it in the normal timescale for construction operations. A contractant soil tested under such a procedure would shown an increase in porewater pressure, and a dilatant one a decrease. Naturally, measurement of the porewater pressure in this test would permit the extraction of fundamental shear strength parameters in terms of effective stresses, but the test results may equally well be analysed in terms of the total stresses acting on the specimen. If these are similar to those likely to act in the field, and are applied in a similar sequence (i.e. the correct *stress path* is followed) then the strengths are applicable without the intermediate vehicle of the effective stress shear strength parameters.

Other tests, intermediate between the two extremes are possible. These include pre-setting the state of effective stress in a sample by a drained stage (*consolidation*—section 7.8) followed by shearing without drainage. This would yield undrained strengths for a more complex stress path. An embankment, for example, which was completed and allowed to consolidate, but then loaded rapidly, might be analysed for stability under this rapid loading with the aid of such *consolidated undrained* test data.

Several samples preset to the same effective stresses would be insensitive to the exact combination of total stress (cell pressure in a triaxial test) and pore-fluid pressure. If they were saturated, the application of further all-round total stress would generate additional pore-fluid pressures, at the end of which the initial effective stresses would be retained unaltered. Subsequently, with the application of shearing stresses in an undrained fashion, the stress paths followed by all of the specimens (this including the generation of undrained pore-fluid pressures) will be nearly identical, and the specimens will yield identical strengths. Hence, a strength which is apparently independent of total stress (cell pressure or confining stress) which implies a zero angle of shearing resistance, and some value of cohesion.

If the soil specimens were partly saturated, the application of additional confining stress will change the initial effective stresses in the specimen, and there is little reason to suspect that they will follow the same stress path when sheared. Plotting the strengths obtained from this sort of test would then show some apparent angle of shearing resistance. The parameters obtained from undrained tests without a consolidation stage would normally be referred to by means of the symbols c_u and ϕ_u, and those from a consolidated undrained test as c_{cu} and ϕ_{cu}. In loosely applied terminology, the shearing strength obtained from an undrained test or consolidated undrained test on a natural soil specimen (or any other that was fully saturated) might be referred to as 'the undrained strength, c_u' even though strictly c_u is a *parameter* and not the undrained strength. Some authors use s_u for drained strength in an attempt to impose discipline in this.

It can therefore be demonstrated that the undrained strength of a soil is more a function of the porewater-pressure response than of fundamental effective stress shear strength parameters. Furthermore, the generation of porewater pressures, and the consequent undrained shear strength, is related to the stress path followed, and will be quite different for tests under increasing stresses than for those where average stresses decrease.

3.5 Progressive failure

What happens during the growth of a slip surface or slip surfaces as a landslide develops? It would pay to examine this in the context of the building of an embankment. To simplify the concepts, we will first consider the embankment and its foundation to be composed of materials which have identical physical properties. To simplify things further, the construction will be assumed to be proceeding at a rate slow enough to allow the dissipation of all porewater pressures, so that the argument can be pursued in terms of effective stresses but without necessitating simultaneous appreciation of porewater pressures. It would lead to identical conclusions, however, and to a broadly similar argument, if the construction were completely undrained. The treatment would then, of course, rely on the undrained stress–strain curves, undrained strength, and undrained brittleness.

Starting from a levelled platform, the construction of the embankment imposes stresses in the foundation. There are direct stresses, but the stresses that concern us directly are the shear stresses. As each lift of fill is placed, it also increases the shear stresses in earlier lifts adjacent to the sloping face of the embankment, as well as in the soil forming the foundation. These applied shear stresses represent the tendency of the soil slope to slide under the influence of gravity.

For low fill heights, the shear stresses are much less than the shear strength of the fill and of the foundation soil, and shear deformations are small. As the fill is raised, the ratio of shear stress to strength rises, and due to the non-linear relationship between stress and strain (Figure 3.1) deformations increase faster than the rate of mobilization of the available shear strength. Furthermore, because the stresses are not uniformly distributed throughout the soil, neither are the strains.

At some point in the construction sequence, increasing fill height will cause the application of stress equal to the shear strength in a part or parts of the soil. From then on, the behaviour of brittle and non-brittle soils diverges. Take first the non-brittle soil: the performance of this is less complex. Additional stress cannot be carried by soil elements already stressed up to their full strength—it must be carried by adjacent soil. This then accelerates the displacements in the slope. Observations of toe heave, or lateral displacement of the slope face will then reflect the shape of the stress–strain curve for this non-brittle soil. Ultimately, the whole of the soil mass will have mobilized its shear strength along a potential failure surface. Extra loading at the crest of the slope will then merely cause movement until the rising of the toe, and settlement of the head of the slide, bring the soil mass back into equilibrium with the shear strength acting along the sliding surface. It may prove impossible to effect an increase in the height of the embankment and the result of placing more fill at the top is simply to cause a lateral spreading of the embankment toes.

Even small movements will sufficiently flatten the slope to reduce the shear stresses and restore equilibrium. In principle, therefore, one would expect the displacements in a non-brittle soil to be small in magnitude. However, further fill placement will trigger additional movements, the magnitude of which will naturally reflect the added weight of fill, or the degree of attempted overstress.

Where brittle soils are present, a more involved process takes place. The bringing of any soil element to the peak of the stress–strain curve, or the mobilization of its peak strength, does not simply mean that that soil element cannot sustain further load: it means that that soil element must shed part of the load it is presently carrying, if it is to be strained further. This load shedding may cause adjacent elements to mobilize all of their strength in turn, and to the spread of a zone or zones within which the soil loses its load-carrying capacity.

Now that further movements must take place, additional losses of strength occur, and failure progresses until *sliding surfaces* have formed. It will be seen later that these sliding surfaces are formed where flat mineral flakes align in the direction of shear. The shear strength on the sliding surface is the residual strength and the soil mass must move into a position which brings equilibrium between the gravity forces in the moving mass of soil or rock, and the residual strength along the sliding surface. If the soil is of appreciable brittleness, relatively large displacements will occur.

The brittleness can be quantified by means of the *brittleness index*, I_b

$$I_b = \frac{S_p - S_r}{S_p} \tag{3.5}$$

as the drop in strength from peak (S_p) to residual (S_r) expressed as a ratio to the original peak strength (Bishop, 1967).

What is more, the rate at which these displacements take place must depend to a large extent on the shape of the load–deformation curve between peak and residual. A rapid drop in strength must induce more rapid displacements than a gradual reduction for instance.

Similar types of failure, either brittle or non-brittle, may take place in an existing slope where the porewater pressures increase, even though shear and total normal stresses remain constant.

3.6 Growth of slip surfaces: particle alignment

In many clay soils, shear surfaces may develop in response to shear strains and deformations. When completely formed, these slip surfaces are quite thin, perhaps a millimetre or so thick, and when separated present a polished, fluted or striated, *slickensided* appearance. This degree of polish (Figure 3.4)

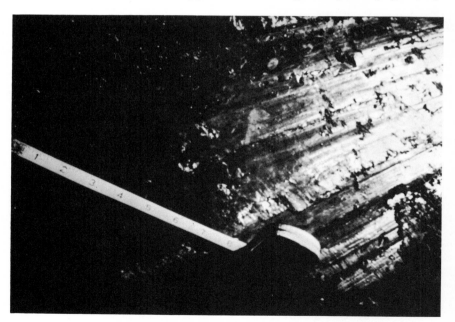

Figure 3.4 Sevenoaks shear surface exposed in a trial pit. The freshness of this shear surface is self-evident from its high degree of polish. This is often dulled in older slip surfaces (i.e. those which have not moved recently) by moisture migration.

is due to the reorientation of individual clay platelet-shaped particles in the direction of shear, and is quite clearly seen in thin petrographically examined sections when viewed under polarized light, although measurement of the orientation ratio is difficult because of the extremely localized concentrations of oriented particles. Such slip surfaces are quite readily produced in the laboratory: they begin to form at strains which are achievable in the triaxial or shear box test, but for full development need greater deformation, as in the ring shear test.

When found in the field, slip surfaces take two main forms. The first of these is typical of slides along bedding planes where there is a particularly weak horizon for instance, or where a change in the stiffness of the soil or rock causes a concentration of strain. These slip surfaces may be relatively planar

Figure 3.5 Herne Bay Miramar shear surface in sample. Even had this slip surface been lost between two sample tubes, the difference in texture above and below the slip surface would have made its location evident. Sliding was along the bedding.

over a large extent, and tend to be isolated, that is they occur singly. Of course, where such a bedding-dominated slip surface is formed, movements along it may be sufficiently great to bring down much disturbed and sheared debris, so that this *basal* shear surface appears to be accompanied by many minor shears, but close examination of these reveals that they have little to do with the major movement that is occurring. Figure 3.5 shows this in detail for a split 100 mm-diameter open-drive sample taken through the basal slip surface of the Miramar landslide in the London Clay of the North Kent coast. Barton (1973, 1977, 1984) describes a number of such instances in his accounts of the landslides which affect the clays of the Barton series in south Hampshire. Some shear surfaces and shear zones are illustrated in Figures 3.6 and 3.7.

The second form occurs where a thick bed of more uniform material is involved in shearing. Many slip surface elements can then be observed, each of which has the same general trend and is separated from its neighbour by anything from a few millimetres to several centimetres. In the London Clay landslides of the North Kent coast, for instance, the remnant shear surface which is left on the rear scarp of a fresh landslide is merely the surface on which *most* movement has taken place: many other minor slip surfaces can be found if sheets of clay are pulled from the face. These separate readily

Figure 3.6 Shear at Warden Point. This is a typical slickensided shear surface exposed by erosion in the toe of a mudslide.

Figure 3.7 Bedding-plane shear, at the level of the knife blade. Such shear surfaces are often quite thin, and usually strongly planar.

along the subsidiary slickensided surfaces, and the whole *shear zone* may be a metre or more in total thickness.

Such shear zones occur in many clays, but were observed and described in detail where they occur in the tectonically sheared clays of Northern India (Skempton, 1964). When seen in section, the individual shear surfaces run into each other and separate many times, separating the clay into lenticular masses. Such shear surfaces have undulations on a major scale, as well as on a minor scale like the striae of a bedding plane slip surface.

An interesting example of the second form of slip surface was encountered in a railway cutting at Uxbridge (Watson, 1956). At this site, a deep cutting into London Clay was excavated in 1902 for a suburban railway. In 1937, the cutting was widened to accommodate more tracks, and a gravity-retaining wall was built. In 1956, following disturbance to these tracks that suggested that a slide involving the retaining wall was under way, the slope behind the wall was excavated to restore stability temporarily while the wall was reconstructed and strengthened, and the faces of this excavation logged. A number

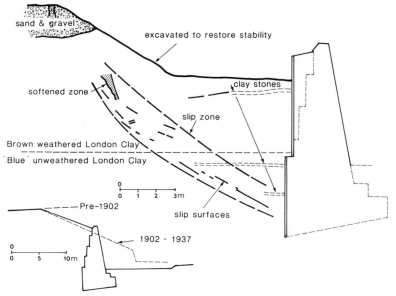

Figure 3.8 Growth of slip surfaces—Uxbridge cutting and retaining wall section. The growth of slip surfaces was arrested by the remedial measures undertaken to stabilize the wall. It appears that had the wall not been constructed, a slide in the cutting, down to the level of the weathering contact, might have occurred.

of shear surfaces were found trending in the same direction, but none were continuous (Figure 3.8). The implication is that the prompt action of the railway engineers had prevented the growth and coalescence of these individual surfaces into a single major slide surface along which more rapid movement could take place.

The Uxbridge cutting penetrated the weathered (oxidized or 'Brown London Clay') into the unweathered material beneath ('Blue London Clay'). Elsewhere in the London railway network (Gregory, 1844; Henkel, 1957) slope failures in a cutting similar to the original one at Uxbridge have main slip surfaces which go down to, and run along, this weathering junction. This phenomenon occurs perhaps in response to the change in stiffness of the clay at this level. It is tempting to see in the Uxbridge data some slip surface segments in the weathered clay which appear to be seeking out the weathering junction. It is possible that these were developing in response to the original cutting excavation. The construction of the retaining wall when the cutting was widened effectively prevented this mode of failure from happening, so failure was delayed while slip surfaces grew under the foundations of the new wall.

3.7 Residual strength

The formation of a slip surface involves the orientation of platy-shaped clay mineral particles in the direction of shear. This orientation can be seen in thin sections under a polarizing light in an optical microscope, both in laboratory sheared specimens and in those taken from the field. It also eliminates some of the attributes of the soil: the effects of previous stress history, for example. The formation of a shear surfaces is the creation of an entirely new fabric in the soil; other, pre-existing, fabrics are wiped out. Later, we will see that this has an important bearing on the way that shear strength tests to measure the residual strength may be carried out, but for the time being, it is an important fact to be kept in mind.

What is essential in the formation of these shear surfaces is that the strong orientation of flat particles at a voids ratio in equilibrium with the applied stress brings the sides of these particles into frictional contact with each other. The strength along such a surface now truly represents interparticle friction, in a way that the simple theories have always argued. It is for this reason that angles of shearing resistance for the residual state in clays are so much lower than the peak angles. In clays even a small percentage of the particles aligned across the direction of shear will have a disproportionate influence on the apparent strength of the soil mass.

It is often found that the residual strength envelopes of clay soils are slightly curved, convex upwards on the conventional τ–σ' plot. There may also be a slight residual cohesion implied by the test, but this could well be simply a failure to take the tests down to a low enough normal stress, or to tiny frictions in the apparatus.

With some soils too, the residual shear strength can be sensitive to the rate at which the soil is sheared. It would seem that the viscous drag forces on the particles as they are sheared tend to prevent their complete alignment. Lupini *et al.* (1981) term this 'turbulent shear' to distinguish it from the 'sliding shear' at low strain rates. There is a threshold for the strain rate below which the effect is negligible, and it is the low strain rate residual strength which is most applicable to slope stability problems. Almost all soils will show this effect at extremely high rates of shear, which is an immediate explanation for the absence of slip surfaces down the sides of piles driven into clay soils.

Furthermore, slip surfaces can only form in soils which have a sufficient proportion of plate-shaped particles to allow the formation of a highly orientated zone. Experiments conducted using mica–quartz mixtures and silt–clay mixtures show that there is a critical percentage of 'clay' particles below which slip surfaces cannot form. Where a slip surface cannot form, the strength remains that of the remoulded, normally consolidated, soil.

3.8 High-sensitivity soils

Some soils demonstrate significant *sensitivity* or undrained brittleness when sheared. Perhaps the classic examples of such soils are the quick clays of Scandinavia and the eastern provinces of Canada. These comparatively recent marine clays, laid down under water during recent glaciations and uplifted as the earth's surface responded to the removal of the ice load have been subject to the leaching out of their salt content by fresh groundwater flows.

Now under saline conditions, the individual clay particles are not separate, but are attached to each other forming *flocs* or large aggregations of particles, the effect of salinity is to encourage the bonding of the particles in this way. Subsequently, the leaching process removes the bonding agency, but leaves the particles in their original structure (Bjerrum, 1954). Naturally, without the interparticle bond, such a structure is like a 'house of cards', needing only a slight stimulus to collapse. The soil then behaves much as a fluid, perhaps of very low viscosity, and it can flow for long distances at comparatively high speed. Some thixotropic regain in strength is possible if the now-fluid soil comes to rest, particularly if the salt content is increased artificially. This is due to the presence of some particle bonding via the thin layer of water molecules adsorbed on to the surface of each clay mineral particle.

Work in Scandinavia has shown that the quick characteristics of these recent marine sediments only occur at salt contents of less than one gram per litre of the pore water, and so the effects of leaching of the pore fluid by natural flows of ground water can be significant in generating localized zones of quick clay in otherwise merely sensitive recent marine sediments.

Many recent sediments do exhibit some sensitivity or loss in shear strength when remoulded. It is now quite conventional when measuring the shear strength of such soils *in situ*, with a vane test for instance, to fail the soil completely with several revolutions of the vane, and then measure its re-moulded strength. The mechanism for this loss in strength, and its eventual partial regain, is similar to that discussed above for quick clays, but is of lesser magnitude.

The collapse of a quick clay may be seen as analogous to the liquefaction of loose non-cohesive soils under shear or vibration. Loose structures in such soils may be the result of placement: spoil heaps of mica and quartz arising from the china-clay industry in Cornwall have a structure which closely mirrors the fabric of the marine silty clays of Scandinavia, but do not require any leaching of the pore water to give them a metastable structure. Other waste tips constructed of loose debris also exhibit metastable fabrics. Examples cited by Bishop (1973) include fly ash, colliery discard and limestone waste.

Some natural soils can be subject to liquefaction under repeated loading, seismic shaking being an important cause of this. They are predominantly loose granular soils of sand size. Larger-sized debris from mountain avalanches and rockfalls can also become fluidized and behave as a soil of extremely low shear strength (Voight, 1978; Shreve, 1968; Kojan and Hutchinson, 1978).

Hsu (1978) cites the 'grain flow' mechanism of Bagnold (1954) to account for the mobility of avalanche flowslides which he terms 'sturzstroms', using a term coined by Heim (1932). In this mechanism, the individual particles are sustained in motion not in a state of zero effective stress by a pressurized pore fluid, but rather by a continuous set of impacts from other surrounding particles. This is rather like the picture we have of atoms in a fluid, but on a macroscopic scale. Just as in the pore-fluid-supported mechanism, there is a threshold point at which the mechanism is no longer self supporting: in this case it is when the kinetic energy transferred by the impacts allows some particles to come to rest, and the chain reaction stops.

3.9 Effect of discontinuities on shear strength: soils

The most important discontinuity in a soil is obviously a shear surface. On such a surface, the shear strength is appreciably reduced, and the reduced strength plays a dominant role in the future performance of that soil. There are, however, other important discontinuities such as fissures and joints which affect the bulk shear strength. It is possible to make a simple checklist of these discontinuities in the following way:

Macro feature	Scale of hand specimen
Fault	Shear surface
	listric surface
Joint	Fissure
Stratigraphic unit	Bedding, lamination
	varve

Engineers are often extremely casual in their use of these varied terms.

In a series of careful tests on soils containing discontinuities of various sorts, Skempton and Petley (1967) showed that the strength along joints and fissures which do not show evidence of previous shear deformation closely approximated to the shear strength of normally consolidated specimens of the same soil. In effect, the formation of the discontinuity had robbed the soil of the cohesive element of its shear strength; the angle of shearing resistance being relatively unaltered by either the overconsolidation or the formation of the discontinuity. In contrast, the bulk strength might show a considerable cohesive element, although the magnitude of this would reflect the degree of overconsolidation, and the spacing of the discontinuities.

Skempton (1970) uses this finding to argue a justification for the empirical observation that the lower-bound field-mobilized shear strength in a number of back-analysed case records of first-time failure in London Clay is described well with this strength envelope. In line with Terzaghi's description of the softening of stiff, fissured, overconsolidated, clays, Skempton terms it 'the fully softened strength'. Although he cautions against a wider application of the rule that the long-term lower-bound strength in any deposit might also be taken as the fully softened strength, or in simple terms c' between zero and about $1\,kN/m^2$, and ϕ' equivalent to ϕ'_{peak}, this finding is supported by other studies, in the Lias for instance.

The findings of the tests on sheared discontinuities, both by Skempton and Petley, and much later work, showed that these tended towards the residual strength. In his Rankine Lecture of 1964, Skempton put forward his 'residual factor' approach. This is much more in line with a progressive failure concept, since the residual factor is closely allied to Bishop's 'brittleness index', and this provides a sounder basis for an understanding of the pheno-menon than does an empirical finding, regardless of the usefulness of the latter for geotechnical engineering. The residual factor is the percentage drop in field-mobilized strength from the laboratory peak (measured on small, intact, specimens) to the residual, and reflects the proportions of each mobilized along the whole sliding surface at the instant of failure.

3.10 Effect of discontinuities: rocks

If anything, discontinuities in rocks play an even greater role than they do in soils. With the exception of *slip surfaces*, the influence of most disconti-nuities in soils can be adequately represented by taking the appropriately modified bulk parameters. In rocks, however, this is rarely the case, and the discontinuities each have their own individual effect.

This poses considerable problems in continuum analysis: stress or seepage, but comparatively little in limit equilibrium analysis (most stability analysis techniques). Once the critical discontinuities in a rock slope have been identi-fied, and the influence of these on the modes of failure which need to be checked, the procedures are identical to those used in the stability analysis of soil slopes (Chapter 5).

In view of the special significance of discontinuities in rocks, and their effect on shear strength, it is worth reviewing these properties. Factors which will influence the strength along a discontinuity in a rock mass, and its influence on stability, are as follows:

The planarity and smoothness of the surfaces of the discontinuity. A smooth, planar surface will obviously yield a much lower strength than an irregular and rough surface

The orientation of the discontinuity to the horizontal, and to the rock slope face

The openness of the discontinuity—whether it is a *joint* or a *fissure*, the weathering of the discontinuity surfaces and the possible infill of the joint with weaker material (clay gouge, etc.). The effect on strength may be direct, with the infill *cementing* the discontinuity, or replacing its shearing resistance by that of a soft clay for example; it may also be indirect. An example of indirect influence is the control on the seepage pattern, good or poor drainage and so on, and the effect of the resulting joint-water pressures on strength

The interaction of a particular discontinuity with others of a similar or different kind. That is, with other discontinuities of the same *set*, or of other *sets*. The commonplace stability problem of two intersecting joints daylighting in a rock face is an example of this, as also is the joint-controlled toppling of columnar rocks and unravelling of slopes.

3.11 Other factors affecting shear strength

Those changes in shear strength observed when soils are sheared at different strain rates are the result of two separate effects. Firstly, at high strain rates, drainage may be impeded, and the most significant effects result from pore-fluid-pressure changes in the soil. These often blanket the effects of strain rate on the interaction between the particles, which is most often what is meant when strain rate effects are mentioned.

The decrease of shearing strength at very low strain rates is often of little importance, and is catered for in routine engineering design by the adoption of appropriate factors of safety overall.

In the measurement of shear strength, the sample size can also have a major effect. The usual argument put forward to explain this is that a large specimen is more likely to contain joints or fissures than a small one. Even if the discontinuities are closely spaced, so that a small sample is intrinsically as representative as a large one, the effects of sample preparation in the laboratory will break up the smaller samples so that they cannot be tested, and only those samples which are unrepresentative in being stronger than average will survive to be tested.

Both drained and undrained strengths of normal sedimentary soils are anisotropic. In the simplest analysis, this is because the effective stress fields are directional, and thus the normal stresses on failure surfaces are different on planes in the soil at various inclinations. This is an effect which is additional to any basic anisotropy resulting from the bedded structures in the soil.

Finally, the shear strength of a soil depends to a large degree on the effective stress path followed to failure. Specimens where the mean effective stress increases have shear strengths larger than those where it decreases.

All of these factors need to be taken into account when applying laboratory measured shear strengths to field conditions.

4 Measurement of shear strength

4.1 Measurement of the peak strength of soils and rocks

Problems involving initially unsheared soil or rock must be approached using peak shear strengths where this peak shear strength is the maximum shear stress that can be sustained under a given set of conditions. Those conditions, however, are usually far more important than intrinsic properties of the soil or rock material!

Typical of the factors to be considered in soil is the movement of moisture during shear, and the consequent behaviour of the pressures in the soil pores. In rocks, the presence and orientation of joint patterns, their condition, and their spacing is a similarly important factor: the bulk strength of a rock is of much less importance than the strength of its joints.

Consider first the shear strength of soil, and the effect of porewater pressure. The shear strength could be measured under drained or undrained conditions, and so an apparatus which permits the control of drainage is essential. Either the provision of physical barriers to the flow of the pore fluid, or the provision of a means of accurately setting the deformation rate, can constitute such control. (Setting a strain rate faster than one threshold will ensure that failure takes place without significant drainage; setting it slower than a second threshold will have the opposite effect, and drainage will take place as fast as the excess porewater pressures are generated.) Secondly, there is the choice between forcing shear to occur on certain surfaces in the specimen, or in allowing it to fail on its weakest surface. Such a freedom may be more apparent than real, if the stress conditions applied do not permit more than a restricted range of failure mechanisms: by way of example, a cylindrical specimen with axial compression could not fail on a planar surface at right angles to the axis, no matter how much weaker this plane was than the rest of the soil sample.

There are two main types of apparatus available for routine use in the soil mechanics laboratory which meet some or all of these requirements. They are the shear box (or direct shear apparatus), and the triaxial apparatus. Other devices (laboratory vane, cone penetrometer, ring shear machine, plane strain and independent stress control 'triaxial' cells) are also used.

Soil mechanics shear boxes can be used to measure rock joint strength, but most soil mechanics apparatus is too flimsy to allow intact rock specimens to be used. A special triaxial cell developed by Hoek and Franklin (1968) is widely used for this purpose.

4.2 Laboratory strengths

In the soils laboratory it is conventional to measure the peak shear strength of soils in either a shear box (direct shear device) or in the triaxial apparatus. The former apparatus has the advantage that it is simple in concept and the test is easy to perform, although it is difficult to control drainage during the test except by judicious choice of a strain rate, and the stress conditions at failure are complex. The triaxial apparatus, on the other hand, is excellent for its control of the flow of water into or out of the specimen, but is relatively complex and much more subject to experimental problems during the test.

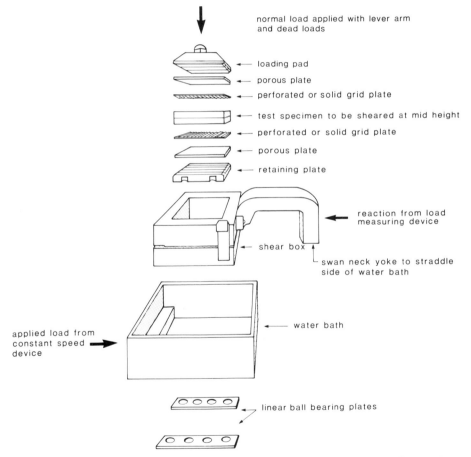

Figure 4.1 Laboratory techniques for peak strength: shear box. In this exploded diagram the principal features may be clearly seen. In the apparatus manufactured by Messrs Wykeham Farrance, this is mounted on a frame similar to that of the ring shear machine shown in subsequent illustrations, and incorporating a counterweighted lever loading device to provide the normal loading.

Figure 4.1 shows an exploded cut-away diagram of the principal features of a laboratory shear box apparatus. These are supplied in a variety of sizes, the commonest being 60 mm square in plan, with 100 mm-square boxes available, often for use on the same loading frame. Shear boxes of up to 300 mm square are used on occasion, but they normally require much larger loading frames and are comparatively rare in practice. The depth of specimen which can be accommodated is of the order of 25 mm in the smaller boxes, limiting the particle size which can be allowed in the test specimen. 300 mm shear boxes have the capacity to accommodate gravelly soils.

Normal loads on the shearing surface are applied through a lever loading system, or direct through a weight hanger, and shearing forces through a motor-driven ram. The shear force transmitted through the specimen is reacted by a proving ring or load cell, and the deformation of this is a source of two problems. Firstly, the strain in the specimen is reduced from the ram movement to a lesser value, which slows the test (possibly allowing some drainage in an undrained test), and secondly, allowing a lever loading system to move from the vertical with a consequent application of a component of this 'normal' force in a direction which helps the specimen resist the applied shearing loads. A direct hanger system (or the hydraulic 'bag' sometimes used in the 300 mm shear box where the normal load is too large to be transmitted through the knife edges required by a lever system) is preferable in this respect.

Volumetric changes during shear can be measured approximately by recording vertical movement of the loading platen.

Since shearing is constrained to occur along the plane of separation of the two box halves, the apparatus is ideally suited to testing specimens containing discontinuity surfaces if these can be aligned in the box. Drained tests are made by running the apparatus slowly, and undrained tests are normally only possible in clay soils where drainage is slow and can be largely prevented in a quick test.

The reader is referred to the many books on soil testing, which include Head (1985), for more details of the test procedure. The triaxial apparatus commonly used in British laboratory practice takes one of two forms: the conventional triaxial cell (Bishop and Henkel, 1957) and the later Bishop–Wesley stress path cell (Bishop and Wesley, 1974). The loading in the former is by a ram at a constant rate of displacement which drives the cell upwards with the axial load reacted via a 'frictionless' ram by a proving ring, or directly by a submersible load cell. As in the shear box, the true rate of deformation is affected by the flexibility of the load measuring device, and mechanical arrangements to allow for this have been made.

In the Bishop–Wesley cell, axial loads are supplied by a hydraulically powered ram in the pedestal of the cell, which does not therefore require an external motorized loading frame. The supply of fluid to the ram can be controlled to allow constant rate of strain or load stress paths to be followed, or for that matter, any combination of axial and cell pressures desired.

Lateral loads, applied by the fluid in the cell, simulate the lateral restraint in the ground.

Drainage in these tests is from the end(s) of the specimen, with porous disks connected via leads to outside pressure source. Tests can therefore be run against a back pressure to keep the soil saturated. Measurement of pore-fluid pressures is by connections similar to those for drainage, or with miniature probes inserted directly into the side of the specimen. The latter give more rapid porewater-pressure response. The soil moisture is kept separate from the cell fluid by means of rubber membranes. Additional drainage by filter paper drains up the sides of the specimen and connected to the porous end filter drain, can accelerate the drainage.

Again, the details of test procedure are extensively covered in the literature, notably by Bishop and Henkel (1957), and also by Head (1984). An unconfined test apparatus, testing axially loaded soil cylinders without the lateral restraint of fluid pressure in a cell, is sometimes used, but its main use is to measure the undrained strength of soil specimens in the field.

Two other devices are used to measure the undrained shear strength of cohesive soils. These are the laboratory vane, and the penetrometer ('fall cone test' or 'pocket penetrometer'). The vane is a multibladed device, pressed into the soil and rotated. Shearing takes place around a cylindrical surface, and the maximum torque (measured at the instant of failure) can be correlated with the undrained shear strength of the soil. Penetrometers operate in a way analogous to bearing capacity tests. The fall cone relies on the penetration of a pointed and weighted cone into the soil. The distance that the cone falls under gravity can be correlated with the shear strength in weak soils—this test is therefore much used in Scandinavia to obtain the shear strength of soft, sensitive sediments. The pocket penetrometer is a spring-loaded piston pressed against the soil surface. The load carried by this at the instant of failure (recorded either as a maximum load, or the load at a set penetration) also gives an indication of the shear strength. The pocket penetrometer, and a hand version of the laboratory vane, are used in routine classification tests: the former in stiff, and the latter in soft, soils. Such simple aids are useful both in the laboratory and in the field.

A major problem in soil testing is to obtain the effective stress dependency of the shear strength. This is in the form of the parameters controlling the shape of the shear strength–effective stress envelope. Tests therefore have to be carried out at different normal stress levels to 'probe' this envelope. The test procedure can either utilize the *same* specimen for all of these probes, or fresh specimens for each. The former procedure is termed a *multistage* test, and will only yield reliable results if the material has a *non-brittle* stress–strain relationship. In the application of multistage test procedures to *brittle* soils and rocks, the full mobilization of shear strength at one normal stress damages the load-carrying capacity for other normal stresses, and an incorrect envelope is obtained.

The other major difficulty in soil testing is the provision of appropriate,

D

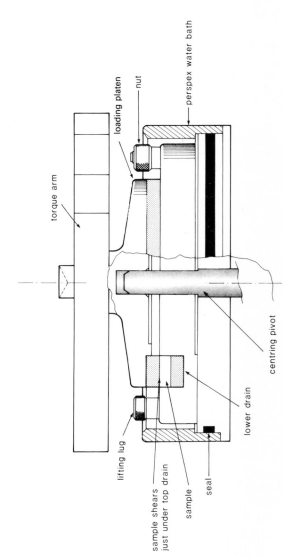

Figure 4.2 Ring shear apparatus: detail of test cell. Shearing takes place through the sample immediately under the top loading platen.

low, effective stresses on potential failure surfaces. This is particularly important where the failure envelope is inconveniently not a straight line on the $\tau - \sigma'$ plot. In the triaxial test, the provision of low fluid pressures is particularly difficult. Loads for low-stress tests can be course be measured with sensitive load cells, but careful laboratory management is required if these are not to be inadvertently overstressed. Larger test specimens, although difficult to handle and manipulate in the laboratory, are a partial answer to the problem, as the loads transmitted through these increase proportionately to the sample size. The major use, then, of the large-size shear box test (s) is in examining low normal stress behaviour.

Normally, in tests on soil, loading is *strain controlled*, and is measured by the reactions to the load transmitted through the soil. This permits the test to follow (albeit approximately in many cases) the post-peak behaviour of stress–strain or load–displacement curve. Stress-controlled tests are possible, but are much rarer. In test to determine the peak strength of rocks, the opposite holds true. The reasons for this are twofold. In the first place, the peak strength and stiffness of intact rock specimens is likely to be high, so that the mechanical arrangements for constant rate of deformation are difficult to make, and the second the rapid drop of strength post-peak and the high brittleness render the precise shape of the stress–strain curve of purely academic interest and little practical significance.

Rocks can be tested unconfined in concrete compression machines to obtain the uniaxial strength. There is an analogue to the soil mechanics triaxial test, the Hoek–Franklin cell in which a confining load can be applied, and also a shear box where shearing and normal loads are applied with hydraulic jacks.

The strength of rock *discontinuities* can be obtained in either type of apparatus: in the Hoek–Franklin cell by aligning the discontinuity obliquely to the specimen axis, and in the shear box parallel with the shearing direction.

4.3 Field tests

Field tests can be divided into classification and identification tests, and those intended to provide an *in-situ* estimation of the strength.

Simple identification and classification tests use the hand vane and penetrometer apparatus already described. Variations on these devices are used for field tests in weaker soils.

A vane apparatus is often used, either in boreholes, or merely jacked into the ground. In the case of more sophisticated devices, the vane is retained within a protective sleeve, and is jacked into the ground either without a borehole, or ahead of the end of the borehole. The cruciform vane blades are pushed from the end of the sleeve to make the test, and are withdrawn on completion. The sleeve not only protects the vane blades, but also alleviates

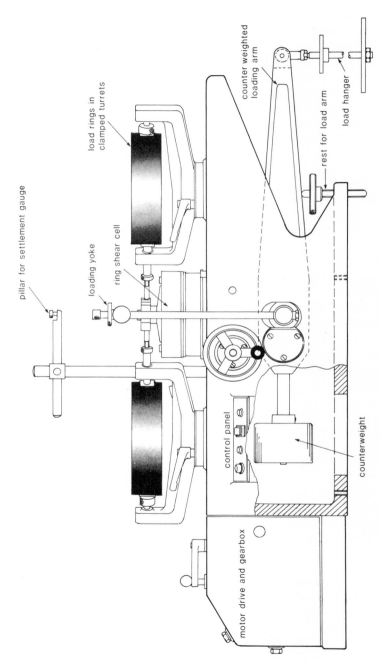

Figure 4.3 Ring shear apparatus: elevation and general layout. Ordinarily, this apparatus is mounted on a combined stand and small table, but is equally at home on a workbench.

friction on the rods if a borehole has not been used, Other, simpler, borehole vane apparatus omits the sleeve.

Some static and dynamic penetrometers can be interpreted in terms of a field peak shear strength. The standard penetration test, and the blow count for any open drive sampler is, in principle, convertible to a peak, undrained, shear strength in a cohesive soil, and some correlations with ϕ' in sands also exist. The interpretation of such data is subject to a considerable latitude, and may only be used as an index, or as a last resort.

Pressuremeter and *in-situ* plate-bearing tests are commonly used to assess soil stiffness ('elastic' modulus). The pressuremeter relies on the lateral expansion of a fluid-filled rubber sack in a borehole to provide a stress–deformation picture, and the plate-loading test relies on the settlement of a vertically loaded plate in a pit or at the base of a borehole. Either can yield a shear strength when interpreted using the correct theory, but neither is *commonly* used for this purpose in slope stability investigations in Britain.

Field shear boxes are sometimes used in studies of the shear strength of the surface layers of a soil where sample disturbance and size effects could be crucial. One such apparatus is described by Chandler *et al.* (1981). The procedures, although of necessity more involved than the laboratory test, follow much the same general principles. Other field shear boxes are described by Hoek and Bray (1980) and are used to measure in *in-situ* strength of rock joints and intact rock masses.

4.4 Measurement of residual strength

Formation of a shear zone or shear surface implies disruption of an existing soil fabric, which perhaps reflected the deposition or weathering of the soil initially, and its replacement by a fabric dominated by the effects of shear. In clay soils, for instance, the shear fabric will contain a very large proportion of clay particles oriented with their long axis in the direction of shear. This results in the development of fluted, polished or *slickensided* surfaces.

Naturally, the alteration of fabric causes changes in the strength properties of the soil, usually a loss in strength. The strength of the soil with its original fabric is termed the *peak* strength, and its strength under large deformation conditions in a shear zone or on a shear surface is referred to as the *residual* strength. This latter term can cause problems in comprehension in those parts of the world where deeply weathered rocks give rise to *residual soils*, and alternative use of the terms *ultimate* or *large-deformation* strength is advantageous. Soils with a large loss of strength from peak to residual are termed highly brittle (Bishop, 1967). In accordance with established UK practice, the term residual strength will be used in this book despite the possible source of confusion.

The measurement of residual strength is directly of significance to soil stability problems where the existence of shear zones is known or suspected.

and an appreciation of the residual strength of a particular soil is also of importance where the possibility exists of non-uniform strain mobilization causing progressive failure (Bishop, 1967). Tests involving the residual strength demand the use of specimens containing shear surfaces, or a laboratory technique which forms them during the test.

It is my experience that the measurement of residual strength is poorly understood in the geotechnical engineering community at large, whereas the measurement of peak strength is much more widely appreciated. The remainder of this chapter will therefore concentrate on the measurement of the residual strength of clay soils, which is perhaps the least well understood of all soil-strength measurement techniques.

4.5 Strategies for obtaining the residual strength

There are two main decisions to be made in planning a programme of measuring residual strength:

whether to use selected specimens containing naturally formed shear surfaces, or to form the shear surface in the laboratory

whether to examine the relationship between normal stress and shear strength on several individual specimens, or to carry out a multistage test.

There are thus four permutations of the selections from these two choices. It will be found that the experimental technique will be different for each combination selected. To a certain extent, too, the results in terms of the measured shear strength will be dependent on the selection. Note that the problems of obtaining representative specimens, of selecting an appropriate strain rate both to ensure full drainage and to eliminate viscous strength, and to guarantee sufficient accuracy in the test procedure are all additional factors, but then they are common to all shear-strength testing.

The formation of a shear surface in a soil replaces the original fabric, with all its locked-in 'stress history' with a new fabric. This new fabric of strongly aligned clay mineral particles will, if fully formed, be unaffected by further strain or deformation. Multistage tests, or 'probes' at different normal effective stresses, are then not an undesirable expedient as in the case of peak strength measurement, but are perfectly admissible. Indeed, by doing a multistage test, it is possible to ensure that the same material is tested at each normal stress. In turn, a more regular residual shear-strength envelope is usually obtained.

4.6 Specimens with natural shear surfaces

Specimens containing shear surfaces can be sampled from pits, shafts, tunnels, open faces and boreholes. Borehole samples are the least satisfactory for testing: without elaborate orientation procedures it is not always possible to be certain of the correct direction of shear, and splitting a sample to log it usually destroys its value for testing.

A technique which may usefully be employed in the sampling of slip surfaces in clay soils is to drill one borehole with continuous samples, split and examine each specimen, and then redrill alongside the first borehole to the requisite depth specifically to sample for the slip surface. Where the slip surface is horizontal the technique is simple, but where the slip surface is inclined, allowance must be made for its dip, or the second borehole located along the strike of the slip surface from the first. This can be done approximately by reference to the morphology of the slope, or in a slightly more exact way by locating the slip surface in three closely-spaced sampling holes initially, and then treating the finding of its dip as an application of the three-point problem of structural geology.

Even when care is taken with the above procedures, it is possible to miss the shear surface in the subsequent sampling. It is extremely frustrating to sample a perfect specimen in the sample tube used for examination, and then to be unable to repeat the process to obtain a specimen for testing. A possible technique to economize on drilling costs is to examine only the ends of each sample still in its tube to decide the most probable location of slip surfaces. Then samples from above and below the selected one(s) are extruded, split and examined. On the basis of the findings of that examination, specimens are retained for extrusion in the laboratory.

Without some clear indication, such as a colour or lithological change across a slip surface and showing in opposite ends of the sample tube, it is difficult to be sure that the desired feature is present. It is equally difficult to find it for testing. The sample must be extruded, and thin slivers cut from its sides. A gentle shearing action on these slivers with the fingers can sometimes reveal the slip surface location. Test specimen(s) can then be cut from the body of the sample.

Shear surfaces are a soil fabric element and are as liable to disruption during the sampling process as are all other fabrics. Sample disturbance (e.g. turned-over edges where the sample was cut) not only affect the flatness of the surface, but also smear the feature and make it hard to find subsequently. This is of importance in shear box testing where the shear surface needs to be accurately aligned with the box separation plane, or in triaxial testing where the shear surface orientation must be known if the stresses on it are to be evaluated. It is noteworthy that the alignment of the shear surface with the joint in the shear box has to be so after the normal stresses have been adjusted to the test values, and allowance must be made for consolidation or swelling that may take place.

The use of the triaxial test with oriented slip surfaces is discussed by Chandler (1966), and the shear box techniques by Skempton and Petley (1967). The triaxial test technique does have the advantage that precise orientation is less critical, and the exact orientation of the slip surface can be measured retrospectively in the failed specimen.

A further difficulty with borehole samples is that it is almost impossible to orientate the slip surface correctly in the apparatus. The exception to this rule is where the slip surface has a steep dip, and the relative directions of shear are obvious. Where the slip surface is sub-horizontal, the best that can be done is to get the striations lined up with the direction of shear, and to take pot luck with whether or not the direction of shearing in the apparatus is the same as in the field.

Samples taken from trial pits or shafts do not suffer from this problem. There is normally so much of the shear surface exposed (contrast, for example, Figure 3.4, 3.5) that its position in the block sample can be clearly identified and marked with short slivers of wood—orange-sticks are ideal. Then the top of the specimen is marked, and with it the direction of shear, both before and after coating the specimen with wax to keep its moisture in during transport back to the laboratory. Alternate layers of paraffin wax and mutton cloth for reinforcement are ideal, although the resulting cover is difficult to remove in the laboratory. An aluminium foil sheet closest to the specimen makes the cover easy to remove. It is a mistake to have the wax too hot: just above melting point is best so that the wax congeals immediately it is in contact with the block sample.

4.7 Laboratory-formed shear surfaces

It is axiomatic that the formation of a shear surface destroys all earlier fabric in the soil. This is a remoulding action, and it is therefore illogical to demand tests on 'undisturbed' specimens, preparation of which can be arduous. Remoulded specimens are adequate if a residual strength alone is required. However, if the stress–strain behaviour as the soil passes from peak to residual is to be observed, remoulded specimens will not do. In this case only a 'remoulded' peak strength could be measured, and an underestimate of the full brittleness would be obtained.

A shear surface may be preformed (often termed 'precut plane' tests) or it may be formed by the test procedure itself. Shear boxes and triaxial apparatus can only give small strains and the preforming of a shear surface is to overcome this limitation, rather than to offer some other, positive, advantage.

Shear surfaces have been preformed by cutting a sample with cheesewires or knives, the separated halves being polished on glass plates to simulate slickensiding. Inevitably this cannot be completely effective, and even the most carefully prepared specimens do display some brittleness in shear as the plate-shaped clay mineral particles on the slip surface adopt a more complete orientation. Ideally, the test specimen would be consolidated to the desired normal stress in the shear box, and then be removed to form the slip surface. This then means that when reconsolidated prior to shearing, the shear surface is located as closely as possible to the plane of separation of the shear box halves. It will be found, however, that for practical reasons it is only worth following this sequence when testing soft and compressible materials.

Shear surfaces do form even at comparatively low strains. These have a low level or particle alignment, however, and truly residual conditions are not reached in most test procedures. Using a shear box, the two specimen halves can be racked backwards and forwards past each other until some lower bound strength is achieved. Observations of shear load v. displacement for each pass do tend to show both decreasing brittleness and decreasing strength with increasing number of reversals of strain direction, but it is not always easy to decide when a 'residual' has been reached. Arbitrarily defined values of total movement and/or percentage change in shear strength in consecutive reversals are often used as a criterion on which to judge the attainment of residual conditions.

With this reversal technique, there are several variants in the experimental procedure. These include measuring the load-deflection relationship both on the first pass and on the reverse direction for each stage. Alternatively, measurements of shear strength can be made in one direction only, bringing the box halves back to the starting position by hand, or at a relatively fast rate using the drive of the machine. My personal preference is for the latter method, with time allowed for re-consolidation after each reversal. However, reversal tests are rarely carried out in my laboratory as I prefer to use the ring shear apparatus when determining the residual strength of clays.

It is only by the use of a torsion or ring shear device that we can achieve large enough strains to produce real residual conditions in the laboratory, starting from an initially unsheared specimen (Bishop *et al.*, 1971). This facility to achieve almost unlimited strain frees the user of a ring shear apparatus from a number of the constraints that beset soil testing generally. For example, since a test can be of whatever duration the user chooses, and sooner or later it must be fully drained at whatever strain rate is chosen, one of the most tricky aspects of the measurement of peak strength, or even of residual strength in a limited strain apparatus, is eliminated completely.

4.8 Experimental procedures for the ring shear test

Imposing a shear surface on a clay soil completely changes its initial fabric in the vicinity of the shear surface. There is therefore little or no merit in attempting to preserve such an initial fabric or structure in the test specimen, *unless it is specifically desired to examine the process of shear surface formation*. Remoulded specimens are quite adequate for residual strength determination alone.

It is usually convenient to remould soil specimens at moisture contents of the plastic limit or less. After all, the shear surface formation process is a result of soil brittleness: and this may only be manifest at moisture contents lower than the plastic limit. When wetter, the samples can extrude very easily from their container under consolidating loads. Bishop *et al.* (1971) have described a mechanically elaborate ring shear machine in which the gaps between the upper and lower rings may be opened and closed. This can accept fairly wet specimens. Shearing in this device takes place through the mid-height of the specimen. In contrast, the simple ring shear device described by Bromhead (1979) needs a drier sample to prevent early extrusion, and shears close to the upper loading platen. It is often found that smaller strains are required to develop the residual strength in this machine, since shearing takes place through the strongly remoulded upper part of the specimen.

The following description is specifically directed to the use of the simpler device, which is in daily use at Kingston Polytechnic (Figure 4.2), and in a number of academic and commercial establishments elsewhere in the United Kingdom. There are rather fewer examples abroad, although there are users throughout much of the English-speaking world.

Soils may be kneaded into the sample container with the fingers or rammed into position with a wooden spatula. A short length of dowel makes a convenient rammer. Final trimming flush with the surface of the container is done with a palette knife. This has the added benefit that it begins the process of orientation of the mineral particles close to the eventual shearing surface. It is not detrimental that the beginnings of an orientated zone close to the top of the specimen are made, so that an undisturbed strength determination is ruled out, since all ring shear devices are totally useless for measuring the peak strength of the soil. This is because with the different strains at the inside and outside radii of the specimen, a form of progressive failure across the specimen takes place. This just about defies analysis, and all that is obtainable by way of a peak strength from this apparatus is some weighted average between a partly initially sheared remoulded strength and the residual.

Initial shearing of the specimen may be performed by setting the machine to a high rotation speed or by use of the manual control. Such high rates of shear cause substantial extrusion in the simple device (and also in the Imperial College/Geonor device if the confining ring gaps are left open), and although a shear surface is usually formed by this rotation, it may not have the desired properties as a result of undrained porewater-pressure generation, or because viscous interparticle drag forces have prevented the formation of an ideal, strongly particle-orientated, low-strain-rate, shear surface. A period of slow shearing is usually required to form this feature correctly.

It has been found convenient for routine tests at Kingston Polytechnic to set up a test in the late afternoon, and to observe only the early stages of this shearing, leaving the sample to shear unattended overnight. If the test

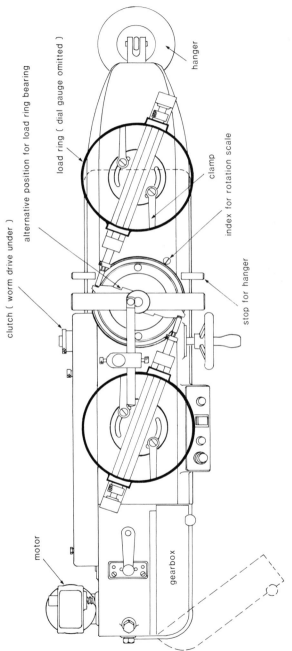

Figure 4.4 Ring shear apparatus: plan (Bromhead, 1979). The torque is measured by the proving rings. These are mounted in turrets so that they can be swung out of the way for sample preparation (the test cell is removable). Two stops are provided on the torque arm to allow easy setting of the radius at which the proving rings act.

Figure 4.5 Ring shear apparatus. The simplicity of the design is evident from this view.

rate is set to allow perhaps three revolutions to take place, then residual conditions will be approximately achieved. A number of tests have been automatically recorded throughout this shear surface formation stage, and they all show that towards the end of the allotted time, the decrease of torque through the specimen with further deformation slows down to an almost imperceptible level. One is left with the problem of ascertaining that a true residual has been reached; solutions to this are discussed below. Also, the selection of a strain rate for the remainder of the test is different to that for other soil tests: the procedure is explored in the following section.

4.9 Selection of a strain rate

When testing clay specimens it is customary to perform *drained* tests. A drained test is one in which any porewater pressures generated by the application of the shearing load are allowed to escape from the test specimen. It does not necessarily imply zero porewater pressure in the sample. The time over which drained tests must be run is normally calculated according to a theory proposed by Gibson and Henkel (1954). In this theory, the shearing load, or rather the resultant generated porewater pressures, are introduced at a constant rate. Concurrently, water escapes from the specimen via the drains, and at the end of the loading period, only $x\%$ of the total applied porewater pressure remains in the worst place (usually the middle) of the specimen. For example, where $x = 5\%$, this time for a doubly-drained soil sample is given by

$$t = \frac{20h^2}{3c_v}$$

Constants equivalent to the $\frac{20}{3}$ factor for other drainage conditions and degrees of remaining porewater pressure are given by Bishop and Henkel (1971).

A strain rate is then chosen such that the test is run for this time to achieve the estimated failure strain. With the small specimen thickness of the simple ring shear apparatus, this yields very short test times, for instance with a c_v of 1 m²/year and a total specimen thickness $2h$ of 5 mm, t is about 22 minutes. Accordingly, our standard procedure for measurement of a residual strength demands that the torque transmitted through the specimen remain sensibly constant for at least this amount of time. When it has done so, it is possible to be sure that the drainage process is complete. It will normally take about half an hour to mobilize the reaction on the proving rings of the apparatus, so that for load stages after the first (in which the shear surface is formed initially) about an hour in total is required.

The virtually unlimited 'strain' capacity of a ring shear device permits a completely novel approach. When the shear-induced porewater pressures dissipate, they cause changes in the shear strength of the soil. This reflects in the torque transmitted through the apparatus. Thus when a constant torque is found, for example over a period of half an hour or so, or for a time equivalent to t in the above formula, this demonstrates the complete elimination of transient porewater-pressure effects. A similar argument may be applied to the dissipation of normal load-induced porewater pressures: these may as well be allowed to escape during the shearing stage. The use of separate consolidation stages gives little additional benefit. Indeed, our standard procedure is to start the shearing as soon as the normal load is applied.

4.10 Achieving adequate total deformation

A major source of uncertainty is in ensuring that large enough strains have been obtained for the complete development of the residual strength. Bishop *et al.* (1971) recommend plotting the load–deformation behaviour on a semi-logarithmic base (deformation on the logarithmic axis), since this is a severe test of the data. However, for rapid determinations, for example on a commercial basis, the time requirements for this are excessive. A more satisfactory alternative is to strain the sample until it *appears* that the residual has been reached, and then to carry on with the next normal load stage. When a full sequence of normal loads has been applied, the total load is reduced and the strength is re-evaluated at the initial normal load. Provided that the strength is comparable with the original measurement, it is safe to assume that the additional deformation of the later load stages has not further reduced the soil strength, and by inference, residual *was* achieved earlier. More than one point may be so checked.

Early in the development of the simple ring shear device, it became obvious that there was a conflict between several of the requirements that the apparatus had to meet. For instance, the specimen thickness had to be small, so that drainage during shear was rapid, and test times could be reduced. However, the small thickness of the sample meant that the monitoring of any consolidation stage undertaken was inaccurate. The procedure for observing the torque and using that as an indirect measure of the progress of consolidation was adopted to overcome what was seen to be a problem. It was only in retrospect that it proved to be a positive feature of the technique. Similarly, the early tests were done utilizing the semi-logarithmic plot of torque *v.* time. It was found that the time taken in the first load stage, and the attention it demanded, ruled out the measurement of residual strength with the ease and rapidity intended in the design.

The idea of returning to the first point on the residual strength envelope was then tried. This was principally to see if there was an overconsolidation effect on the residual strength. It was found that the results were frequently erratic, but, with further strain, the earlier residual strength could be obtained. In the course of exploring the erratic behaviour, the reasons for it were discovered, and a systematic experimental procedure found which was more in line with the initial concept of a simple and rapid method of residual strength determination.

4.11 Effect of strain rate on residual strength

Since in the ring shear test it is possible to achieve full drainage at whatever strain rate is chosen, merely by extending the test, it is practical to assess the influence of strain rate on the drained residual shear strength. The effect has been explored by Lupini *et al.* (1981) who show that increases in strain rate can cause increased strength in the soil, with some brittleness becoming apparent when the strain rates are subsequently reduced. This is explained as a result of the disruption by viscous drag forces of the strongly orientated zone produced under slower shearing conditions. At strain rates below a threshold value the influence of strain rate is negligible. This strain rate threshold has been found to correspond to a speed of $1°/min$ in a $100\,mm$-diameter shear apparatus for most clay soils, and a much lower speed of typically $0.048°/min$ provides a safety factor on this as well as allowing a convenient test programme schedule in the laboratory.

High rates of shear may be applied deliberately or inadvertently in ring shear testing. Should the normal load be reduced, for instance, the energy stored in the torque measuring system will cause rapid deformation of the specimen with consequent changes to the nature of the shear surface. This rapid deformation was taking place in the direction of shearing, and so is not related to the effect of a reversal where a change in the direction of shearing upsets the particle alignment. By way of analogy, this can be visualized by thinking of stroking a cat's back. Do it repeatedly in the same direction, and the fur becomes orientated and smooth. However, if the fur is then stroked the 'wrong way', it stands on end. That is the effect in a reversal test. Consider then the effect of stroking the cat the 'right way', but at high speed. Static electricity is then generated so that the fur refuses to lie flat.

It was this effect that was causing the erratic behaviour of test specimens when unloaded to try to reproduce the residual strength of the first load stage. Reduction in torque is therefore essential before relieving the normal load. Notwithstanding this, some brittleness may still be introduced into the slip surface by unloading, but not so much that the routine of about an hour's further straining is not adequate to re-establish the drained residual strength to an appropriate experimental degree of accuracy. The mechanisms for this secondary effect are not fully understood.

A rapid assessment of the strain rate sensitivity of the soil can be made by switching off the drive. If the specimen can hold the applied torque, then the test has been made at a rate to which the soil is insensitive, and may be deemed satisfactory. A significant loss of torque should give rise to concern that the strain rate was too fast. About an hour is ample time to check this.

Typical test results from a ring shear test are given in Figure 4.6.

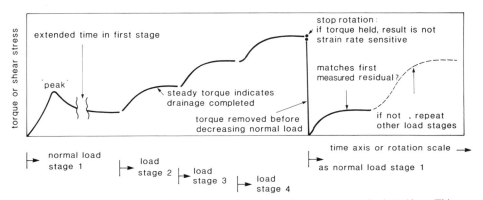

Figure 4.6 Ring shear test record: torque v. rotation. A complete test sequence is plotted here. This procedure enables checks to be made to ensure that the test is strain-rate insensitive, fully-drained, and to a high enough total strain.

4.12 Choice of testing method

Both direct shear tests on naturally occurring slip surfaces and ring shear tests have been shown to produce residual strength parameters comparable with the results of back analysis (e.g. Bromhead, 1978; Hutchinson *et al.*, 1980; Chandler, 1984; Bromhead *et al.*, 1985) and with each other (e.g. Bromhead and Curtis, 1983; Hawkins and Privett, 1981; Newbery and Baker, 1981; Warren, 1985). However, both techniques appear to yield lower results in general than, for instance, reversal or precut plane tests. Some apparent correlations of these with field data appear in retrospect to have been merely fortuitous. All this data contradicts Skempton's results (1985), which would seem to indicate perfect agreement between tests on natural shear surfaces

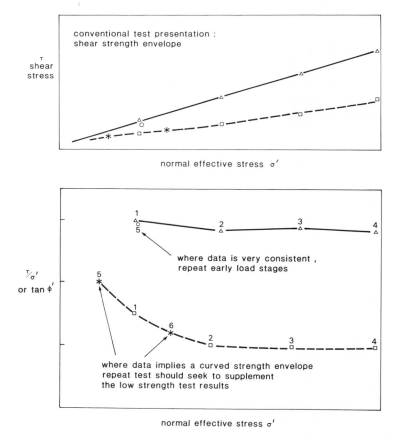

Figure 4.7 Ring shear test of Figure 4.6 plotted as a conventional residual strength envelope. It is often found that the plot of ϕ' against normal stress shows up poor data better than the conventional Mohr–Coulomb plot.

and back analysis, with the ring shear tests giving $1-1\frac{1}{2}°$ degrees lower for the residual angle of shearing resistance.

Strength tests on natural slip surfaces will always be expensive, as it is difficult and costly to obtain the test specimens, but it will always be worth the effort and expense to obtain them for the confidence that they bring. On the other hand, there is little doubt that ring shear testing is more convenient and cost-effective if a number of specimens need to be tested. My belief is that, rather than view ring shear testing as a cheap alternative method, it would be better to see it as a way of producing more data within a given budget. It is never satisfactory to have a mere handful of laboratory results; for confidence one needs to have tests in all the major clay strata on a particular site, and the only way to do this both systematically and economically is with ring shear testing.

It is often stated by the ill-informed that 'smear' types of apparatus do not measure the correct strength, because the top platen skids over the top of the soil. This is not so. All that is required is for the slip surface to form one or two clay platelets below the platen soil contact, and the residual strength will be measured. Provided therefore that the platen is 'rough' compared to a clay mineral particle, the residual will be measured. This is amply proved through comparisons of the results with other apparatus and with field measurements. One might also have some reservations about its use with larger particle sizes, but the trials seem to indicate that quite good results can be obtained. We have for instance tested crushed mica schist and found agreement with shear box tests, and have also tested sand and gained the appropriate ϕ_{cv}.

4.13 Discussion of the applicability of laboratory-measured strengths

Earlier in this chapter it was remarked that peak strength and its measurement were far better understood in geotechnical circles than residual strength. This is due to a number of factors, including:

apparatus for measuring peak strength is commonplace in industrial laboratories as well as research establishments, and is in daily use

many peak strength tests are *undrained* and hence are quick and cheap, leading to their routine use

residual strength as a concept is relatively recent, and is viewed as being applicable to 'landslides' and not as an aspect of behaviour with far wider ramifications.

Furthermore, the view is prevalent that simple peak strength measurements in the laboratory genuinely represent field strengths, whereas residual strength measurement is in some way still 'experimental'. In fact, the opposite is true!

The essence of peak strength measurement in the laboratory is to test a representative soil sample which preserves its fabric. Peak strength testing is beset with major problems:

disturbance, which usually *decreases* the measured strength

failure to follow the correct stress path in the test, which can either decrease or increase the measured strength relative to the field

unrepresentative sampling, which *increases* the strength relative to that which is likely to be operative in the field

poor testing technique, leading to partial or inadequate drainage in 'undrained' or 'drained' tests respectively, so affecting the measured strengths

failure to take into account progressive failure in brittle soils (e.g. by doing multistage tests on them) so that their behaviour is modified by systematic changes in the soil fabric as the test progresses

errors in choosing strain rates, which cannot later be rectified.

In comparison, residual strength tests are measuring a soil property which is largely independent of the stress path followed, and in the ring shear apparatus it is possible to allow for, or rectify, many of the factors listed above because of the unlimited strain capacity of the machine.

5 Principles of stability analysis

5.1 Assessment of stability

The forms taken by landslides are many and varied, so much so that an attempt to predict mathematically the nature of developing instability and its progress is beyond present-day capabilities. What we do observe, however, is that many forms of instability commence, or progress, by sliding along surfaces within the soil or rock mass. Many big rockfalls start with an initial shear failure and sliding phase before the debris entirely parts company with the parent rock mass; many landslides remain in contact even though they move considerable distances. Thus it is that for most engineering purposes (and indeed for many others), simple sliding models can suffice. They even appear to be applicable to slow-moving flows of debris, although not perhaps to flowslides and avalanches.

It is not only as a result of the computational difficulties that attend attempts to model real soil and rock behaviour that engineers have to resort to highly simplified sliding models for slope stability: even if it were possible to use a complete theory, and developments in electronics made the analysis by computer almost without cost, the many independent material properties required to define such behaviour would be extremely costly to obtain from the field, and this factor alone would rule out the technique for much routine work.

Simple sliding models fall into the category of limit equilibrium methods. Take, for example, a single mass of soil or rock resting on an inclined plane. This plane in the rock mechanics case could be a joint, or a fault. It will have some frictional strength properties, and in addition could have some strength independent of the normal load (cohesion). In the soil mechanics case it could be a bedding plane, or even just some arbitrary plane on which it is suspected that failure will take place. Resisting this destabilizing force is some part of the available shear strength on the sliding surface. Unless the block is on the verge of sliding (or is in fact doing so without acceleration), then the disturbing force is not identical with the shear strength. The force that resists motion, however, is equal to the destabilizing force. It is termed the *mobilized* shear strength. The ratio between the actual strength available, and that mobilized, gives an index of relative stability called the *factor of safety*.

The main force which tends to cause movement of the block is principally the resolved component of the weight of the block acting down the plane.

This could be supplemented by other forces from a variety of sources: structural loads, inertial loads from blasting or seismic shocks, and so forth. Forces which resist movement may also arise from resolved weight components, structural loads, etc., but are mainly due to the cohesive and frictional strength elements of the soil. In order to evaluate the frictional strength of the soil, an estimate of the normal stresses on the sliding surface must be obtained. This is simple for the single sliding block, as the normal force on the slip surface can be calculated along with the other unknown forces by simple resolution. With the more common, curved, slip surfaces this is no longer possible, and the so-called *method of slices* is used.

In the method of slices, the soil mass above the slip surface is divided into wedges or slices. Conventionally, this subdivision is by vertical lines, but this is by no means vital for the method, nor is it necessary to have the slices all of the same width. The simplest approach would then be to take each slice in turn, and to calculate its factor of safety in isolation: the factor of safety for the whole series of slices could then be the average. Numerical problems of sign would arise if the slip surface had a rising portion at the toe; to avoid this all the resisting forces are summed, and divided by the sum of all of the disturbing forces. This gives a better average value for the factor of safety, and resolves the problem. Indeed, in some cases there are rational arguments for doing it this way round.

Take for example a slip surface that forms the arc of a circle in section. Instead of defining disturbing and resisting forces for each slice, it would be better to define moments about the centre of rotation of the sliding mass, and to term these disturbing and resisting moments. These moments can be simply shown to be proportional to the summed disturbing and resisting forces assumed previously. Other theoretical simplifications arise from the use of circular arc slip surfaces (or 'slip circles') and in the following, all of the ideas will be developed in terms of slip circles. In much of the literature, the moment of the disturbing forces is termed 'the overturning moment'. That convention will be followed here.

5.2 The $\phi = 0$ and conventional methods of analysis

Simplest of the slip circle methods is the $\phi_u = 0$ method (Skempton, 1948). In this, the soil strength is assumed to be purely cohesive. This simplifies the calculation of the maximum available resisting moment since it can only be the sum of the cohesive strengths multiplied by the areas or lengths over which they act and the radius of the slip circle (Figure 5.1), viz.;

$$M_r = R\Sigma l c_u \tag{5.1}$$

Note that the strength need not be the same on every segment of slip surface: if the slip circle passes through two or more strata, then the slices should

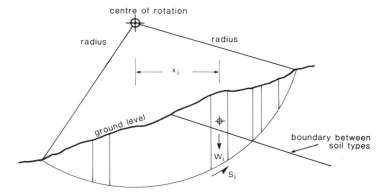

Figure 5.1 The $\phi_u = 0$ method. The distribution of shearing resistance around the slip surface is defined by the shear strengths available, and hence the analyst need assume nothing about the internal force distribution.

preferably be arranged so that each segment of the slip surface passes through but one of these layers.

In comparison, the overturning moment is likewise simply defined: it is the sum of slice weights multiplied by the distance of their centroids from the centre of the slip circle, measured horizontally:

$$M_o = \Sigma W x \qquad (5.2)$$

These two can be expressed as a ratio to yield the factor of safety:

$$F = M_r / M_o \qquad (5.3)$$

It will be seen that dividing through by R, the radius of the slip circle, gives the common form of this expression (since $x/R = \sin \alpha$):

$$F = \Sigma c_u l / \Sigma W \sin \alpha \qquad (5.4)$$

Now it is worthy of note that this equation does not retain any explicit trace of its slip circle origins. Indeed, it is identical (but for the summation signs) to the equation for a single slice sliding down an inclined plane with only a cohesive resistance. This might well then be used as an argument for the use of this expression for slip surfaces of all shapes. The writer has found this to be a reasonable suggestion when the results of this equation are compared to more complex methods, not inducing relative errors in the calculation of factor of safety by more than 5–8%.

Also no account needs to be taken of the forces between the slices, since these are dependent variables. The distribution of shear stress on the slip surface is a simple function of the distribution of shear strength: the *interslice* forces cannot influence this in any way. With frictional shear strength,

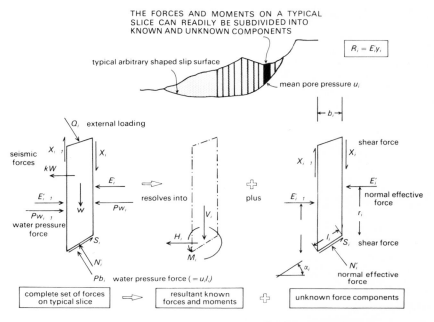

Figure 5.2 Forces on a typical slice for an effective stress analysis.

however, the converse is true, and a correct estimate of the interslice forces is essential to a correct determination of the normal force on the slip surface, and hence on the shear strength for each slice.

This is harder to do in practice than it seems at first. Shown in Figure 5.2 are the forces that act on a single typical slice. In order to solve the problem, it is essential that the analyst understands the internal structure of the slope, the soil strength properties and unit weights of the different strata, and the shape of the slip surface. He also needs to be able to define the porewater-pressure distribution, at the very least along the slip surface, but ideally everywhere in the soil mass, if the shear strength parameters are defined in terms of effective stress. It is also likely that external loadings will also be known (many structural loads will be negligible compared to gravity loading on the soil mass and can safely be ignored).

Thus the weight, the external loading and the water pressure resultants, both on the slip surface and on the sides of the slices, are known quantities. They could, if desired, all be resolved into two orthogonal forces and a single moment. (This is done in one of the later methods to be discussed below in the context of arbitrarily shaped slip surfaces.) Removal of the known forces, and their known lines of action, leaves as unknown only the following, which for a set of n slices have in total $6n - 2$ values:

normal effective force on base of each slice	n
shear force on base of each slice	n
position of line of action of normal forces	n
interslice normal effective forces	$n - 1$
interslice shear forces	$n - 1$
position of line of action of normal forces	$n - 1$
factor of safety	1
total	$6n - 2$

In contrast, there is for each slice a relationship between the normal force and the shear strength, and the three equations of equilibrium (sum of horizontal forces, sum of vertical forces, and sum of moments all equal to zero). There are thus $4n$ equations.

With only one slice, the problem is trivial; with more than one slice, it is indeterminate. All of the methods currently available introduce assumptions to counter the indeterminacies. Sometimes, with the introduction of too many assumptions, the problem becomes redundant, and additional unknown parameters are brought in to compensate. The simplest of the methods merely ignores the interslice forces, removing $3n - 3$ unknowns from the balance. It must therefore also violate some of the equations of equilibrium, and thus the number of equations balances the number of unknowns.

In this *conventional* method, sometimes referred to as the *US Bureau of Reclamation method*, no account is taken of the interslice forces. It thus provides at least a datum against which more satisfactory methods may be measured.

The normal effective force on the base of each slice can be found by resolution to be $W \cos \alpha$, the subtraction of the porewater pressure resultant along the slip surface (equal to the mean porewater pressure multiplied by the length of slip surface, ul) gives the normal effective force N'. Hence, the maximum available strength on this section of slip surface, S, is

$$S = c'l + (W \cos \alpha - ul) \tan \phi' \qquad (5.5)$$

The maximum available resisting moment is the sum of these S forces multiplied by the radius R, and the overturning moment is as before, so expressing the factor of safety in ratio form and dividing by the radius:

$$F = \frac{\Sigma\{c'l + (W \cos \alpha - ul) \tan \phi'\}}{\Sigma W \sin \alpha} \qquad (5.6)$$

This equation is fairly simple to solve, but has been found to yield conservative results (lower than actual factors of safety) especially where the slip surface is deep, or where the porewater pressures are high. In both of these cases the fault lies with the neglect of the interslice forces. For deep slip surfaces the rising portion at the toe of the slip would give additional support to the back of the slide if these were considered properly: in the case of high pore-water pressure, a loss of effective stress, and hence of strength in part of the slip surface, arising from the high porewater pressures, would transfer additional load to another part of the slip surface.

5.3 Bishop's method

The problems with the conventional method were first described by Bishop (1955), who brought recognition to an improved method. Referring to Figure 5.3, which shows a polygon of forces for a typical slice, resolving the forces *vertically* and then rearranging the equation, the normal effective force N' is found to be

$$N' = \frac{W + \Delta X - l(u \cos \alpha + c' \sin \alpha/F)}{\cos \alpha + (\tan \phi'/F)\sin \alpha} \tag{5.7}$$

It should be noted that the resolution is done in this way to avoid explicit reference to the horizontal components of the interslice forces (E), for reasons

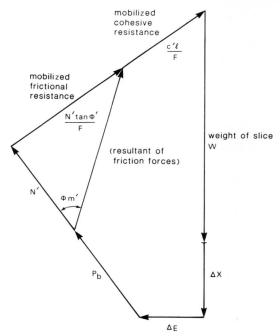

Figure 5.3 Polygon of forces on a slice for an effective stress analysis. ΔX is the difference between the interslice shear forces, ΔE is the difference between the normal components.

that appear later. This normal effective force can replace that assumed in the conventional method in the computation of the maximum available strength on the appropriate segment of the slip surface (equations (5.5) and (5.6)). Now, this resulting equation can be simplified, principally by bringing all the terms in the resisting force summation over the common denominator, but also by dividing through by the radius to yield

$$F = \frac{\Sigma\{c'b + (W - ub + \Delta X)\tan\phi'\}\dfrac{\sec\alpha}{1 + \tan\alpha\tan\phi'/F}}{\Sigma W\sin\alpha} \tag{5.8}$$

Values of F, and of ΔX for every slice, that satisfy this equation, give a rigorous solution to this problem. As a first approximation, take ΔX to be zero for every slice; this leaves only F in the equation unknown. Unfortunately, no amount of algebraic manipulation will bring both F values together; the equation must be solved in some other way. A solution scheme usually used for this is in fact suggested by the very form of the equation.

Examine the form of the expression

$$\frac{\sec\alpha}{1 + \tan\alpha\tan\phi'/F} \tag{5.9}$$

and suppose that it is evaluated using a value of F much larger than applies for this particular slip. Both $\tan\alpha$ and $\tan\phi'$ are likely to be less than 1 in the vast majority of cases. Thus it is the value of 1 that dominates the denominator of this expression, and when the result is evaluated, it will be found to be too large, but not excessively so, despite the starting value of F being much bigger than the correct value. When the F on the left-hand side of equation (5.8) is evaluated it is closer to the correct value than was the starting value. This can be the first of a series of iterations, the result of each being taken as the starting value for the next. A high degree of accuracy can be obtained with only a few iterations, regardless of the starting value.

A similar argument applies when the starting value is taken lower than the correct result, except in cases where the chosen value leads to the 1 no longer being the chief numerical influence on the magnitude of the denominator. Where that is so, numerical instability sets in. The writer's preference is to evaluate the factor of safety for the conventional method, multiply it by about 1.2, and use that as the start point for the iteration: most of the terms have to be evaluated anyway, and this gives a good starting point.

In Bishop's paper, he outlines the method of re-including the X forces. This is based on different additional resolutions from the basic force polygons and the two sums:

$$\Sigma\Delta E = 0 \tag{5.10a}$$

and

$$\Sigma\Delta X = 0 \tag{5.10b}$$

The procedure is to take sets of E and X forces that satisfy these expressions, and then to re-evaluate the factor of safety. Sets of both E and X forces have to be taken, so that for instance, a position of the line of action of the E forces (say at $\frac{1}{3}$ of the slice height) can be obtained. This leads to an order-of-magnitude increase in the complexity of the equations, for surprisingly little refinement in the factor of safety. Bishop noted that feasible solutions were obtainable with markedly different distributions of X forces (i.e. with alternative positions for this line of thrust), but that all seemed to give factors of safety in a range of about 1%. The reasons for this later became clear, but at that early date all that could be said was that there was little merit in the additional complexity of the full method, and accordingly, Bishop recommended the simplified equation (5.8) for routine work. Some writers use the term 'Bishop's simplified method' for this equation, but others apply that description to the conventional expression (which first occurs in UK literature in Bishop's paper). To avoid confusion, the preferred title (indeed, it seems always to have been called this at Imperial College, where the method originated) is 'Bishop's routine method'.

The horizontal interslice forces are implicit in the derivation of this equation; only the vertical force components have been neglected.

5.4 Application of computers

It was readily recognized that even early in the development of modern electronic digital computers, great savings of effort could be made by doing these repetitive computations by machine. In a typical earth dam, several hundred of these slip circle analyses, each with its own iterative solution, might have to be made. Little and Price (1958) report the first of many computer solutions for slope stability. They used Bishop's routine method. At that time, this was thought to be the most complicated program ever written for a computer of British origin.

A blind application of this routine equation may give rise to some numerical inconsistencies in the results. This is pointed out by Whitman and Bailey (1967). For certain combinations of soil properties, slice dimensions, and factor of safety, both the numerator *and* denominator of expression (5.9) may take zero or negative values, giving rise to zero, negative or infinite implied normal effective forces. A typical case arises when a thin layer of highly frictional granular soil overlies a slope of much weaker material, for which the overall factor of safety is low. The conflict in this case occurs in the slice at the toe of the slope (Figure 5.4) where the combination of high magnitude, negative sign, α and high magnitude ϕ' are incompatible with a low factor of safety.

Whitman and Bailey suggest checking for such cases, but it is better to recognize their occurrence when collecting data, and simulate these layers

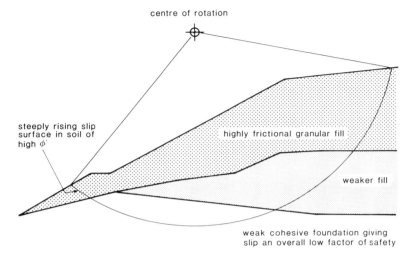

Figure 5.4 Conditions which give rise to problems with Bishop's method (after Whitman and Bailey, 1967.) The rising segment of slip surface at the toe in material with a high ϕ' value is not compatible with the overall low F resulting from the majority of the slip surface lying in a weak $\phi_u = 0$ material.

with soil of a nominal shear strength independent of friction, but commensurate with the levels of effective stress present. Where effective stress tension is implied, it reverses the direction of the shear forces on the slip surface, and leads to errors of a conservative kind in calculating F.

It is essential in this, and in all other methods of slices, to take sufficient slices to adequately represent the shape of the slip surface, and the distribution of forces around it. A minimum of 30 slices is recommended.

5.5 Spencer's method

Although the complete method proposed by Bishop satisfies all of the conditions of equilibrium with respect to forces and moments, and takes into consideration all of the components of the interslice forces, the routine method does not. Why this could therefore be so was eventually explained by a further extension to the theory by Spencer (1967). Taking the resultant of all the interslice force components *on a single slice* to be Z he derived:

$$Z = \frac{c'b/F + (W\cos\alpha - ub\sec\alpha)\tan\phi'/F - W\sin\alpha}{\cos(\alpha - \theta)\{1 + [\tan\phi'\tan(\alpha - \theta)]/F\}} \tag{5.11}$$

in which θ is the angle with respect to the horizontal of this resultant.

For equilibrium of the whole mass, the vectorial sum of the interslice forces and their moments about the centre of rotation on the slide must be zero, hence the following three expressions:

$$\Sigma Z \cos \theta = 0 \qquad (5.12a)$$

$$\Sigma Z \sin \theta = 0 \qquad (5.12b)$$

$$\Sigma Z \cos(\alpha - \theta) = 0 \qquad (5.12c)$$

(the third of these expressions assumes that the slice widths are small, so that the assumption that the force Z acts through the centre of each slip surface segment is approximately true.) With one further simplification, that both α and θ are constants, these three equations degenerate of two, matching the two unknowns F and θ.

The easiest solution scheme is to evaluate F from each equation taking a range of θ values, and to find the intersection point of the two functions. Spencer describes in some detail a method for this.

An important finding of this work was that the factor of safety given by the moment equation varied only a little with increasing θ values. For the $\theta = 0$ case, it is identical with Bishop's routine method, as might be expected. In contrast, the equation derived from force equilibrium was very sensitive to θ. This, then, is the key to the relative accuracy of Bishop's routine method: it is soundly based on the equilibrium of moments. These are much larger in magnitude than the forces, and so satisfying moment equilibrium brings about near satisfaction of force equilibrium. The reverse, however, is not true. If moment equilibrium requirements can be satisfied, even with a simplified assumed interslice force assumption, a better solution is obtained than if force equilibrium alone is achieved.

5.6 Graphical wedge method

So, having seen that the Bishop and Spencer solutions rely heavily on the equations of moment equilibrium to give an accurate result for the factor of safety, it is appropriate to ask of solutions that ignore the equilibrium of moments and concentrate on force equilibrium: what does this alternative offer? What is found is that there is a sacrifice in precision, and an increase in uncertainty about the result, but a gain in calculation simplicity. If the moment condition is discarded, a solution to the equations of force equilibrium can even be obtained graphically. In these days of readily accessible computer power, it is difficult to recall how it was when the average engineer had no computational aid other than a slide rule or a set of logarithm tables. A graphical solution then offered significant advantages over an equation. Even today, a graphical solution can give the engineer greater insight into the meaning of his computations than any tabulated list of numerical results.

Paramount among the force equilibrium methods is the graphical wedge method. The basis for this is the polygon of forces that can be drawn for

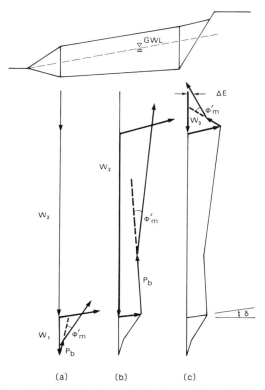

Figure 5.5 The wedge method. The development of the force polygons for the wedges is shown in the diagrams from left to right in very much the same way as they would be produced graphically by an engineer working at a drawing-board.

each slice in turn, once the *directions* of the interslice force resultants are known. Take for example the three-wedge slide shown in section in Figure 5.5. The lowest wedge, at the toe of the slide, has a weight W_1; this is drawn as a vertical vector (part (a) of the figure). There is no interslice force at the very toe of the slope, and so from the bottom of the W_1 vector can be drawn the vector relating to forces on the slip surface. The only one of these that is known at this juncture is the water-pressure resultant, so this is drawn next, parallel to the normal to the base of the wedge and to a scale length.

Suppose now that the direction of action of:

(a) the resultant of the interslice force components

and

(b) the resultant of the shear and normal (effective) forces on the slip surface were both known. Then these two directions could be drawn on the polygon of forces. Where they intersected would define the *magnitude* of both of them, and subsequently, the individual component forces could be recovered if so desired.

Suppose that the inclination of the resultant of the interslice forces is taken as known. Presumably, it must be at some angle lying between 0 and ϕ'. The angle of the resultant of shear and normal effective forces is known also: it is the mobilized angle of shearing resistance. Hence, if ϕ' is known, and a value for F is estimated, it is possible to complete the diagram.

The second polygon of forces, for the middle wedge, is completed in a similar way. This time, however, there is a common interslice force, and the end of this, rather than the bottom of the weight vector, is taken as the starting point for vectors of forces acting on the slip surface ((b) in the figure). It is convenient to draw the two diagrams as one (c) rather than separate, preventing errors in transcribing the interslice force.

When the last polygon of forces is drawn, it will be found that a small 'interslice' force is required at the head of the slide to make the polygon close. This is because the estimation of F, and hence of ϕ'_m, is incorrect. If this is corrected, by trial and error, or systematically, then the factor of safety can be recovered from the final value of ϕ'_m.

A systematic method of altering ϕ'_m is to correct it by an amount $\Delta\phi'_m$, where

$$\Delta\phi'_m = E_n/\Sigma(W - P_b \cos \alpha) \tag{5.13}$$

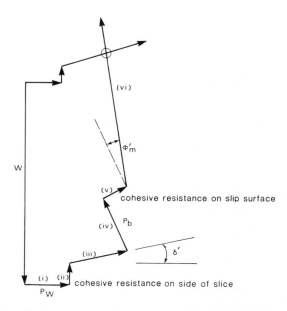

Figure 5.6 Addition of cohesion to the force polygons in the wedge method. This complicates the detail of Figure 5.5, but does not change the principles in any way. Force components equal to the mobilized cohesion forces (using the 'current' factor of safety) are added in the appropriate directions before an attempt is made to close the polygon of forces for each wedge.

To save computations, these two quantities can merely be scaled off the drawings.

The effect of cohesion, on the slip surface or on the interface between wedges, can readily be incorporated (Figure 5.6), as can the effect of defining the mobilized interslice friction angle in terms of effective stress.

One needs to be cautious in the choice of inclination for the interslice forces, because a poor choice here can have a major influence on the calculated factor of safety. The advantages of the method are that virtually no computational aids are required, only a drawing-board; and that the influence of different parts of the slide on its stability can readily be appreciated. Sensitivity studies can be made, altering the inclination of the interslice forces, for little additional effort.

The details of the construction of these force polygons are instructive. It is recommended that those new to the stability of slopes try out this method, or draw up the force diagrams for cases analysed by computer, to benefit from the additional insight that this exercise gives.

5.7 Janbu's method

Following the success of Bishop's theory, and with good results obtainable from the routine method, efforts were made to develop a similar theory applicable to slides with any shape of slip surface. The chronology of the early developments is confused, but appears to have been as follows. Bishop's paper was presented at a conference a year before it appeared in print. Separate researches led Janbu and Kenney to the same result (Kenney working under Bishop at Imperial College). Janbu published first, in 1955, but in an incorrect form, and Kenney's thesis appeared the following year. Subsequently, Janbu republished a corrected form of the equations, but in a form relatively inaccessible to English readers. In the meantime, Bishop had persuaded Price, who had been one of the authors of the first stability analysis computer program (see Little and Price, 1958), to try to program Kenney's equations for non-circular slips. It was then found that the basic equations could give rise to numerical problems when evaluated to high precision that would not appear at slide-rule accuracy: he and Morgenstern (Morgenstern and Price, 1965; 1967) then developed a more sophisticated method (again at Imperial College). This time, they were secure in the knowledge that complexity in computation was no longer a bar to widespread use of a method because of the growing availability of computers. Janbu developed his method further, and published his generalized procedure of slices in 1973, and a number of other methods also appeared in print throughout the late 1960s and 1970s. The key methods are described below, and the salient points of the others are listed at the conclusion of this chapter.

A problem when dealing with a slip surface of arbitrary shape is that the convenience of a single point through which a number of the force components act, and are therefore lost from a moment equation based on that point, is no longer available. Substitution of any other arbitrary point causes great additional complexity. In Janbu's method, therefore, the force equilibrium equation, rather than the moment equilibrium equation, was chosen for the development of a routine method of stability analysis.

Following the general lines of Bishop's method, this finally gives the following:

$$F = \frac{\Sigma\{c'b + (W - ub + \Delta X)\tan \phi'\}\dfrac{\sec^2 \alpha}{1 + \tan \alpha \tan \phi'/F}}{\Sigma W \tan \alpha} \qquad (5.14)$$

Exactly as in Bishop's method, a systematic iterative procedure can be used

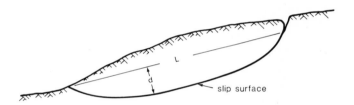

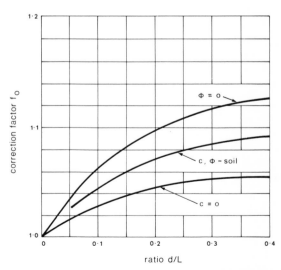

Figure 5.7 Janbu's correction factor chart. This was developed originally for use with low-angled slopes and needs to be used with caution.

to solve this equation. An equally suitable starting point can be found from the evaluation of the conventional method formula, 'plus a little'. The same general observations on likely numerical problems with this equation apply here too.

Remembering that this work predated Spencer's findings as to the reasons for the relative accuracy of Bishop's routine formula, it is perhaps not surprising that Janbu found some inaccuracies in the results of his routine method in comparison to more involved computations. On the basis of a strictly limited number of such calculations, he proposed an empirical correction to be applied to the results of a calculation made using his routine method. This is shown in Figure 5.7. This correction factor is in the nature of an increase in the factor of safety, and depends on the relative depth of the landslide in relation to its length, and on the nature of the soil properties. It has a maximum value of 13% increase in F. Note that the correction should be applied *after* the routine procedure has been followed, i.e. the correction is made to the converged factor of safety, not during the iterative procedure, as follows:

$$F_{corrected} = f_o F \tag{5.15}$$

where f_o is taken from the chart. Narrow slices must be taken when using this method.

5.8 Morgenstern and Price's procedure

In response to the difficulties in programming Kenney's method for a computer, a new method was devised and described with improvements by Morgenstern and Price (1965, 1968), Morgenstern (1969) and Chen and Morgenstern (1984). The essence of the method is to divide the sliding mass into a relatively small number of linear sections or wedges which are vertical-sided in the conventional way. Within each of these sections, which may be many times wider than the slices considered in most other methods, an element (in the calculus sense) can be considered. The conditions of force equilibrium can be considered; taking directions normal and parallel to the slip surface. In the normal direction the equilibrium equation yields

$$dN' + dP_b = dW \cos \alpha - dX \cos \alpha - dE' \sin \alpha - dP_w \sin \alpha \tag{5.16}$$

and in the shear direction,

$$dS = dE' \cos \alpha + dP_w \cos \alpha - dX \sin \alpha + dW \sin \alpha \tag{5.17}$$

These two can be combined with the Mohr–Coulomb failure criterion in terms of effective stresses:

$$dS = \frac{1}{F} \{c' dx \sec \alpha + dN' \tan \phi'\} \tag{5.18}$$

E

infinitesimal element of a larger, linearly bounded, slice

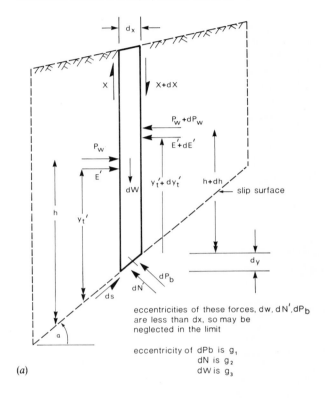

(a)

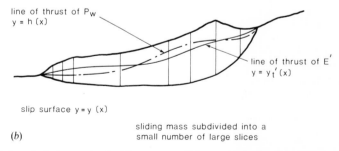

line of thrust of P_w
$y = h(x)$

line of thrust of E'
$y = y_t'(x)$

slip surface $y = y(x)$

(b)

sliding mass subdivided into a
small number of large slices

Figure 5.8 A typical element in the Morgenstern–Price method. The element is an element of infinitesimal width within a larger slice. Quantities of less than the slice width (e.g. eccentricities of forces acting on the element of slip surface) disappear when the element is taken to the limit.

which after proceeding to the limit as $dx \to 0$ eventually yields the following force equilibrium equation:

$$\frac{c'}{F}\sec^2\alpha + \frac{\tan\phi'}{F}\left\{\frac{dW}{dx} - \frac{dX}{dx} - \frac{dE'}{dx}\tan\alpha - \frac{dP_w}{dx}\tan\alpha - \frac{dP_b}{dx}\sec\alpha\right\}$$

$$= \frac{dE'}{dx} + \frac{dP_w}{dx} - \frac{dX}{dx}\tan\alpha + \frac{dW}{dx}\tan\alpha \tag{5.19}$$

The moment equilibrium condition for the element gives the equation that follows once moments are taken about the midpoint of the base of the *element*:

$$E'\left\{(y - yt') - \left(-\frac{dy}{2}\right)\right\} + P_w\left\{(y - h) - \left(-\frac{dy}{2}\right)\right\} - (E' + dE')\left\{y + dy - yt'\right.$$

$$\left. - dyt + \left(-\frac{dy}{2}\right)\right\} - X\frac{dx}{2} - (X + dX)\frac{dx}{2} - (P_w + dP_w)$$

$$\times \left\{(y + dy) - (h + dh) - \frac{dy}{2}\right\} - dP_b g_1 - dN'g_2 + dWg_3 = 0 \tag{5.20}$$

When this equation is simplified, and after proceeding to the limit as $dx \to 0$, bearing in mind that the three eccentricities, g_1, g_2 and g_3 are all much less than dx, it can be shown that

$$X = \frac{d}{dx}(E'yt') - y\frac{dE'}{dx} + \frac{d}{dx}(P_w h) - y\frac{dP_w}{dx} \tag{5.21}$$

Solving the pair of simultaneous partial differential equations (5.19) and (5.21) is by no means a simple task. There are two important starting assumptions: first that the E' and X forces can be related simply to one another, and secondly that slices are chosen in such a way as to make the individual differential coefficients simple to evaluate. The choice of the relationship between E' and X forces is perhaps the most distinctive feature of the Morgenstern–Price method, and certainly gives the most difficulties to users of the method.

If X and E' are related by a single constant, as in Spencer's method, then the interslice forces will all be inclined at a uniform angle to the horizontal. Some variation in these inclinations can be introduced by using a range of constants for different positions through the landslide. Unfortunately, to do this overconstrains the solution. A way to avoid such a difficulty is to take a relationship between the two components of the interslice force which is partly made up of a coefficient defined *a priori*, which specifies the *relative* inclination of the interslice forces, together with a coefficient determined as part of the solution process, which then fixes the *absolute* inclination values. The predefined coefficients must be in the form of a continuous function so that the differential equations can be solved.

At first sight it might seem that this approach is unduly prescriptive, but with an almost infinite range of possible functions available, it is not found to be a restriction at all. The predefined interslice inclination function must be defined for all x and so will be termed $f(x)$ in the following. λ is the scaling constant. Hence

$$X = \lambda f(x)E' \tag{5.22}$$

A great deal of simplification in the differential equations can be made by choosing to subdivide the landslide mass into wedges or thick slices such that the lines that subdivide the slope into soil zones are all linear (or approximately so) in each of the chosen slices. This linearity must also apply to the ground level, slip surface, and piezometric line. Choosing for example

$$y = Ax + B \tag{5.23a}$$

$$\frac{dW}{dx} = px + q \tag{5.23b}$$

$$f(x) = kx + m \tag{5.23c}$$

Suppose that the water pressure force P_w on the side of each slice is defined by a hydrostatic variation of pressure down from the piezometric line to the slip surface. A linear piezometric line and slip surface would then make the P_w force vary quadratically across the slice. A similar argument can be applied to the lateral body forces arising from seismic acceleration. Thus there are the additional relations:

$$\frac{dP_b}{dx} = rx + s \tag{5.23d}$$

and

$$\frac{dP_w}{dx} = tx^2 + ux + v \tag{5.23e}$$

Using these simple relationships in the force equation (5.19), it can be shown that this takes the form

$$(Kx + L)\frac{dE'}{dx} + KE' = Nx + P \tag{5.24}$$

where K, L, N and P are functions of the coefficients in equations (5.23). In order not to interrupt the development of the theory, the exact definition of these expressions is relegated to Appendix A.

This equation can be integrated between the limits* of $x = 0$ and $x = b_i$:

---

*Conventionally, the summation of terms relating to slices i, where i ranges from 1 to n, is represented by showing the limits of summation on the summation symbol, and terms which vary from slice to slice are represented with the subscript i. For clarity, the ranges have been omitted, and in the foregoing, so have the subscripts. However, in the following derivations, terms relating not only to i, but also to $i + 1$ or $i - 1$ are required. Terms in equations are fully subscripted where this is required to make the meaning clear.

$$E'_{i+1} = \frac{1}{(L_i + K_i b_i)} [E'_i L_i + P_i b_i + \tfrac{1}{2} N_i b_i^2] \tag{5.25}$$

Applying this to each slice in turn, starting from $E' = 0$ at the toe of the slide, the complete set of E' forces can be obtained for any assumed factor of safety and λ values. The last of these forces at the crown of the slide should be zero, but unless a fortuitous (or correct) pair of values have been selected, will not be so. Even values for F and λ which satisfy this condition will not be a complete solution unless they also satisfy the moment condition of equation (5.21).

Suppose the moment of the E' force about the level of the slip surface is defined as R, so that

$$R_i = E'_i(yt_i - y_i) = E'_i r_i \tag{5.26}$$

Referring to the moment equation (5.21), it will be seen that the change in this moment is given by

$$\int (X + yE')dx + Q \tag{5.27}$$

where Q is a moment containing the following elements:

the moment of the known porewater pressure forces on the sides of the slice

the moments arising from seismic accelerations on the soil mass.

Thus the moment of the horizontal effective interslice force at one side of a finite slice, R_i, could be computed from the moment at the other side of the slice, R_{i-1}, and the physical properties of the slice itself, as follows:

$$R_i = R_{i-1} + \int_0^{b_i} (\lambda(kx + m) + y)E'dx + Q_i \tag{5.28}$$

Then this can be integrated across each slice in turn, starting from the toe of the slope and working to its crest, in the same way as was done earlier for the E' forces. This gives a last moment which is unlikely to be zero. A systematic adjustment procedure, altering both F and λ is then used to reduce these out-of-balance forces and moments. Ideally, when the final moment and interslice force R_n and E'_n are both equal to zero, the values of F and λ which lead to this result are the solution to the problem, and may then be used to derive a whole range of subsidiary results such as stresses acting on the slip surface. These quantities can be derived from the individual force equilibrium equations for each slice. In practice, convergence is assumed when the errors in *either* R_n and E'_n, *or* in F and λ, are sufficiently small.

Suppose that the errors in E' and R at the end of the slip surface are $\delta E'$ and δR. A fresh estimate for the values of F and λ can then be obtained by adding δF and $\delta \lambda$ respectively to the starting estimates, where δF and $\delta \lambda$ are evaluated by a two-variable Newton approximation method:

$$\delta F = \frac{\delta R \dfrac{\partial E'}{\partial \lambda} - \delta E' \dfrac{\partial R}{\partial \lambda}}{\dfrac{\partial E'}{\partial F}\dfrac{\partial R}{\partial \lambda} + \dfrac{\partial E'}{\partial \lambda}\dfrac{\partial R}{\partial F}} \tag{5.29a}$$

and

$$\delta \lambda = \frac{\delta E' \dfrac{\partial R}{\partial F} - \delta R \dfrac{\partial E'}{\partial F}}{\dfrac{\partial E'}{\partial F}\dfrac{\partial R}{\partial \lambda} + \dfrac{\partial E'}{\partial \lambda}\dfrac{\partial R}{\partial F}} \tag{5.29b}$$

All the necessary differential components can be obtained from differentiating equations (5.25) and (5.26) with respect to F and λ and then using the resulting equations across each slice in turn to find the values of the differentials at the crest of the slope. Formulae for these differentials are listed in Appendix A along with the details of the integration of the moment equation, which is a fairly complex procedure in its own right.

5.9 Maksumovic's method

It is difficult to include such things as external concentrated forces in Morgenstern and Price's method. Furthermore, the computations are lengthy, even taking into account developments in computing technology, so the method is unsuited to routine stability-calculation work. It does have the advantage of allowing wide slices to be used, so that a smaller number are required to represent any slip surface, but this advantage is rarely realized in practice, the reason for this being that to obtain slices with genuinely linear characteristics, it is necessary for the computer program to insert slice boundaries at every x coordinate in the cross-section where an interface between strata changes slope. In a dam cross-section, this can make the *natural* subdivision into slices very fine anyway.

Suppose that this advantage was disregarded. Could there be advantages in taking small slice widths? It would be possible to use Morgenstern and Price's approximate integration formulae with some confidence there. However, the problems of applying external forces still remain.

In a method devised by Maksumovic, this problem is removed. In Figure 5.2, it was pointed out that the known force and moment components acting on a slice could all be resolved into two orthogonal forces and a single moment, leaving a far less complicated picture (Figure 5.9). Writing the equations of horizontal force equilibrium

$$E'_i - E'_{i+1} - N'_i \sin \alpha_i + S_i \cos \alpha_i - H_i = 0 \tag{5.30}$$

individual trapezium–shaped
soil zones within a slice subdivided
into triangles to simplify calculation
of force and moment contributions

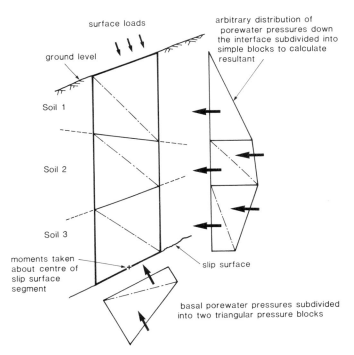

surface loads

arbitrary distribution of
porewater pressures down
the interface subdivided into
simple blocks to calculate
resultant

ground level

Soil 1

Soil 2

Soil 3

moments taken
about centre of
slip surface
segment

slip surface

basal porewater pressures subdivided
into two triangular pressure blocks

Figure 5.9 Resolution of known forces in Maksumovic's method. By dividing the water pressures on the side of each slice into separate 'triangles', it is easy to evaluate the force and moment components arising from them. Similarly, a fine subdivision into triangular segments within each slice makes it simple to evaluate the self-weight and seismic forces and moments.

vertical force equilibrium

$$X_i - X_{i+1} + N_i' \cos \alpha_i + S_i \sin \alpha_i - V_i = 0 \tag{5.31}$$

and moment equilibrium

$$E_{i+1}'(r_{i+1} + \tfrac{1}{2} b_i \tan \alpha_i) - E_i'(r_i + \tfrac{1}{2} b_i \tan \alpha_i) - \tfrac{1}{2} b_i (X_{i+1} - X_i) + M_i = 0 \tag{5.32}$$

together with the Mohr–Coulomb failure criterion

$$S_i = \frac{1}{F}(c_i' b_i \sec \alpha_i + N_i' \tan \phi_i') \tag{5.33}$$

and Morgenstern and Price's relation between X and E' forces of equation (5.23)

$$X_i = \lambda f(x)E'_i.$$

It is possible to devise recurrence formulae for E'_i and r_i in terms of the known parameters of the slice and E'_{i-1} and r_{i-1}. The intermediate steps in this derivation are omitted for space considerations, but the resulting equations are the two following:

$$E'_i(1 - f_i a_i) = E'_{i-1}(1 - f_{i-1} a_i) + \frac{b_i c'_i}{F}(1 - \tan \alpha_i a_i) + a_i V_i - H_i) \quad (5.34)$$

and

$$E'_i(r_i + \tfrac{1}{2} b_i \tan \alpha_i) = E'_{i-1}(r_{i-1} - \tfrac{1}{2} b_i \tan \alpha_i)$$
$$+ \tfrac{1}{2} \lambda b_i (f_i E'_i + f_{i-1} E'_{i-1}) - M_i \quad (5.35)$$

in which

$$a_i = \frac{\dfrac{\tan \phi_i}{F} - \tan \alpha_i}{1 + \dfrac{\tan \phi'_i \tan \alpha_i}{F}} \quad (5.36)$$

Maksumovic originally suggested using these two equations on each slice in turn, from the toe of the slip to its head, and then adopting a trial and error procedure, to find the values of F and λ which would give negligible force and eccentricity errors at the head of the slide. His procedure was based on *regula falsi* (the intersection of chords procedure), with fairly large differences in F and λ initially used. Elsewhere, (Bromhead, 1984), I have proposed using the whole of Morgenstern and Price's solution procedure, substituting Maksumovic's recurrence equations for their partial differential equations. There is a slight problem in that whereas the E' force is the dominant term in Maksumovic's method, it is the M term which is of larger magnitude in Morgenstern and Price's theory. To make the two methods more compatible, it is suggested that the eccentricity (r) equation be replaced by a moment equation, in which $R = E'r$:

$$R_i = R_{i-1} - b_i \frac{\tan \alpha_i}{2}(E'_i + E'_{i-1}) + \frac{b_i}{2}(f_i E'_i + f_{i-1} E'_{i-1}) - M_i \quad (5.37)$$

The two equations (5.34) and (5.37) can then be treated exactly as the equations in Morgenstern and Price's method, using the two-variable Newton iteration procedure with all the convergence controls applied. The two equations may readily be differentiated with respect to F and λ. For reference, the four partial differential equations are listed in Appendix B.

5.10 Sarma's method

In his paper of 1973, Sarma adopts a radically different approach to the calculation of the stability of slopes. He finds the critical value for a uniform horizontal acceleration which will just cause failure of the slope. This method can be adapted to find an ordinary static factor of safety by merely factoring the soil strength parameters until a zero horizontal acceleration is required for failure. Sarma, and later on, Sarma and Bhave (1976) argue that this critical horizontal acceleration is as meaningful an index of relative stability as is the conventional definition of a factor of safety; although they do concede that sometimes the surface which has the lowest static factor of safety may not necessarily yield the lowest critical acceleration, and vice versa.

The derivation of the equations is on relatively conventional lines initially, although the reader is advised to follow this derivation rather than the original which is full of typographic errors, adopts unconventional forms of presentation and muddles computer symbology with mathematical expressions. As an example of the latter, both D and X are used separately and in combination as DX to represent something completely different!

The horizontal and vertical equilibrium conditions give the following pair of equations:

$$N_i \cos \alpha_i + S_i \sin \alpha_i = W_i - \Delta X_i \tag{5.38}$$

and

$$S_i \cos \alpha_i - N_i \sin \alpha_i = kW_i + \Delta E_i \tag{5.39}$$

Taking the first of these in conjunction with the Mohr–Coulomb failure criterion which relates the N and S forces:

$$S_i = (N_i - P_{b_i}) \tan \phi_i' + c_i' b_i \sec \alpha_i \tag{5.40}$$

it becomes possible to derive expressions for both N and S for substitution into the second equation, i.e.

$$N_i = \frac{(W_i - \Delta X_i - c_i b_i \tan \alpha_i + P_{b_i} \tan \phi_i' \sin \alpha_i) \cos \phi_i'}{\cos \alpha_i \cos \phi_i' + \sin \phi_i' \sin \alpha_i} \tag{5.41}$$

(The sharp-eyed will note the near-identity of this expression with that derived for Bishop's method in equation (5.7).)

$$S_i = \frac{(W_i - \Delta X_i - P_{b_i} \cos \alpha_i) \sin \phi_i' - c_i' b_i \cos \phi_i'}{\cos \alpha_i \cos \phi_i' + \sin \phi_i' \sin \alpha_i} \tag{5.42}$$

yielding an expression for the maximum horizontal seismic force that can be sustained kW_i:

$$kW_i = D_i - \Delta E_i' - \Delta X_i \tan(\phi_i' - \alpha_i) \tag{5.43}$$

in which D_i stands for the expression

$$D_i = W_i \tan(\phi'_i - \alpha_i) + \frac{c'_i b_i \cos \phi'_i \sec \alpha_i + P_{b_i} \sin \phi'_i}{\cos \phi'_i \cos \alpha + \sin \phi'_i \sin \alpha_i} \tag{5.44}$$

Now, the items in this latest expression can be summed over all the slices in order to find the total horizontal seismic force $k\Sigma W_i$. When this is done, it is found that the sum $\Sigma\Delta E$ has to be zero. $\Sigma\Delta X$ would also be zero, but when the individual terms are multiplied by different coefficient, a non-zero result is obtained. Hence,

$$\Sigma\Delta X_i \tan(\phi'_i - \alpha_i) + \Sigma k W_i = \Sigma D_i \tag{5.45}$$

This is the main force equilibrium equation of the method.

Moment equilibrium must also be satisfied, and the choice of an appropriate point about which to take moments can simplify computations enormously. Sarma chooses the centre of gravity of the sliding mass so that the sums of W and also of kW vanish. Furthermore, the interslice forces do not provide any net moment, so that

$$\Sigma(S_i \cos \alpha_i - N_i \sin \alpha_i)(y_i - y_g) + \Sigma(N_i \cos \alpha_i + T_i \sin \alpha_i)(x_i - x_g) = 0 \tag{5.46}$$

In this, x_i and y_i are respectively the coordinates of the midpoint of the base of the slice, so it will be seen that the simplifying assumption that normal forces act through the middle of this slip surface segment has been made. This in turn demands the use of narrow slices.

The moment condition can be rewritten in terms of the main force equilibrium expression (equation (5.45)) together with the Mohr–Coulomb relationship, to give rise to the following equation:

$$\Sigma\Delta X_i\{(y_i - y_g)\tan(\phi'_i - \alpha_i) + (x_i - x_g)\} = \Sigma W_i(x_i - x_g) + \Sigma D_i(y_i - y_g) \tag{5.47}$$

Sarma chooses to define each ΔX in the form of

$$\Delta X = \lambda \psi_i \tag{5.48}$$

in which ψ_i is known, and where $\Sigma\psi_i = 0$. This then makes the two key equations become

$$\lambda\Sigma\psi_i \tan(\phi'_i - \alpha_i) + k\Sigma W_i = \Sigma D_i \tag{5.49}$$

and

$$\lambda\Sigma\psi_i\{(y_i - y_g)\tan(\phi'_i - \alpha_i) + (x_i - x_g)\} = \Sigma W_i(x_i - x_g) + \Sigma D_i(y_i - y_g) \tag{5.50}$$

These two may be solved simultaneously to obtain λ and k.

$$\lambda = \frac{\Sigma W_i(x_i - x_g) + \Sigma D_i(y_i - y_g)}{\Sigma\psi_i\{(y_i - y_g)\tan(\phi'_i - \alpha_i) + (x_i - x_g)\}} \tag{5.51}$$

$$k = \frac{(\Sigma D_i - \lambda \Sigma \psi_i \tan(\phi_i' - \alpha_i))}{\Sigma W_i} \qquad (5.52)$$

Subsequently, a complete range of derived results including the actual values of E, X, N, and so on can be obtained by considering the force and moment conditions for each slice in turn now that the k value and the X forces are known.

There remains but one problem: selection of the individual ψ values. If a computer solution is used in which the slice subdivision is automatic, then it may be difficult to select ψ values *a priori*. Sarma considers the equilibrium along interslice boundaries. If we take the normal effective force E_i' on this interface to be made up of 'earth pressures', or to be evaluated from the equilibrium of stresses in the soil mass at this particular location, then a good assessment of the likely value for the X force might be

$$X_i = \frac{\lambda(E_i')}{\text{local } F} \qquad (5.53)$$

(local F against shear on the interslice boundary). This is similar to the Morgenstern–Price assumption (equation (5.22)), sufficiently much so for us to write f_i for the inverse of the expected local factor of safety. In turn, this then gives a set of X forces for use in the main equation, and of course, the set of X forces chosen must satisfy $\Sigma \Delta X = 0$. When the solution is complete, however, an equilibrium set of E' forces are obtained, which then define the local factor of safety on an interslice boundary, which may be very different from the inverse of f_i because the normal components of the interslice forces may be dissimilar to those assessed initially. In Sarma's method, then, the specification of f_i values for all x coordinates is nowhere near so prescriptive of the eventual interslice force inclinations as are the values in Morgenstern and Price's method.

The procedure follows a line of argument broadly similar to that advanced by Rowe (1970), for the prescription of $f(x)$ values for use in the Morgenstern and Price method, and is amplified considerably by Sarma in the source reference.

5.11 Miscellaneous other methods

Other methods of stability analysis have been developed, and the details published. Sarma (1976), for instance, following the general line of his earlier work, points out that it is by no means essential to use vertically bounded slices in stability analyses, and that there are, indeed, some advantages in adopting a series of wedges. Not least of the advantages is that this more accurately models the mechanics of some slides, notably graben slides, than does a vertical slice analysis.

Most of the authors referred to in the above have made more than one attempt at the problem. Spencer, for instance, published further thoughts on the thrust line criteria for stability analysis in 1973; Janbu wrote a series of papers culminating in his definitive work of 1973; and Morgenstern extended his method in 1969 and again in 1984, latterly concentrating on the validity or otherwise of certain $f(x)$ distributions. Normally, these 'second thoughts' merely amplify points raised earlier, but not covered in depth. Occasionally, though, they do produce valuable new data, and represent a genuine advance in our understanding of the implications of analytical techniques presented earlier.

An alternative procedure for analysing non-circular slip surfaces is presented by Bell (1968). Although he cannot reproduce the results quoted by Whitman and Bailey (1967), who use the Morgenstern and Price method, this does not invalidate his technique, since neither can Sarma (1973). The problem here is a misunderstanding of precisely which slip surface is referred to in the Whitman and Bailey paper, a matter which is discussed further in section 5.14.

Lowe and Karafiath (1970) and Seed and Sultan (1967) present variants of the wedge method for analysing sloping-core earth dams. Their methods are capable of being interpreted directly in terms of models or techniques outlined above, and therefore it is not necessary to continue in depth with them here.

Early analyses of non-circular slides have been made with composite slip surfaces made up of two circular segments, one at each at each end of a block which moves linearly. These segments are respectively an 'active' and a 'passive' zone. Similar analyses substitute the active and passive thrusts for the circular segments. This is particularly useful where the active thrust comes from a weak but thin vertical core, where a slip surface is not an appropriate model, or in the initial and approximate sizing of the berm width for a large embankment on soft ground.

5.12 The infinite-slope procedure

Some interesting results can be obtained from the basic theory if a few simplifications are made. Take for example the equation of the conventional method. This must be exact for a single slice, because there are no interslice forces to consider. Now consider an elongate landslide, where the influence of the toe and head portions of the slide are negligible (Figure 5.10). This may be represented as a single slice. When the slip surface is approximately parallel to the ground surface, the equation for factor of safety may be written:

$$F = \frac{c'l + (\gamma z b \cos \alpha - ul)\tan \phi'}{\gamma z b \sin \alpha} \tag{5.54}$$

(in which z is the depth of the slip surface, γ is the soil unit weight, and u is the mean porewater pressure acting on the slip surface).

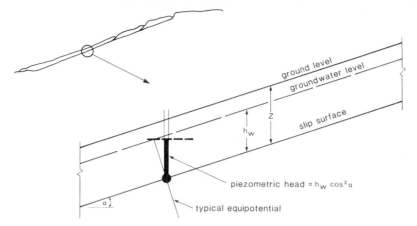

Figure 5.10 The infinite slope method. Both sliding and seepage are taken as acting parallel to the slope face (Haefeli, 1948).

In such an extensive slip, typical of the movement types experienced where a weathered mantle of material, or a fossil solifluction sheet, moves over much more competent material at depth, it is likely that flow is parallel to the slope surface. It the groundwater table is at an elevation of h_w above the slip surface, the porewater pressure u can be expressed as

$$u = \gamma_w h_w \cos^2 \alpha \tag{5.55}$$

Note that the trigonometric function here derives from the geometry of the flow net corresponding to the assumptions.

Thus the final expression for factor of safety becomes

$$F = \frac{c' + (\gamma z - \gamma_w h_w) \cos^2 \alpha \tan \phi'}{\gamma z \sin \alpha \cos \alpha} \tag{5.56}$$

and this equation is referred to by Skempton (1957) as the 'infinite slope expression', because strictly, it refers to failure parallel to the surface of a slope of infinite extent. Indeed, in its original form, it was presented as a method of analysis for slopes acted upon by seepage parallel to the slope face (Haefeli, 1948). It can be re-cast in terms of the pore-pressure ratio, r_u, viz.

$$F = \frac{c'/\gamma z + (1 - r_u) \cos^2 \alpha \tan'}{\sin \alpha \cos \alpha} \tag{5.57}$$

Where c' is zero, or small relative to γ_z, it may safely be neglected in the above expression. This then becomes

$$F = (1 - r_u) \frac{\tan \phi'}{\tan \alpha} \tag{5.58}$$

Since the realistic range of r_u values lies between 0 and 0.5 (corresponding to a groundwater level at or below the slip surface for $r_u = 0$, and the groundwater level at ground level for $r_u \approx 0.5$), it may readily be seen that slope angles for natural slopes subject to shallow slides can be predicted. For slopes without groundwater pressures, the limiting slope angle will be ϕ', and slopes where the groundwater table reaches ground level will have slopes such that $\tan \alpha \approx \frac{1}{2}\tan \phi'$. Where the slope angle is small (under residual strength conditions, for example, where ϕ' is the residual value), it may be permissible to express this latter relationship as $\alpha = \frac{1}{2}\phi'$.

The first of these two results is the reason behind the commonly observed phenomenon of a loosely heaped pile of dry granular soil adopting a slope angle equal to its angle of shearing resistance, and the second (see section 2.5) accounts for the normal range of natural slope angles for degraded slopes in clay strata in temperate climates.

For very shallow, translational landslides, the infinite-slope expression is as useful as any other method of slope stability analysis, and yields extremely good results. The engineer should not be put off by its simplicity. However, it is necessary to resort to more sophisticated methods where localized effects are to be considered.

5.13 Toppling modes of failure

Not all modes of slope failure can be analysed using simple sliding theories, and probably the other main area which is amenable to analysis is the onset of toppling modes of failure. These can be stimulated by a large horizontal water-pressure thrust in an already open joint or crack behind the face of a steep slope (Figure 5.11) causing a large overturning moment to be exerted. The depth of the joint and the narrowness of the block are both features which will contribute strongly to this mode of failure. However, it does presuppose that there is no outlet for the water at the bottom of the joint, so that winter conditions where seepages on the cliff face freeze and seal the outlets while unfrozen groundwater accumulates inside the joint will be especially conducive to the build-up of high water pressures. Comparatively small quantities of water are needed to fill nearly closed joints, but even a small opening leads to considerable drops in the total thrust as the water level falls, and when the joint is wide open (see for example Figure 2.20) further filling with water is improbable. Water thrusts of this sort are not therefore a good mechanism to account for all toppling failures.

An alternative mechanism can be the progressive weakening by weathering or otherwise of a massive bed underlying jointed rock. This leads to a loss of elastic modulus and a consequent rotation of a joint-bounded block. One alternative to weathering induced changes in the properties of the massive bed could be 'swelling' due to ingress of moisture drawn in by suctions which were

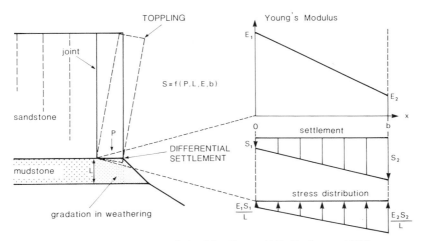

Figure 5.11 Toppling analysis. (After Evans, 1981; Dunbaven, 1983.)

initially induced by stress relief, with the consequent loss of suction-maintained effective stresses. The term *swelling* could be misleading here, if it were taken to imply a rise in the top surface of the mudstone; of course, the volume changes accompanying this drawing in the moisture take place entirely with outwards movement, and the top surface of the mudstone will in all probability settle.

Evans (1981) and Dunbaven (1983) consider this mode of failure. It is obviously progressive, since the initiation of the toppling movement transfers load to the front of the toppling block, increasing the settlement and accelerating the topple. Furthermore, the deterioration in stiffness of the underlying incompetent stratum will be encouraged by the rise in stress close to its unsupported face.

Other modes of toppling failure can be induced by undercutting (stream erosion, seepage erosion, or marine attack) or by thrusts or drags from associated slide movements. Many of these modes of failure defy simple analysis, or require specific techniques unrelated to the limit equilibrium models of this section (Goodman and Bray, 1976).

5.14 Mobilization of shear parameters

The methods of slope stability analysis presented above all make fundamentally similar assumptions at the outset. First of all, they all assume that whatever strength is available in the soil, it is mobilized uniformly along a potential sliding surface. For example, where the calculated factor of safety is 2, one half of the available strength is utilized in resisting the sliding at the toe of the slope, along the main part of the slip surface, and at the head of the slide. This takes no account of the stress–strain behaviour of the various soils and rocks involved in the slide, and leads to some absurdities. Take the case where two slip surfaces share a common segment. Slip surface *A* has a higher computed factor of safety than slip surface *B*. The degree of mobilization of shearing resistance along the common section is calculated to be *higher* for slip surface *B* than for slip surface *A*, and yet this segment of slip surface is common to both slides, and rationally can have only one set of stresses on it, and one degree of mobilization of its shear strength. It is only possible to resolve the problem using proper continuum mechanics theories. At present, such methods are too complex for use in routine engineering analysis and design.

A useful concept in the application of limit equilibrium methods for slope stability analysis and design is the idea of *mobilized shear strengths* and *mobilized shear strength parameters*. Since the definition for the factor of safety against sliding for a particular slip surface can be expressed as the ratio of the available shear strength to the applied shear stress, the mobilized shear strength, which must be the same as the applied shear stress, can be expressed as

$$s_{mob} = \frac{s_{available}}{F} = \frac{1}{F}(c + \sigma \tan \phi) \tag{5.59}$$

In an effort to make the description general, no subscripts have been added to the shear strength parameters c and ϕ, which could equally well be drained peak or residual parameters, or undrained strengths. In any case, however, the appropriate normal stress σ, is necessary.

The analysis would be just the same in terms of computed forces and moments for equilibrium if the actual shear strength parameters were replaced by c_{mob} and ϕ_{mob}, such that

$$c_{mob} = c/F$$

and

$$\phi_{mob} = \tan^{-1}(\tan \phi)/F \tag{5.60}$$

Using these parameters will exactly balance the applied forces and moments with the available resisting forces and moments. It is useful to compute these mobilized shear strength parameters for use in the design of remedial works (section 8.3) or just as an alternative to using the factor of safety F as an index of stability or the relative mobilization of shear strength. This can assist in reconciling the apparent anomalies between the definition of factor of safety computed as in this section, and that computed as a *load factor* along lines conventional in other branches of civil engineering. An example of this might help.

A cutting had side slopes of 1:2.5, and a computed factor of safety of 1.3. This was low, but broadly acceptable in that particular case. At one section, the cutting was to be crossed by an overbridge, but, where the bank seat load was applied, the factor of safety fell to 1.25. One immediate result of this was to cause consternation among the structural designers of the bridge, who demanded a factor of safety of 1.8 or better for the bank seat stability so as to be compatible with the basic design parameters for the superstructure! This then resulted in an outline for the side slopes being flattened to 1:4 in the vicinity of the bridge, and 1:3.5 elsewhere. This was naturally accompanied by an increase in cost for the earthworks and also for the bridge, which had to be extended.

A degree of rationality was re-introduced when it was pointed out that the bridge bank seat load would have to be increased *six* times to cause collapse of the original slope (actually the change in factor of safety with increasing bank seat load was not quite linear), and thus the *load factor* was not 1.25, but 6! The natural response of the bridge section—to demand that the earthworks group immediately adopt less conservative designs (after all, the real factors of safety were in excess of 6, and not the 1.4 claimed initially!)—was countered by pointing out the degree of mobilization of available shear strength. Of the 36.5° of shearing resistance, some 29.6° were being used up, or mobilized to support the self-weight of the slope, leaving 6.9° to cope with other things, such as the load of the bank seat. After applying this load, some 30.6° were mobilized, leaving 5.9° 'in reserve' in comparison to the 6.9° in adjacent sections of the slope.

5.15 Corrections for end effects and curved shear strength envelopes

The methods outlined in the foregoing are all applicable to slides of large lateral extent, since they are all based on a slice taken at right angles to the slope face with no correction for side shear. Some slides are like this, particularly those which are controlled by geological structure such as bedding, but many slides have slip surfaces that are bowl-shaped, or are lobate slides, where the influence of side shear is considerable. Lateral variations in the section geometry need to be taken into account where slip surfaces are bowl-shaped, or where they vary in depth from one side of the slide to another, and side shear effects should be incorporated where the width of the slide in plan is less than perhaps twice the length in plan from head to toe.

The effect of lateral variation of section can be taken into account of analysing different sections through the slide, and using a weighted average of the results. This weighting can be made on the basis of the relative proportions of the slide for which each of the sections are appropriate: weighting coefficients based on the *lengths* of slide, *plan areas*, or even *approximate volumes* of slide debris may all be appropriate in different situations. An example of such a weighting is given by Hutchinson (1969).

Skempton (1985) suggests that the correction to be applied to a stability analysis to incorporate the effects of side shear is in the form of a reduction factor for shear stress:

$$\frac{1}{1 + KD/B} \qquad (5.61)$$

in which D and B are the average depth and width of the slide respectively, and K is an earth pressure coefficient. In most cases the correction is small, although it can become significant for lobate slides. Skempton, for instance, took K to be 0.5, and found corrections of typically 5% in a number of case examples.

It is admissible to make these corrections for side shear when performing back analysis, but caution should be exercised in using them in unslipped slopes. This is because the assessment of the likely length of a failing section is difficult, and the incorporation of a factor which decreases the computed shear stress will be unconservative.

Skempton also considers the effects of curved shear strength envelopes, and comments that it is often useful to take the 'best fit' linear envelope over the range of pressures involved. This range of pressures can be estimated approximately at the outset of an analysis by taking just the overburden stresses on different sections of a slide. When the analysis has been completed, some slight modification of the parameters selected may be in order, but despite calling for a repeat of the analysis with revised parameters, it is preferable to trying to recast the analysis in terms of a curved strength envelope because standard techniques and computer programs may be used.

An alternative procedure is to use a *secant* angle of shearing resistance appropriate to the stress range involved. This is less satisfactory than using chords of the failure envelope, since the individual angles of shearing resistance are applicable for quite small ranges of effective stress, and more adjustment stages are required.

6 Techniques used in stability analysis

6.1 The search for the most critical slip surface

Searching for the most critical sliding surface in a slope can be made easier if slip circles are used. The total quantity of data required to set up a systematic search is relatively small. To perform such a search, first consider a series of slip circles of different radii, but each with the same centre of rotation (Figure 6.1). If the factor of safety for each of these circles in turn is found, and plotted on a graph against radius, then it is likely that at some point a minimum factor of safety will be found. This can come about merely from the shape of the slope, or because that particular slip circle has intercepted a weak soil.

Assume now that a similar range of slip circles are investigated from each of a whole array of centres—a regular rectangular grid would be most convenient. A minimum factor of safety for each centre could then be found, and the centre with the lowest factor of safety would show the factor of safety for the whole slope, provided of course, that sufficient centres, each with sufficient circles, had been investigated. To be sure of this, it is necessary to have a fairly large number of circles, with a big spread of radii from each centre, and quite a large grid of individual centres.

It is often convenient to 'contour' the values of factor of safety over the grid of centres: a closed minimum-value contour probably indicates that at least the location of the worst slip circle has been identified, whereas an open contour (Figure 6.2) almost certainly indicates that it has not. Such a contour plot will also indicate (from the spacing of the contours) whether or not further analyses are required.

Slope stability analyses would indeed be simple if this was all that was required, but unfortunately it is not. Even in a slope with a relatively simple shape in section, there may well be more than one 'critical' slip surface, each of which is the worst for a different mode of slope failure. One could be for shallow failures of an embankment fill, and another for a slip surface penetrating deep into the soil under the foundation of the embankment. The contour plot of factor of safety over a grid of slip circle centres in this example would have two low points. Each additional feature in the slope profile, extra berms for instance, brings extra complexities to the contour plot of factor of safety.

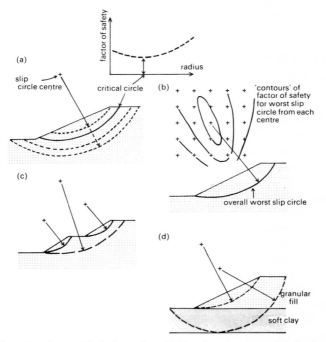

Figure 6.1 Searching for a critical slip surface. Taking first a variety of circles from the same centre, a minimum is found. This is then compared to the minimum factors of safety for other centres to find the most critical one.

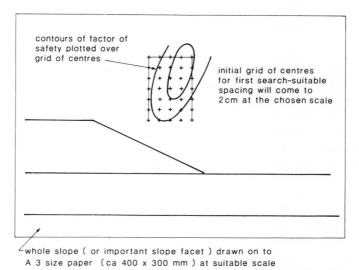

Figure 6.2 Techniques for systematic searching. Guidelines on dimensions for a check plot, and the location of a grid of centres for a systematic search, are given in the text.

Further intricacies are introduced if the soil is not homogeneous; the presence of weaker strata, or even zones of high porewater pressure, can also contribute their own low point to the plot. In view of this, the use of automated search procedures in a computer program can prove unsuccessful. For a start, the programming complexities are immense, and even when a minimum factor of safety is found, there is little or no guarantee that it is in any way related to the absolute minimum for the whole problem. I will refer to these factors of safety for modes of failure that are not critical in the overall sense as *local minima*.

There are also cases where it can be shown that the minimum factor of safety lies in a local minimum which is not within a closed system of contours. An example of this is the shallow mode of failure of a slope in granular soil. If a search procedure utilizing slip circles is followed, then the centre of the critical circle (which has infinite radius!) is found to lie an infinite distance from the slope face—resulting in a shallow, skimming mode of failure equivalent to the 'infinite slope' assumptions. It can readily be shown for this (section 5.12) that the limiting factor of safety is given by

$$F = (1 - r_u)\frac{\tan \phi'}{\tan \alpha} \tag{6.1}$$

—a value that is approached, but never reached, by the slip circle analysis.

It is recommended, therefore, that the first step in searching for such a critical slip circle is to make a preliminary assessment of just how many individual modes of failure there might be. Then the results of a search analysis with a grid of slip circles can be examined to see whether all the local minima have been obtained. Sometimes it will even be found that local minima for two or more different modes of failure are for slip circles with a common centre! This search analysis then throws up additional modes of failure (particularly in slopes with complicated internal details) and gives the analyst an insight into how the slope will behave.

It is important to understand why each local minimum factor of safety exists: what factors in the slope have given rise to it. This is an essential prerequisite to the making of systematic changes in a design to improve performance, for instance. In addition, some modes of failure are inconsequential, and low factors of safety can be accepted in design, whereas they could not be accepted for more serious modes. A simple example is of the granular fill embankment on a soft cohesive foundation. Shallow failure modes, such as those discussed above, really have little serious impact on the performance of such an embankment, although they can be unsightly and would be avoided if possible. Deep-seated modes of failure are much more critical to performance, but the designer is often far more concerned about factors of safety against this mode of failure than he is in respect of shallow modes of surface instability.

It is scarcely worth wasting computer time to demonstrate the validity of the limit given by equation (6.1), but there are a number of other simple cases that are easily found. Take, for instance, an embankment of a relatively strong soil built on a layer of weaker material. This case often arises with embankments on alluvial plains. The most critical slip surface then tends to penetrate almost to the base of the weak soil, and a more cursory check of shallower failure modes may be all that is required.

Where very deep deposits of alluvium are concerned, the deep slip circles which correspond to the above may not be realistic, as the natural increase in strength with depth that comes about through the effect of consolidation under higher pressures tends to localize the zone in which the slip surface can reasonably lie. Furthermore, the finite size of the embankment in cross-section will limit the maximum slip-circle radius which it is reasonable to analyse.

6.2 Tension cracks

Tension cracks at the head of a slide are an important indicator of the onset of instability. They are sometimes considered in slope stability calculations, and in such cases are usually considered to be full, or partly full, of water, and the resulting destabilizing force added in to the analysis. Just how deep such a tension crack might be is the subject of a number of differently held views. If the soil strength is purely cohesive, as in clay soils in the undrained state, then a depth in the range from 2 to 4 times c_u/γ is indicated. This can be very deep even for the weakest of soils. Another suggestion is that the tension crack might penetrate to the 'water table', or at least to the mean seasonal zero-pressure line, and stop there. The pressures arising from water filling such a tension crack must therefore be seen as transient effects arising out of isolated climatic events.

The appearance of a tension crack at the head of a slope, always assuming that is that the crack is genuinely tension-induced, and not merely the result of desiccation shrinkage, probably makes stability analysis a fairly academic exercise: failure is imminent, and the factor of safety close to 1. There are, however, instances where the tension crack concept is useful (Figure 6.3).

One of these instances is related to the formation of desiccation shrinkage cracks in dry weather. These can rapidly fill with water, and exert high lateral thrusts. Although only of short duration, these lateral thrusts can cause shallow slide movements to occur.

Another situation where the tension crack concept can be helpful is where soil strengths are high at the head of a slide. This causes numerical problems in many of the methods of slope stability analysis, and arises quite simply from one of the basic assumptions of limit equilibrium theory: that there is a single factor of safety equally applicable all round the slip surface. If the strengths are

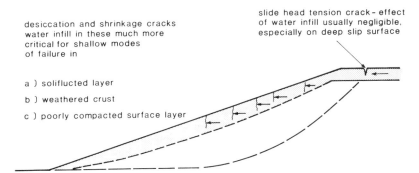

desiccation and shrinkage cracks
water infill in these much more
critical for shallow modes
of failure in

slide head tension crack – effect
of water infill usually negligible,
especially on deep slip surface

a) soliflucted layer

b) weathered crust

c) poorly compacted surface layer

Figure 6.3 Tension cracks. Not usually important in stability analysis, but can become important in some special cases.

Figure 6.4 Tension crack. Is this the result of past instability, or the cause of future movements?

high at the head of the slice then this concept of a uniform factor of safety implies tension between the slices. Although in itself undesirable, this can then make the normal forces on the steep segments of slip surface become negative, which is the source of the worst numerical problems of all. A tension crack judiciously inserted can overcome these difficulties at a stroke. On the other hand, a steeply rising section of slip surface at the head of a slice will be subject to small normal stresses, and in a frictional material, the computed shearing resistance will be low, and a tension crack facility may not be necessary in order to obtain realistic results.

Numerical difficulties of this sort are most frequently experienced in those methods where the order of the slices and their interaction is significant, for example in Morgenstern and Price's method, and least common in the methods where the forces are merely summed over all the slices. Of course, in many of the latter methods, the treatment of the interslice forces is so cursory that the effect is unlikely to be detected anyway. The writer has often found it necessary to use tension cracks when analysing slopes with Sarma's method. This may well reflect the net horizontal forces injected into the slide via k_{crit}.

The general cautions of section 8.4 on the inclusion of horizontal force components in a stability analysis should also be taken into consideration here. These water forces in upslope tension cracks not only increase destabilizing forces, but can also lower the normal stresses in adjacent parts of the slip surface. This makes their effect especially potent, and the neglect of one of the components yields a result which is on the *unsafe* side.

6.3 Partly submerged slopes

Slopes that are partly submerged present some additional problems to the analyst in the treatment of the applied loading. The vertical components of the water load could simply be added to the appropriate slice weights, but it is in the correct application of the horizontal force components that difficulties occur.

If the method will allow it, it is best to apply the force and moment components that arise from the water pressure to each slice. This is a consistent treatment that applies to both 'upstream' and 'downstream' water loads (Figure 6.5, slip XY) but many of the stability analysis techniques in common use will not allow the application of the horizontal force and moment terms without further development to the basic theories. Bishop's routine method and Janbu's method are cases in point.

Bishop (1955) recognized this problem, and developed a technique for overcoming it. Examine the downstream slip circle in Figure 6.5. Ideally, the part submerged analysis would be carried out without recourse to calculating the forces due to the water pressures on the slope face. Is there then

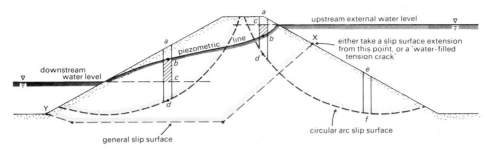

Figure 6.5 Submerged and partly-submerged slopes. Bishop's technique for dealing with these is poorly understood, and often misapplied.

another set of forces equal to those external forces, but of an opposite sense, that could also be omitted from the analysis? Both sets of forces could then be left out, and the effect of the water forces would be simulated.

There is indeed such a set of forces. If, for example, in Figure 6.5, only submerged weights were to be used in the calculation of the slice weights beneath the projection of the external water level into the slope, and if, in addition, only heads of porewater pressure in excess of those required to reach this external water level are used, then inspection will show that forces exactly in equilibrium with the external water loading have been removed. In the figure, it means using submerged unit weights in segment *cd* of the slice, and only using a part (*cb*) of the porewater-pressure head. For an 'upstream' slip surface such as in Figure 6.5, the subtraction of head *cd* from *db* may leave an apparently net negative head *cb* to be dealt with by the analysis.

Bishop's technique only caters for water loading on the downslope side of a slide. Returning to slip surfaces such as *XY* in Figure 6.5, it may be possible to use this approach on the downstream water load, but the upstream water load must be simulated *either* by taking the appropriate vertical loading and inserting a 'water-filled tension crack' at the position of *X or* by taking any extension from *X* up to the upstream external water level as though this segment of slip surface was in a strengthless soil with a unit weight equal to γ_w, and with the appropriate porewater-pressure conditions.

Both upstream and downstream water bodies can be treated as extra soil types with hydrostatic porewater pressures based on the level of the external water surface and no strength, if desired.

A word of caution to the users of standard commercial computer packages for slope stability calculations is probably in order here. There is no consensus about the meaning of the pore-pressure ratio r_u when Bishop's technique for dealing with part submergence is being used. Some programs may well operate on the reduced, or partly buoyant, weight of the slice in order to evaluate porewater pressures, others utilize the full weight (with or without reservoir loading) in order to do this calculation. The results can be radically different, not only in comparison to each other, but also in comparison to the analyst's intentions.

When an external water load is applied to a slope, it tends to increase the factor of safety against sliding *if the porewater pressures remain the same* (the modification procedure for porewater pressure outlined above is simply a method for eliminating the external water loading from consideration in the analysis, and should not be confused with any porewater pressures induced in the slope by the increased total stress, or through a change in the seepage pattern). The effect of such partial submergence can be quite considerable, and caution should be exercised before quoting the changes in factor of safety computed in this way, as they are sensitive to alterations in the porewater pressures in the slope. There may be situations in which the application of the loading is more than offset by these changes in porewater pressure.

Application of the above technique will always yield a higher factor of safety for the partly submerged slope than for the same slope without an external water level *provided that the internal conditions in the slope remain unaltered*. In an undrained case, this is equivalent to there being no change in the undrained shear strength; in an effective stress analysis, to there being no change in the porewater pressures present in the slope. The external water is an undrained loading, and changes in porewater pressure under its effect are to be expected. Thus, in the early stages of impounding of a dam, a slight worsening of stability may result.

Since one is impelled to use a conservative estimate of the pore pressure parameter $\bar{B}$ to calculate these increments of porewater pressure, worse changes in stability are normally predicted than are actually experienced. It is therefore essential to monitor the first impounding of a reservoir to examine the pore-pressure response actually encountered, lest this in combination with an end-of-construction condition gives rise to stability problems.

6.4 Back analysis and internal force distribution

Stability analyses are performed not only to provide a factor of safety once the soil properties are known, but also to establish field shear strengths from the study of failures. It is therefore necessary to do some analyses in reverse, to obtain soil properties from a known factor of safety (known to be 1 at the instant of failure). This process is usually termed *back analysis* (Chandler, 1977). An additional requirement in such an analysis is a determination of the mean levels of effective and shear stresses on parts of the slip surface—a knowledge of these, however imprecise, can be important in the selection of appropriate stress levels for laboratory testing.

Repeated analyses with different soil properties to find the set for which $F = 1$ is unnecessary.

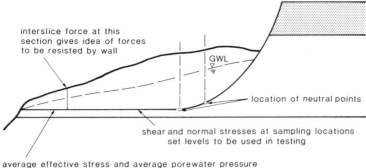

interslice force at this
section gives idea of forces
to be resisted by wall

GWL

location of neutral points

shear and normal stresses at sampling locations
set levels to be used in testing

average effective stress and average porewater pressure
immediately indicate responsiveness to drainage

Figure 6.6 Back analysis and internal force distributions. Not only does a back analysis provide information on field shear strengths, but it can also have some important by-products, such as the magnitude of likely forces on walls.

In the computation of the internal stresses a geotechnical engineer may not have access to a program which calculates these explicitly. Alternatively, he may have neither the skill nor the inclination to include this facility for himself. Indeed, if he is using a standard program, it is likely that its originators would have taken steps to prevent such unauthorized tampering.

Some methods of analysis will also give rise to difficulties in calculation of internal stresses explicitly from the equilibrium of forces on each slice. Bishop's (1955) method is a case in point. This arises out of the simplifying assumptions in the method itself which, while not preventing the calculation of an adequate factor of safety overall, may give rise to incorrect or even nonsensical stress predictions locally (Whitman and Bailey, 1967).

It is, however, possible to estimate average shear and normal effective stresses on the slip surface without recourse to the internals of the method of stability analysis actually being used.

In 'slip circle' analysis, some programs may output an overturning moment M_o or resisting moment M_r in the tabulated results. If this is so, then the average shear stress τ_{av} is given by

$$\tau_{av} = \frac{M_o}{RL} = \frac{M_r}{FRL} \tag{6.2}$$

Alternatively, an analysis of the slide with a single cohesion value c can be used. The average shear stress τ_{av} can be recovered from the resulting factor of safety F_c from

$$\tau_{av} = \frac{c}{F_c} \tag{6.3}$$

In this equation, τ_{av} represents the *mobilized* cohesion (section 5.14). It could equally well be represented by c_{mob}.

A repeat analysis is then carried out using any combination of c' and ϕ' parameters (usually $c' = 0$ will be chosen), and the average normal stress can be obtained from the resulting factor of safety $F_{c,\phi}$

$$\tau_{av} = \frac{1}{F_{c,\phi}}(c' + \sigma'_{av} \tan \phi') \qquad (6.4)$$

Equations (6.2) and (6.3) give an exact and correct answer when applied to slip circles, but give a small error when used for slip surfaces of arbitrary shape.

Much has been written on the subject of the admissibility or otherwise of interslice forces and their line of action ('the thrust line'). This literature contains a fair proportion of nonsense. Such statements as 'the thrust line must lie within the middle third of the slice depth to prevent tension' are as patently false as is 'soil cannot sustain tension'. There is nothing intrinsically wrong with tension in soils or rocks.

Any set of forces between the slices which is self-consistent (i.e. are in equilibrium) gives a lower bound to the collapse load according to the *lower bound theorem* (stated simply in many works on plastic theory, for example by Atkinson (1981)). Hence a conservative factor of safety is predicted. Whatever factor of safety is obtained from a limit equilibrium analysis on a given slip surface is less than or equal to the actual factor of safety for that surface, and so if a method can be found which improves that factor of safety, it must by definition be closer to reality.

One cannot obtain complete equilibrium without interslice forces. Methods which rely on the equilibrium of forces tend to be more conservative than those which are based on moment equilibrium, and so the latter are to be preferred. In most cases, use of stability analyses based on moment equilibrium reduces the likelihood of computing improbable interslice forces: the freedom to have some component of interslice shear is an important factor in this, however small the shear component is. I have found that the use of *parallel* interslice forces (Spencer's method, or Morgenstern–Price-like procedures with $f(x) = 1$) gives not only entirely admissible interslice forces, but a highly satisfactory solution in all but the most extreme cases. These extreme cases usually turn out to be the 'multistorey' landslides of section 1.3. At the narrow neck between different slide elements the forces may be computed as large tensions or compressions. The calculation of tension implies that the use of a uniform factor of safety overall is not valid, and that the mechanism of failure of the whole is more probably movement of the lower block first, with undermining of the upper one(s) and their movement subsequently. Large compressive interslice forces in this critical section indicate the reverse sequence of events.

The prediction by a slope stability analysis of an improbable stress state is more likely to be a demonstration to the analyst that he had chosen an unlikely failure surface, or that his simplifying assumptions are unreasonable, than that he should discard the analysis and try some other set of interslice forces.

6.5 Seismic slope stability analyses

It has already been seen that Sarma's (1973) method leads to the calculation of a horizontal acceleration coefficient which, when uniformly affecting a slope, just causes failure along the analysed slip surface. After a search has been made for the most critical slip surface the *critical* seismic coefficient, k_{crit}, for the slope is obtained. In the next section, the use of this parameter in obtaining probable displacements of the slope under a design earthquake will be reviewed. However, regardless of the utility of k_{crit} as a design parameter, it is an unfamiliar parameter to the majority of engineers. It also requires an analysis technique not readily available in standard computer programs, and hence it is perhaps less well used than it deserves.

The alternative method is to calculate the 'static' factor of safety that would apply if the design earthquake acceleration was applied as a set of horizontal forces. In the Morgenstern–Price and Maksumovic methods the necessary force components may readily be applied. Indeed, different accelerations may be utilized in different parts of the slope. This would probably be implemented in a computer program by assigning an acceleration coefficient to each soil zone. Furthermore, changes in the porewater pressures within the slope accompanying the shaking could be considered simultaneously, although these would have to be evaluated outside the slope stability analysis itself.

Those methods which do not allow the inclusion of horizontal force components can still be used if the acceleration field is simulated by simply tilting the slope. This allows any analysis technique, for example an existing slope stability analysis computer program which otherwise has no seismic facilities, to be used, but of course, only works for uniform acceleration fields.

Each part of the slope is rotated by an angle θ, where

$$\theta = \tan^{-1} gk_h/(1 + gk_v) \qquad (6.5)$$

in which gk_h and gk_v are respectively the horizontal and vertical components of the seismic acceleration. This method is mentioned *en passant* by, for instance, Binnie *et al.* (1967), but a clear account of the procedure is not readily available. In effect, the method returns the line of action of the resultant of the seismic and gravitational forces to the vertical and increases all the unit weights by a factor ρ where

$$\rho = \sqrt{[(1 + k_v)^2 + k_h^2]} \qquad (6.6)$$

The need for the increases in the unit weights does not always appear to have been understood in the past, and the unit weights have frequently not been amended as above. In frictional soils the error has no effect, because both resistance and driving force are proportional to the unit weights, but failure to include this second part of the method can be significant in cohesive soils. When the error exists it becomes increasingly significant with larger seismic coefficients such as are required when attempting to discover a value for the critical acceleration.

Most stability analysis is now computer-based, and it is convenient to arrange for the data to be 'rotated' automatically. Some additions to, or deletions from, the data set may be needed to ensure continued compatibility with the particular program being used. Such problems can constitute major pitfalls for the unwary analyst.

It should further be noted that errors in the porewater pressures will arise if a piezometric line is rotated, or if r_u values based on the original, untransformed, section are used. The errors in rotating the piezometric line may be

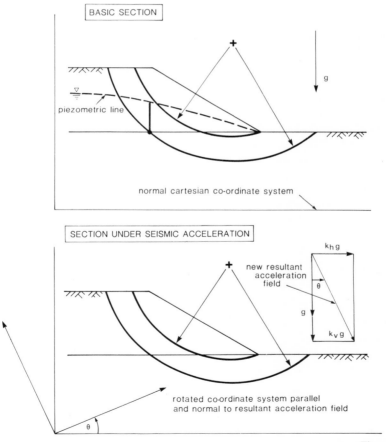

Figure 6.7

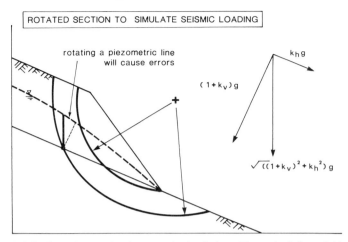

ROTATED SECTION TO SIMULATE SEISMIC LOADING

rotating a piezometric line
will cause errors

$k_h g$

$(1 + k_v) g$

$\sqrt{((1+k_v)^2 + k_h^2)} \, g$

Figure 6.7 Seismic analyses—the 'slope rotation' technique. The net body force field when a slope is shaken is inclined to the vertical. For the simple case of a uniform seismic acceleration, the resultant field is inclined to the vertical, and the combined effect can be simulated by simply rotating the slope.

insignificant for small values of θ, but will become substantial with larger seismic coefficients.

The principal advantage of the tilted-slope method is that it allows the effect of seismic acceleration to be readily visualized, and it is easier to see why the critical modes of failure for static and seismic stability may be different. It is useful to do this section rotation even if the former techniques are used. Furthermore, it may conceivably be more economical to establish critical accelerations with repeated trials using a tilted-slope model if this means that existing or standard computer programs can be used.

6.6 Rates and magnitudes of slope movements

The engineer has to consider a number of different deformation magnitudes and their rates in slopes problems. These include:

compressions in fill and strains during placement and compaction, and allied rebound deformations following unloading by excavation

consolidation or swelling

movements in brittle soils and rocks during a 'first-time failure'

creep in landslides, hillwash or non-brittle materials caused by repetitive loading. This loading may be cyclic applied loading, seasonal effects or small seismic effects

movement during earthquakes

run-out of falls and flows.

Many of the problems are simply not amenable to a theoretical or numerical treatment. Indeed, however appealing the treatment of some of the problems might be to the natural scientist, the rarity of some of the events, combined with their extreme violence, makes it improbable that a complete treatment would be useful to the ordinary practising geotechnical engineer or engineering geologist as a design tool. Others in this list fall outside the strict range of a work on the stability of slopes, and are routinely dealt with in soil and rock mechanics using simple theories: consolidation and swelling deformations being cases in point.

Consider first the general problem of deformation of a natural slope at the point of failure in a 'first-time slide', and having a brittle stress–strain relationship. Intuitively, once sliding has started, the rate of acceleration will depend on the delicate balance between the loss of strength as the soil mass moves (the gradient of the post-peak element of the stress–strain curve) and the decrease of the destabilizing forces as the slide progresses along the newly forming slip surface. At the end of the process of forming the slip surface, the soil mass may not be in equilibrium with the residual strength along the sliding surface, and so movement must continue until this is so. Depending on the kinetic energy of the slide, there may be an 'overshooting' effect.

Efforts to provide general solutions to the slide deformation problem lack general applicability and will not be described here. Rather, an empirical approach will be followed.

Take first the total movement of the slide in a first-time movement. This is obviously related to the brittleness index I_b. I propose an empirical factor to relate the movement to the slope height, $viz.$

$$\delta = I_d H I_b \qquad (6.7)$$

in which H is the slope height, I_d is the new factor, and δ is the displacement of the slide. There is some difficulty in identifying a reasonable brittleness index so that I_d can be calculated. A figure of I_d of about 0.9 to 1 would seem to be indicated for the majority of cases, and this in turn suggests that as a rule of thumb that the brittleness index itself is a useful index of deformation magnitudes under brittle failure. It will be noted that the total deformation, expressed as movement along the slip surface, is related to the scale of the problem: big slides move farther relative to their size.

The rate of deformation during a first-time failure is also of general interest, but has to be treated just as empirically. There are few records of failures in earth slopes where the failure has been observed in progress. However, such evidence as there is suggests that the rates of movement are quite scale-dependent: small slips in highway cuts of up to 15 m deep in London Clay take a day or so to develop, whereas slides in the coastal cliffs of the same material 45 m high move more rapidly, taking only a couple of hours. Thus not only do big slides suffer larger deformations, they also occur more quickly.

The general problem of seismic deformation is treated by Ambraseys and Sarma (1967) following the approach of Newmark (1965).

Consider a single block, with the potential for sliding down a plane inclined at an angle α to the horizontal. At first, and with no earthquake loading, the block is stable, but during shaking, the horizontal acceleration exceeds the critical acceleration (section 5.10) and the driving forces exceed those available to resist. Initially, the driving force is given by $W \sin \beta$, but during the shaking, this force is supplemented by the seismic force $kW \cos \beta$. This gives a driving force of

$$D = W \sin \beta + kW \cos \beta \qquad (6.8)$$

The resisting force, under the seismic loading, is given by

$$R = c'l + (W \cos \beta - kW \sin \beta - P_b) \tan \phi' \qquad (6.9)$$

When the seismic acceleration kg begins to exceed the *critical* acceleration $k_c g$, the maximum available resisting force becomes fully mobilized, so that it is no longer possible to write

$$D = R/F \qquad (6.10)$$

and the mechanics of the problem are now governed by the *net* driving force, $D - R$. Using Newton's law, the acceleration will be proportional to this net driving force, and it is then easy to show that, for accelerations of greater than the *critical* acceleration

$$\frac{W}{g} \frac{d^2 s}{dt^2} = D - R = g(k - k_c) \cos \frac{(\phi' - \beta)}{\cos \phi'} \qquad (6.11)$$

This equation may be integrated for simple acceleration records to obtain the displacements, s.

For real slip surfaces, this simple single-block model is still adequate as a first approximation. What needs to be done is to evaluate the *equivalent* single block. This is done by taking the following resultant forces:

the resultant of all shearing resistance terms acting on the bases of all of the slices in the problem.

the resultant of all of the normal forces acting on the bases of all the slices.

The directions of these two resultants, which will be at right angles to each other, then give the inclination of the base slope of the equivalent single block, β; the resultant of the normal forces is at right angles to it, and the resultant of the shearing forces is aligned along it.

Dynamic changes in porewater pressure as a result of the shearing during the earthquake are taken into account by the method of evaluating k_{crit}, that

F

is, the analysis for k_{crit} should be done with 'earthquake-induced' porewater pressures.

Sarma (1976) has evaluated the displacements for single pulses of the following kinds:

Rectangular-form wave, of duration $T/2$, and amplitude $k_m g$

Triangular-form wave, of duration $T/2$, and amplitude $k_m g$

Half sine wave, of duration $T/2$, and amplitude $k_m g$

and these are plotted in non-dimensional form in Figure 6.8. In a given 'design earthquake record', there may be several pulses for which the maximum acceleration, $k_m g$, exceeds the critical acceleration, $k_{crit} g$. In this case, the individual displacements are summed.

Naturally, this method relies on a number of implicit assumptions, for example that the displacements will be small relative to the overall slope geometry during the earthquake, so that β does not change. This then demands that the soils involved are non-brittle, and that deformations initiated during the shaking rapidly cease once it is over. Soils with a brittle stress–strain behaviour could continue to fail in a progressive manner after the earthquake is over, although it is worth noting that the residual strength of clay soils may well be significantly higher at the high rates of stress application in an earthquake that at drained laboratory test rates. This could

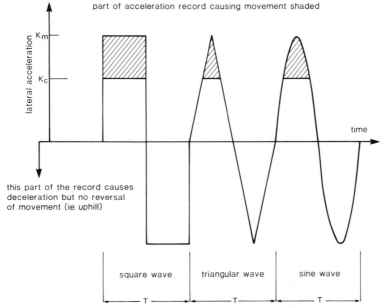

Figure 6.8 Seismic analyses—deformation prediction. Sarma gives the deformations for each of three simple pulses which exceed the critical lateral acceleration.

account for the observation that, generally speaking, failures of earth dams during earthquakes involve those dams which are constructed of non-cohesive soils, rather than clay fills.

6.7 Stability charts and their use in undrained analyses

Even early in the application of calculation to the solution of slope stability problems it became obvious that these could involve a substantial amount of arithmetic, all of it prone to error, and most of it an unproductive use of engineer's time: a series of analyses which showed an unacceptable factor of safety would have to be repeated for a modified design, for instance.

Many slope stability problems could also be reduced to a much simpler design idealization, particularly in the early stages of the evolution of a design, and this meant that the same slopes would have to be repeatedly analysed unless there was a corpus of established solutions. An attempt on this problem was made by Taylor (1937). He noted that for a slope of simple profile, and in the undrained case where the soil properties could be represented by c_u, ϕ_u and γ, and the slope by β and H, that the whole range of combinations of these parameters could be shown on just one graph (Figure 6.9).

Basically, Taylor analysed a range of slopes of different heights and slope angles, using a variety of combinations of soil parameters and his graph is presented in the form of a *stability number v.* the slope angle β. This stability

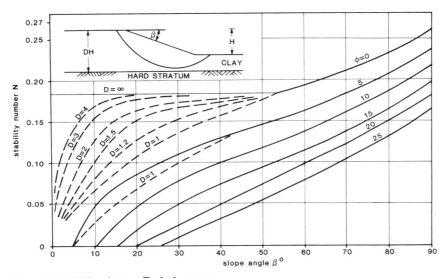

Figure 6.9 Stability charts—Taylor's curves.

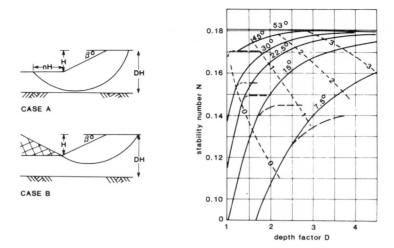

Figure 6.10 Stability charts—continued. n is the break-out factor (T in Figure 6.14).

number is the dimensionless term

$$N = \frac{c}{F_c \gamma H} \qquad (6.12)$$

A range of lines on this plot for different values of the *mobilized* angle of shearing resistance of the soil (see section 5.14) enable the effects of undrained 'friction' to be included.

Taylor found that when the soil strength was dominantly cohesive, failure surfaces tended to be deep seated, as a result of which the presence of a 'hard stratum' at shallow depth, or of a restriction on the position of the toe of any deep slip circle, could increase the factors of safety, and this is taken into account with further lines on the chart. Where the soil is predominantly frictional, however, these effects become insignificant.

The mode of operation of a stability analysis using Taylor's chart is as follows.

Take the slope shown in Figure 6.11. The slope angle is arc tan (1/2.5), or approximately 22°, and reading from the chart for a ϕ_u of 0 gives a stability number of approximately 0.18. A simple calculation shows that this corresponds to a factor of safety of 1.44.

Suppose this slope was underlain by a 'hard stratum' (shown dotted in the figure) at a depth of 12 m. This corresponds to a depth factor D of 1.5 with a slope height of 8 m. The corresponding stability number is 0.16, and hence the factor of safety is raised to the value 1.62.

Suppose now the slope had a more complex shear strength, being represented by a c_u of 20 kN/m² and ϕ_u of 15°. Inspection of the chart will show

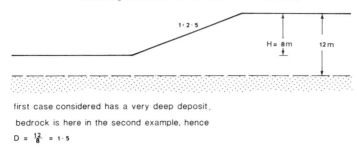

first case considered has a very deep deposit,

bedrock is here in the second example, hence

$D = \frac{12}{8} = 1 \cdot 5$

Figure 6.11 Example of the application of Taylor's stability chart.

that the depth factor is immaterial, and hence the critical mode of failure must be a shallow one. The solution proceeds iteratively. First assume that all the 'friction' is mobilized. Hence F_{mob} is equal to 1, and ϕ_{mob} is equal to ϕ_u. Referring to the chart, it will be seen that this combination corresponds to a stability number of 0.025, and the factor of safety with respect to cohesion, F_{cu}, is 3.

The factor of safety, which is the same for both cohesion and friction, must be somewhere between these two limits. Choosing 2 for F_ϕ gives a ϕ_{mob} of 7.6°, a stability number of 0.07, and F_{cu} of 1.78. Repeating the calculation with F_ϕ of 1.90 matches F_{cu} to F_ϕ.

More detailed work for purely cohesive soils, in which the effect of a restriction on the toe breakout position exists, can be made with an auxiliary chart (Figure 6.10) on which the data is plotted to a larger scale and different axes. The solid lines give the stability numbers if the toe breakout position is totally unconstrained, and the dashed lines show the effect of restricting the critical slip surface to a toe circle. Also shown on the diagram is another series of faint dashed lines, which show the critical toe breakout factor: this is where the constraint arises from the depth limitation and not the toe limitation (or vice versa).

Several other workers have also contributed to this problem, notably Gibson and Morgenstern (1962) who analyse the case where the undrained shear strength of a soil increases with depth. Broadly, however, the approach of Taylor is followed.

6.8 Stability charts: effect of porewater pressures

Porewater pressures can be included in stability charts, provided that some gross simplifying assumptions are made. These usually take the form of assuming a particular distribution of water pressures in the slope, either by taking a water table of some specific shape in the slope, below which the porewater pressures increase hydrostatically with depth, or by the use of an

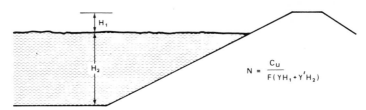

Figure 6.12 Treatment of partly submerged slopes when using stability charts.

r_u value. This concept is explored in depth in the next chapter, principally with respect to its use in computer based analyses of slope stability, where its *over-simplified* description of porewater pressure distribution is a grave disadvantage. However, in stability chart work, it is convenient to have fewer parameters to deal with.

The porewater-pressure ratio r_u used in slope stability charts is the ratio of porewater-pressure head at a point in the slope to the vertical total stress as estimated approximately from the depth of soil above that particular point and its unit weight. In all but the most extreme cases, this cannot possibly be the same at all points in the slope, but may take only a small range, so that the calculation of a mean value implies only a relatively small degree of approximation. Hence the introduction of porewater pressures brings only one extra parameter to the problem, which is accommodated in one of two ways.

The first of these is the approach followed by Spencer (1967). This requires a set of charts, each broadly similar to Taylor's chart, but calculated for a single r_u value. (Spencer chooses 0, 0.25 and 0.50.) An estimate for the factor of safety of the slope is made using each chart in turn, and then the factor of safety for the slope is interpolated for the actual r_u value that applies for the slope in question.

As an example, take the slope in the previous case, but with the following shear strength parameters:

$c' = 10 \, \text{kN/m}^2$

$\phi' = 20°$

$\gamma = 20 \, \text{kN/m}^3$

$r_u = 0.2$ (the evaluation of r_u coefficients is introduced in section 5.12, and is explored in detail in sections 7.12 and 7.14.)

First of all, a procedure identical for that with Taylor's curves is followed with r_u equal to 0. Initially estimating F_ϕ to be 2.5,

$$\phi'_{mob} = \tan^{-1} \tan(\phi'/2.5) = 8.4°$$

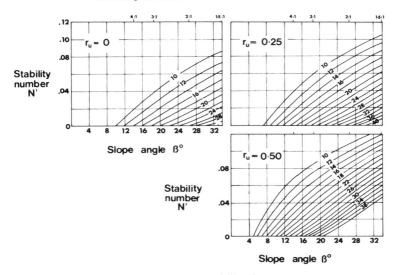

Figure 6.13 Spencer's stability chart.

From the chart (Figure 6.13), the stability number N' is about 0.07. Thus F_c works out at about 1.12 – obviously not the same as F_ϕ, so another iteration is required. For this, we start with a value intermediate between 2.5 and 1.12; a value of 1.8 is convenient. Using the same procedure, ϕ'_{mob} is 11.4°, N' is 0.045, so F_c is 1.74. It is not necessary to make a further calculation to settle on a value of 1.77.

The second and third stages are to repeat the above procedure for r_u values of 0.25 and 0.50, giving factors of safety of 1.37 and 1.00 respectively. Then, a graph (Figure 6.17) is drawn showing the relationship of the factor of safety to r_u, from which the factor of safety corresponding to the r_u of 0.20 in the present case can be obtained. This is approximately 1.53. Note that although not a straight line, the plot could be represented as one to an adequate first approximation. This is the basis of the other main method of dealing with porewater pressures in stability charts, that of Bishop and Morgenstern (1960).

Bishop and Morgenstern's charts are especially intended for earth dam design, and therefore deal with soils which have relatively small cohesion. For these, they note that the factor of safety for a given slope is decreased with increasing r_u, such that

$$F = m - nr_u \qquad (6.13)$$

They then tabulate and graph the m and n values. Since both m and n are functions of c', ϕ', and slope angle β, a series of charts are required. In the original paper, charts were prepared for the three $c'/\gamma H$ ratios 0, 0.025

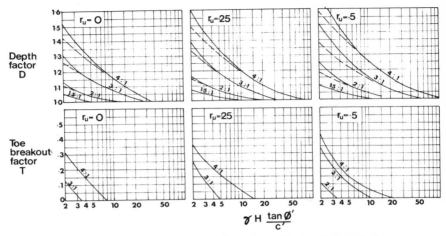

Figure 6.14 Location of critical slip surface with Spencer's method.

and 0.050, so that factors of safety could be obtained using the above formula to use in an interpolation on the basis of the actual cohesion.

Additional pairs of graphs were required for the cohesive cases to cope with depth factors D other than 1. There was a chart for $D = 1.25$ for the intermediate case, and for $D = 1.25$ and 1.50 for the most cohesive case. A primary interpolation on the basis of depth factor would then seem to be called for before the secondary interpolation on the basis of $c'/\gamma H$.

Treating the same example as for Spencer's method, the following results can be obtained. This is set out as in Table 6.1 since the principles are elementary.

The actual $c'/\gamma H$ value is 0.625. This is only a small extrapolation, which is probably valid (Figure 6.17), giving a factor of safety of about 1.4, for the

Table 6.1 Stability chart: Bishop and Morgenstern's method.

$\dfrac{c}{\gamma H}$	m	n	(D)	F
0	0.90	0.05	1.00	0.69
0.025	1.35	1.17	1.00	1.12
	1.45	1.32	1.25	1.19
0.050	1.61	1.25	1.00	1.36
	1.65	1.54	1.25	1.34
	1.85	1.60	1.50	1.53

(Values from the collected charts, Figure 6.16)

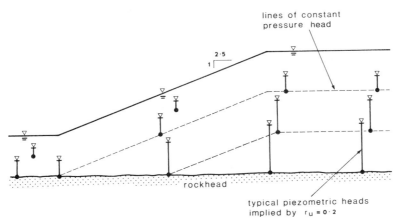

Figure 6.15 Effective stress problem solved with stability charts.

critical case, which is that of a depth factor of 1.25. Plotting a further graph, this time of F v. depth factor for each $c'/\gamma H$ case, and taking the minimum in each case could lead to a slight refinement in this value, but the effort is not thought worth while in the present case.

The use of Bishop and Morgenstern's charts is impaired by the small range of cohesion values that are covered. This has been corrected to a certain extent by additional charts published more recently by O'Connor and Mitchell (1977), who enlarge the range of cohesion values covered by the technique, but the charts are used in exactly the same way as the original charts. Larger cohesion values also tend to lead to the critical slip circle being more deeply seated than in cases where the strength is dominated by 'friction', and this accounts for the small difference in factor of safety predicted by Spencer's charts, and by means of Bishop and Morgenstern's charts. The former case allows the critical surface to be as deep-seated as possible, the latter charts constrain it to have a certain limiting depth factor. One can see from the total stress charts (Figure 6.16) that this has a bearing.

As a check on this, examine the charts presented by Spencer for finding the position of the critical slip surface (Figure 6.14). This position is given in terms of the toe break-out factor T, and the depth factor D for that case. Ordinarily, one would find these factors for the three r_u values, and interpolate, but since this is by way of illustration only, we will examine the $r_u = 0.25$ case. The governing variable is found to be $\gamma H \tan \phi'/c'$ which here may be calculated to be 5.8, and the critical depth factor appears to be of the order of 1.1. Toe circles are also indicated by the charts.

Figure 6.16 (*over*) Bishop and Morgenstern's stability charts. From A. W. Bishop and N. Morgenstern, *Géotechnique*, December 1960, reproduced with permission from The Institution of Civil Engineers.

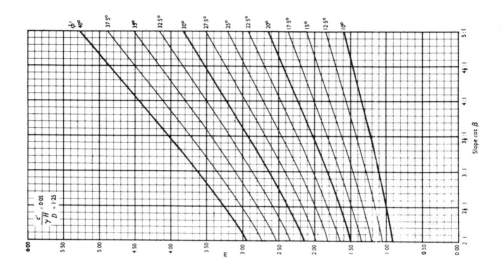

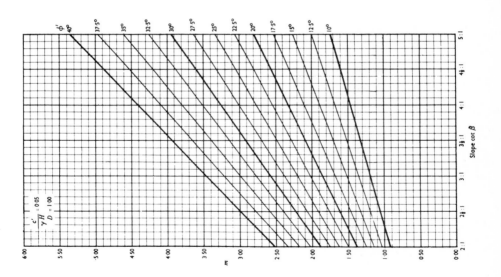

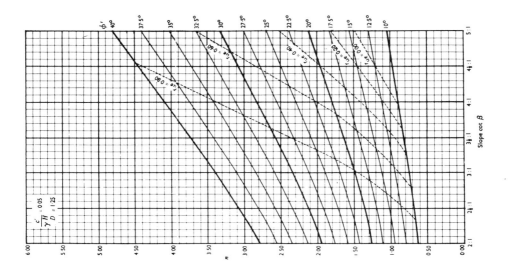

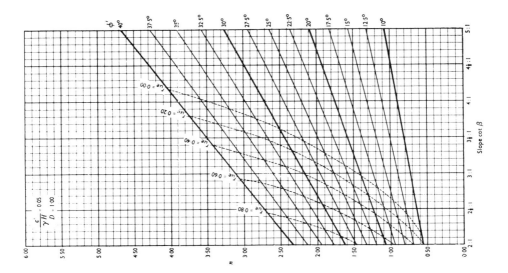

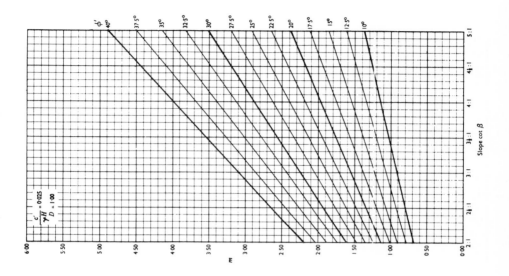

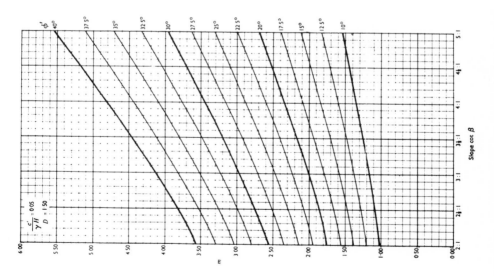

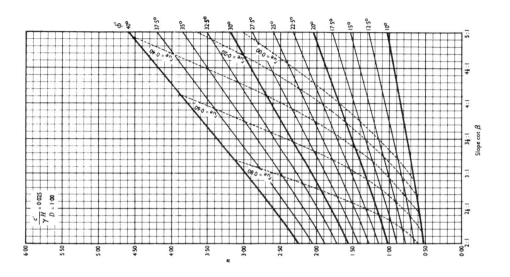

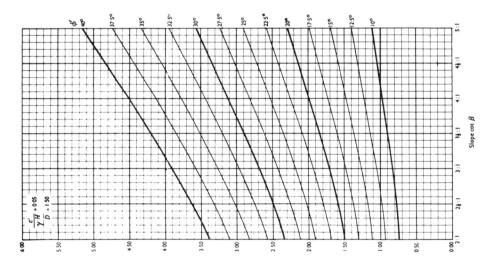

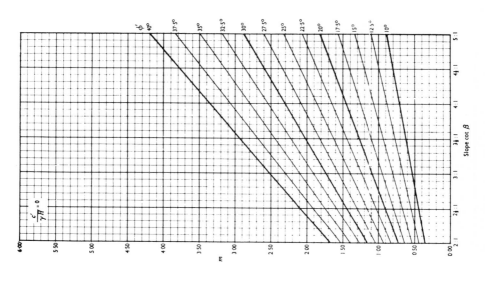

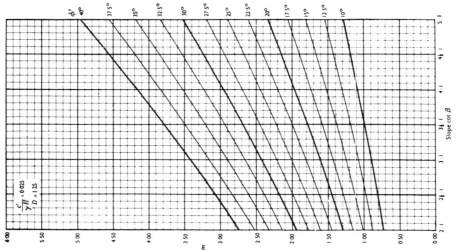

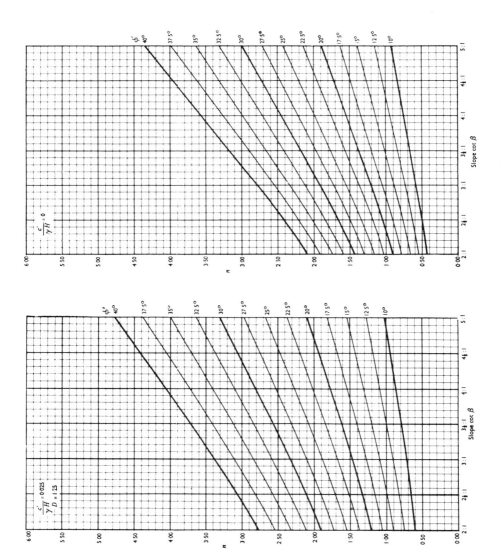

6.9 Additional techniques for use with stability charts

Taylor's charts can be used to provide approximate solutions to a range
of more complicated problems than their designer originally intended. Slopes
with berms, for instance, can be treated by dividing them up into facets, and
applying the charts to each facet in turn. Overall stability must then be
investigated by fitting a single line to the best average slope. Although this

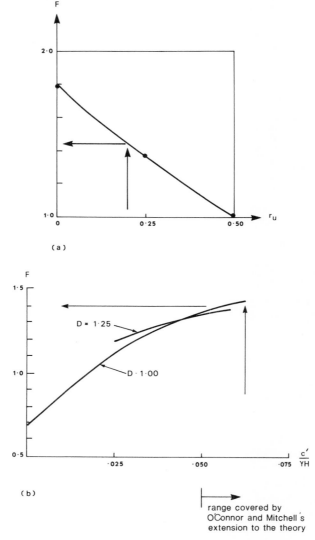

Figure 6.17 Interpolation graphs: effective stress problem.

may seem to be a very *ad hoc* treatment, it does, in fact, simulate the elements of a full analysis, giving safety factors to the different *modes* of failure, both superficial and deep-seated.

Non-homogeneous slopes can also be investigated using simple stability charts: an example of this is the common case of an embankment of non-cohesive fill placed on a cohesive (undrained) foundation. Although detailed solutions are available, for instance those of Pilot and Moreau (1973), their use is not entirely necessary. It will be found that the factor of safety for shallow modes of failure is quite adequately predicted by using stability charts for the fill slope itself, with a hard base assumed at a depth factor of 1; or even more simply by resorting to equation (5.58), i.e.

$$F = (1 - r_u) \frac{\tan \phi'}{\tan \alpha} \tag{6.14}$$

As a second stage, and to a first approximation, the factor of safety of the whole slope can be estimated by treating the whole slope as having the shear strength of the weaker foundation soil, although to keep the destabilizing forces in the analysis correct, the proper unit weight for the fill must be retained. This procedure works because the frictional shear strength on the steeply upward curving part of the slip surface is likely to be small and of minor importance to overall stability.

As suggested previously, seismic loadings can be simulated by the slope rotation technique.

Part submergence of the slope is also accommodated by a simple manipulation. In this case the mean unit weight of the soil is taken to be the average of the submerged and bulk unit weights, weighted on the basis of the heights of slope below and above the external water level.

Where access to a better method is not available, it is possible to resort to Taylor's curves using a modified angle of shearing resistance, ϕ'', where

$$\tan \phi'' = (1 - r_u) \tan \phi' \tag{6.15}$$

A quick trial using this modified angle of shearing resistance will quite readily demonstrate that a factor of safety of about 1.4 is required to obtain the same values for F_c and $F_{\phi''}$.

6.10 Computer program for slope stability calculations using Bishop's method

A short computer program for simple slope stability calculations using Bishop's method is included below. This has been kept as simple as possible, and will run on most computers including micros, although it may need to be rendered into BASIC to be usable on many of the smaller machines. The essential steps in a slope stability analysis can be traced through this program.

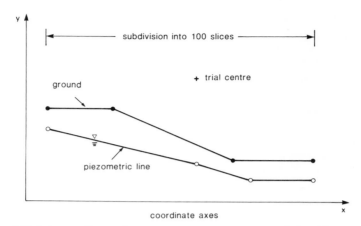

Figure 6.18 Reference diagram for use with computer program utilizing Bishop's method.

First of all, the slope shape must be input to the program. In the demonstration program, the ground surface of the slope is restricted to three straight-line segments, but in general, arbitrary and complicated slope profiles may need to be input, and details of the internal structure of the slope may be needed too. Then the porewater-pressure definition stage follows. Again in the demonstration program this is extremely simple, being again three straight-line segments. These do not necessarily have to follow the ground slope profile (see Figure 6.18).

Inputting soil properties is a simple matter, requiring just c, ϕ, and γ values for the single soil which is all that the program will accept. It would, however, be a trivial matter to make the program take a different unit weight for soil above and below the piezometric line, treating it as though it also formed the groundwater table.

The remaining input stage is the specification of trial surfaces. In its simplest form, and with slip circles, the centre coordinates and radius are all that are required. In more sophisticated programs a grid of centres with many slip circles of different radii from each centre may be input. There are other strategies for the specification of the slip circle geometry, for example forcing all slip circles to pass through a common point, usually the toe, or making them tangent to a given level in the soil profile.

In this simple program, there is an *a priori* subdivision of the slope into constant-width slices. The program then checks to see which of these are intersected by the slip circle, neglecting partial slices at the ends of the slip surface. The errors in so doing, and in assuming the properties of each slice to be located at its centre-line, are small if sufficient slices are taken. A maximum of 100 slices are used in this program, but more can obviously be incorporated with simple changes to the program structure. This 'fitting of the slip surface to the slope' is a calculation repeated many times: it pays the computer programmer to make the code here as efficient as possible. Ideally, this means as few calls to trigonometric functions as possible.

Particularly in more complicated procedures, for example the Morgenstern and Price procedure, where slices of considerable width can be accommodated, the subdivision into slices comes *after* the fitting of the slip surface to the slope. However, some organization of the data to enable this subdivision to be carried out economically will need to be done. That may be analogous to the stage in the demonstration program where the slope and intercept of each of the segments of the ground surface and of the piezometric line are evaluated.

The fitting of the slip circle to the slope in the demonstration program is in two stages. Firstly, it is assumed that no slices more distant from the slip circle centre than its radius can possibly be involved. The first and last of these slices are found (IST and LST in computer code). Secondly, the evaluation of the geometric properties of each slice is done at the same time as the examination of which slices are cut by the slip circle. No such evaluation is made where the slice is not cut, so that a ready check on which were found to be cut is present in the stored coefficients. A subset of slices within the range (IST:LST) can then be found.

In the same way that an initial assessment was made of which range of slices should be examined, later to be refined as more data became available, it is desirable now to have an initial factor of safety. This is obtained for Bishop's method through the conventional method, the equation of which is given above (equation (5.6)). Since the conventional method is well known to be conservative, an arbitrary decision to multiply it by 1.2 has been made. This then gives the starting factor of safety for iteration with Bishop's method.

Following convergence comes the final, output, phase of the program. In this example, the only additional information computed is the overturning moment. Other useful information, such as the effective and shear forces on each slice, could be displayed here, but such code would obscure the logical flow of the demonstration. Furthermore, very few data checks are carried out by this program. Largely, this is to ensure simplicity. So that the potential user has come control over the program, a number of messages about the logical flow of the program are displayed as it runs. As the computer will accept both upper- and lower-case instructions, these messages appear in lower case. They may safety be omitted, or replaced by error-checking code.

The program will quite happily accept slopes defined in such a way as to cause slip circles indicating failure from left to right (as shown in Figure 6.18), or from right to left. In the latter case, slices with a negative value for their base slopes, α, predominate, and a negative factor of safety is computed. However, the magnitude of the factor of safety is correct, and in more complex programs it may be found convenient to leave the sign unchanged, so that the direction of movement of slip circles on the slope section is clearly indicated.

```
       PROGRAM BISHOP
C-------------------------------------------------------------------
C              THIS PROGRAM IN "STABILITY OF SLOPES"
C              BY E.N. BROMHEAD
C              IS A SIMPLE, ILLUSTRATIVE, PROGRAM TO SHOW
C              THE ITERATIVE TECHNIQUE USED IN BISHOP'S METHOD
C
C---------------------MAKE ARRAY DECLARATIONS + FUNCTION FOR X-------
C
       DIMENSION COEF1(100),COEF2(100), GX(4), GY(4), WX(4), WY(4)
       DIMENSION GRAD1(3)  ,GRAD2(3)  , H1(3), H2(3)
       X(I,B)=(I-0.5)*B
C
C---------------------INPUT GEOMETRIC AND SOIL PROPERTY DATA---------
C
       OPEN(5,FILE='CON:',STATUS='OLD')
       OPEN(6,FILE='CON:',STATUS='NEW')
       WRITE(6,500)
       READ (5,*)    (GX(I),GY(I),I=1,4)
       B      =  0.01*(GX(4)-GX(1))
       WRITE(6,510)
       READ(5,*)     (WX(I),WY(I),I=1,4)
       WRITE(6,520)
       READ(5,*)    COHES, PHI, GAMMA
       PHI    =  0.017453*PHI
       PHI    =  SIN(PHI)/COS(PHI)
       GAMMAW =  9.807
C
C---------------------SORT OUT EQUATIONS FOR SEGMENTS OF SLOPE--------
C
       DO 5 I=1,3
       GRAD1(I)=  (GY(I+1)-GY(I)) / (GX(I+1)-GX(I))
       H1(I)   =  GY(I) - GX(I)*GRAD1(I)
       GRAD2(I)=  (WY(I+1)-WY(I)) / (WX(I+1)-WX(I))
5      H2(I)   =  WY(I) - WX(I)*GRAD2(I)
C
C---------------------NOW INPUT THE SLIP SURFACE DEFINITION----------
C
10     WRITE(6,530)
       READ(5,*)    XCEN, YCEN, RAD
C
C---------------------AT WHICH SLICES DOES THE CIRCLE BEGIN/END?-----
C
       XFIRS  =  XCEN - RAD
       XLAST  =  XCEN + RAD
       IST    =  (XFIRS-GX(1))/B + 1
       LST    =  (XLAST-GX(1))/B
       IF (IST .LT. 1)   IST =   1
       IF (LST .GT. 100) LST = 100
         write(6,550) 'first',ist
         write(6,550) ' last',lst
C
C---------------------INITIALISE HELD VALUES TO ZERO----------------
C
       RAD2=RAD*RAD
       DO 20 I=1,100
       COEF1(I) = 0.0
20     COEF2(I) = 0.0
C
```

```
C----------------------NOW SORT OUT THE RELATIONSHIP OF THE------------
C                       SLIP CIRCLE TO THE SLOPE, AND TO THE
C                       SLICES.   THE SUBDIVISION INTO EQUAL
C                       WIDTH SLICES IS IMPLICIT IN THIS.
C
      CTOP   = 0.0
      BOTTOM = 0.0
      DO 100 I=IST,LST
      XCOR  = X(I,B)+GX(1)
      XDIFF = XCEN - XCOR
      IF (ABS(XDIFF) .GT. RAD) GO TO 100
      YDIFF = SQRT(RAD2-XDIFF*XDIFF)
      YCOR  = YCEN - YDIFF
C
C----------------------FIND WHICH SEGMENT OF SLOPE IS ABOVE CURRENT----
C                       POINT.  NOTE IMPLIED EXTRAPOLATION AT START &
C                       END OF THE SEGMENTS.
C
      IF (XCOR .LT. GX(2)) THEN
            J=1
            ELSE IF (XCOR .GT. GX(3)) THEN
            J=3
                  ELSE
                  J=2
                  ENDIF
      YGL = H1(J)+XCOR*GRAD1(J)
C
C----------------------THE NEXT IF IS IMPORTANT.  ONLY IF SLIP --------
C                       SURFACE IS BELOW G.L. DO WE EVALUATE THE
C                       COEFFICIENTS FOR THE NUMERATOR AND THE
C                       DENOMINATOR OF THE EQUATION FOR F.S.
C
      IF (YGL .GT. YCOR) THEN
            WEIGHT = (YGL-YCOR)*B*GAMMA
            SINALF = XDIFF/RAD
            COSALF = YDIFF/RAD
            TANALF = XDIFF/YDIFF
            CBI    = COHES*B
C
C----------------------SIMILARLY, THE POSITION OF THE PIEZO LINE-------
C                       RELATIVE TO THE SLIP SURFACE IS FOUND
C
      IF (XCOR .LT. WX(2)) THEN
            J=1
            ELSE IF (XCOR .GT. WX(3)) THEN
            J=3
                  ELSE
                  J=2
                  ENDIF
            YWL = H2(J)+XCOR*GRAD2(J)
      UBI = (YWL-YCOR)*B*GAMMAW
C
C                       IF RU WAS BEING USED, UBI=WEIGHT*RU
C
C----------------------NOTE A TEST FOR PIEZO LINE BELOW SLIP-----------
C
      IF (UBI .LT. 0.0) UBI=0.0
      COEF2(I)= TANALF*PHI
      COEF1(I)= ((WEIGHT-UBI)*PHI+CBI) /COSALF
      CTOP    = CTOP    + (WEIGHT*COSALF-UBI/COSALF)*PHI + CBI/COSALF
      BOTTOM  = BOTTOM + WEIGHT*SINALF
C
```

```
C----------------------THE FOLLOWING ENDIF IS THE END OF THE-----------
C                        TEST FOR INCLUSION IN THE MAIN LOOP
C
          ENDIF
100   CONTINUE
C
C--------------------NOW ELIMINATE DUD SLICES AT EACH END OF---------
C                        SLIP, CHECK ON INTERSECTION ETC.
C
105       IF (COEF1(IST) .EQ. 0.0) THEN
          IST=IST+1
          IF (IST .LT. LST) GO TO 105
          WRITE(6,560)
          GO TO 10
          ENDIF
110       IF (COEF1(LST) .EQ. 0.0) THEN
          LST=LST-1
          GO TO 110
          ENDIF
C
C----------------------NOTE A SMALLER RANGE OF SLICES NOW--------------
C
      write(6,550) 'first',ist
      write(6,550) ' last',lst
      OTM   = BOTTOM*RAD
      CONVFS= CTOP/BOTTOM
       write(6,540) 'convf',convfs
       write(6,540) 'denom',bottom
      FINIT = 1.2*CONVFS
      NITS  = 1
150   BTOP  = 0.0
      DO 200 I=IST,LST
      BTOP  = BTOP + COEF1(I) / (1.0 + COEF2(I)/FINIT)
200   CONTINUE
      FNEXT = BTOP / BOTTOM
C
C----------------------TEST FOR CONVERGENCE, PRINTOUT OR ITERATE-------
C
      IF (ABS(FNEXT-FINIT) .LT. 0.001) THEN
          WRITE(6,570) XCEN,YCEN,RAD,CONVFS,FNEXT,NITS,OTM
          IF (OTM .LT. 0.0) WRITE(6,580)
          GO TO 10
          ELSE
          NITS=NITS+1
          write(6,540) 'finit',finit
          FINIT=FNEXT
          write(6,540) 'fnext',fnext
          GO TO 150
          ENDIF
C
C--------------------------OUTPUT FORMATS------------------------------
C
500       FORMAT(// ' SLOPE STABILITY ANALYSIS BY BISHOP''S METHOD'/
     1              ' INPUT PAIRS OF COORDINATES AS FOLLOWS:'/
     2              ' LEFT HAND EDGE OF SECTION,'/
     3              ' TOE OF SLOPE, OR CREST,'/
     4              ' CREST OF SLOPE, OR TOE,'/
     5              ' RIGHT HAND EDGE OF SECTION'/)
510       FORMAT(' NOW FOUR PAIRS OF COORDS TO DEFINE THE PIEZO LINE'/
     1           ' IN EXACTLY THE SAME FORM AS FOR G. L.'/)
520       FORMAT(' NOW INPUT COHESION, FRICTION ANGLE, AND UNIT WT'/
     1              ' ALL UNITS TO BE IN kN AND m PLEASE'/
     2              ' DEGREES FOR FRICTION ANGLE'/)
```

```
530        FORMAT(' NOW INPUT THE X,Y COORDINATES OF THE CENTRE,'/
   1                ' AND THE RADIUS FOR THE NEXT SLIP CIRCLE'/)
540        format(' intermediate check ',a5,f13.4)
550        format(' intermediate check ',a5,i10)
560        FORMAT(' CIRCLE DOESN''T CUT SLOPE')
570        FORMAT(' CENTRE:   X=',F10.3/
   1                '          Y=',F10.3/
   2                '          R=',F10.3/
   3                ' FS BY CONVENTIONAL METHOD =',F10.4/
   4                '            BISHOP''S METHOD =',F10.4/
   5                ' NUMBER OF ITERATIONS       =',I5/
   6                ' OVERTURNING MOMENT         =',E15.7/)
580        FORMAT(' --[N.B. MINUS MEANS MOVING IN - X DIRECTION]--'/)
           END
```

6.11 Computer program for slope stability calculations using Janbu's method

This is similar in structure to the program with Bishop's method. The principal difference is that the definition of the slip surface assumes a three-segment surface, starting and finishing where the ground level begins and ends, rather than using a slip circle. Two minor differences will also be found: these are in the Janbu equation, which uses 'tan' in the denominator, rather than 'sin', and 'sec^2' to replace 'sec', and in the function which is used to compute the correction factor. This factor is interpolated from values extracted from Figure 5.7, and requires a little geometric manipulation to obtain the d/L ratio for the whole slide.

In order to avoid lengthy computations dealing with the intersection of the slip surface and the ground surface of the slope, the assumption is made that the slip surface emerges at the endpoints used to define ground level. An excessive number of slices is used, so that the actual position of 'corners' in the various lines can be neglected, since, as with the first program for Bishop's method, the properties of a slice are those that apply on its centre-line. Doing this also removes the need to compute the intersection points of the piezometric line and the slip surface, since the errors are of second- or third-order significance only.

Once again, the result of the calculation made using the conventional method is taken as the starting value for the iterative procedure. Applying the correction factor takes a disproportionate amount of code, even though there are only two points from which it is necessary to calculate the offset from the line joining the two ends of the slip surface. In contrast to the case for slip circles, non-circular slip-surface analysis programs are usually run for just a few slip surfaces at any time. It therefore follows that code executed once per slip surface can be less efficient, if this brings clarity, than in the slip-circle case, and not bring too great a penalty in execution time. However, code executed many times, such as that used once or more per slice, must still be written with an eye to the eventual response time of the program.

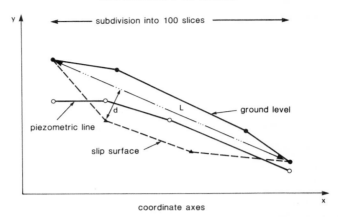

Figure 6.19 Reference diagram for use with computer program utilizing Janbu's method.

Therefore, although it might seem preferable to find a polynomial which fits the data values for the correction factor, and to use that instead of the tabulated values which I have scaled from Janbu's chart, very little benefit will be obtained in practice.

As in the case of the program for use with Bishop's method, this one will accept slip surfaces drawn so that movement takes place from left to right on the section (see Figure 6.19), or from right to left. A similar consideration with respect to the sign of the output factor of safety applies here.

```
      PROGRAM JANBU
C----------------------------------------------------------------------
C              THIS PROGRAM IN "STABILITY OF SLOPES"
C              BY E.N. BROMHEAD
C              IS A SIMPLE, ILLUSTRATIVE, PROGRAM TO SHOW
C              THE ITERATIVE TECHNIQUE USED IN JANBU'S METHOD
C
C----------------------------MAKE ARRAY DECLARATIONS + FUNCTION FOR X-------
C
      DIMENSION COEF1(100),COEF2(100), GX(4), GY(4), WX(4), WY(4)
      DIMENSION GRAD1(3)  ,GRAD2(3)  , H1(3), H2(3)
      DIMENSION GRAD3(3)  ,             H3(3), SX(4), SY(4)
      X(I,B)=(I-0.5)*B
C
C------------------------INPUT GEOMETRIC AND SOIL PROPERTY DATA---------
C
      OPEN(5,FILE='CON:',STATUS='OLD')
      OPEN(6,FILE='CON:',STATUS='NEW')
      WRITE(6,500)
      READ (5,*)   (GX(I),GY(I),I=1,4)
      B       =  0.01*(GX(4)-GX(1))
      WRITE(6,510)
      READ(5,*)    (WX(I),WY(I),I=1,4)
      WRITE(6,520)
      READ(5,*)      COHES, PHI, GAMMA
      PHI     =  0.017453*PHI
      PHI     =  SIN(PHI)/COS(PHI)
      GAMMAW  =  9.807
C
```

```
C-----------------------SCRT OUT EQUATIONS FOR SEGMENTS OF SLOPE--------
C
      DO 5 I=1,3
      GRAD1(I)=  (GY(I+1)-GY(I)) / (GX(I+1)-GX(I))
      H1(I)    =  GY(I) - GX(I)*GRAD1(I)
      GRAD2(I)=  (WY(I+1)-WY(I)) / (WX(I+1)-WX(I))
5     H2(I)    =  WY(I) - WX(I)*GRAD2(I)
C
C-----------------------NOW INPUT THE SLIP SURFACE DEFINITION-----------
C
10    WRITE(6,530)
      READ(5,*)    SX(2),SY(2),SX(3),SY(3)
      SX(1)   =  GX(1)
      SY(1)   =  GY(1)
      SX(4)   =  GX(4)
      SY(4)   =  GY(4)
      GRAD4   =  (SY(4)-SY(1)) / (SX(4)-SX(1))
      H4      =  SY(1) - SX(1)*GRAD4
      DO 15 I = 1,3
      GRAD3(I)=  (SY(I+1)-SY(I)) / (SX(I+1)-SX(I))
15    H3(I)   =  SY(I) - SX(I)*GRAD3(I)
C
C-----------------------NOW SORT OUT THE RELATIONSHIP OF THE-----------
C                       SLIP SURFACE TO THE SLOPE, AND TO THE
C                       SLICES.  THE SUBDIVISION INTO EQUAL
C                       WIDTH SLICES IS IMPLICIT IN THIS.
C
      CTOP    = 0.0
      BOTTOM  = 0.0
      DO 100 I=1,100

      XCOR  = X(I,B)+GX(1)
C
C-----------------------NOW FIND SLIP SURFACE AT CURRENT X--------------
C
      IF (XCOR .LT. SX(2)) THEN
           JSEG=1
           ELSE IF (XCOR .GT. SX(3)) THEN
           JSEG=3
                 ELSE
                 JSEG=2
                 ENDIF
      YCOR = H3(JSEG)+XCOR*GRAD3(JSEG)
C
C-----------------------NOW FIND G.L. AT CURRENT X---------------------
C
      IF (XCOR .LT. GX(2)) THEN
           J=1
           ELSE IF (XCOR .GT. GX(3)) THEN
           J=3
                 ELSE
                 J=2
                 ENDIF
      YGL = H1(J)+XCOR*GRAD1(J)
C
      IF (YGL .LT. YCOR) THEN
           WRITE(6,590)
           GO TO 10
                       ELSE
           WEIGHT=(YGL-YCOR)*B*GAMMA
           TANALF=GRAD3(JSEG)
           ALF   =ATAN(TANALF)
           SINALF=SIN(ALF)
```

```
          COSALF=COS(ALF)
          CBI  =COHES*B
C
C----------------------SIMILARLY, THE POSITION OF THE PIEZO LINE------
C                         RELATIVE TO THE SLIP SURFACE IS FOUND
C
      IF (XCOR .LT. WX(2)) THEN
          J=1
          ELSE IF (XCOR .GT. WX(3)) THEN
          J=3
              ELSE
              J=2
              ENDIF
          YWL = H2(J)+XCOR*GRAD2(J)
      UBI=(YWL-YCOR)*B*GAMMAW
C
C                         IF RU WAS BEING USED, UBI=WEIGHT*RU
C
C----------------------NOTE A TEST FOR PIEZO LINE BELOW SLIP-----------
C
      IF (UBI .LT. 0.0) UBI=0.0
      COEF2(I)= TANALF*PHI
      COEF1(I)= ((WEIGHT-UBI)*PHI+CBI) /COSALF/COSALF
      CTOP    = CTOP    + (WEIGHT*COSALF-UBI/COSALF)*PHI + CBI/COSALF
      CBOT    = CBOT    + WEIGHT*SINALF
      BOTTOM  = BOTTOM  + WEIGHT*TANALF
C
C----------------------THE FOLLOWING ENDIF IS THE END OF THE-----------
C                         TEST FOR INCLUSION IN THE MAIN LOOP
C
          ENDIF
100   CONTINUE
      CONVFS= CTOP/CBOT
        write(6,540) 'convf',convfs
        write(6,540) 'denom',cbot
        write(6,540) 'denom',bottom
      FINIT = 1.2*CONVFS
      NITS  = 1
150   BTOP  = 0.0
      DO 200 I=1,100
      BTOP  = BTOP + COEF1(I) / (1.0 + COEF2(I)/FINIT)
200   CONTINUE
      FNEXT = BTOP / BOTTOM
C
C----------------------TEST FOR CONVERGENCE, PRINTOUT OR ITERATE-------
C
      IF (ABS(FNEXT-FINIT) .LT. 0.001) THEN
        F0   = FOFIT(H4,GRAD4,GX,GY,SX,SY,COHES,PHI)
        FNEXT= F0*FNEXT
          WRITE(6,570) CONVFS,FNEXT,NITS
          IF (BOTTOM .LT. 0.0) WRITE(6,580)
          GO TO 10
          ELSE
          NITS  = NITS+1
          write(6,540) 'finit',finit
          FINIT = FNEXT
          write(6,540) 'fnext',fnext
          GO TO 150
          ENDIF
C
C----------------------OUTPUT FORMATS---------------------------------
C
500       FORMAT(// ' SLOPE STABILITY ANALYSIS BY JANBU''S METHOD'/
     1                ' INPUT PAIRS OF COORDINATES AS FOLLOWS:'/
```

```
      2                      ' LEFT HAND EDGE OF SECTION, (ONE END OF SLIP)'/
      3                      ' TOE OF SLOPE, OR CREST,'/
      4                      ' CREST OF SLOPE, OR TOE,'/
      5                      ' RIGHT HAND EDGE OF SECTION (& OTHER END OF SLIP)'/)
510      FORMAT(' NOW FOUR PAIRS OF COORDS TO DEFINE THE PIEZO LINE'/
     1              ' IN EXACTLY THE SAME FORM AS FOR G. L.'/)
520      FORMAT(' NOW INPUT COHESION, FRICTION ANGLE, AND UNIT WT'/
     1              ' ALL UNITS TO BE IN kN AND m PLEASE'/
     2              ' DEGREES FOR FRICTION ANGLE'/)
530      FORMAT(' NOW INPUT THE X,Y COORDINATE PAIRS FOR THE MIDDLE'/
     1              ' TWO POINTS OF THE SLIP SURFACE'/
     2    ' (THE FIRST AND LAST POINTS ARE THE SAME AS FOR THE G.L.)'/)
540      format(' intermediate check ',a5,f13.4)
550      format(' intermediate check ',a5,i10)
570      FORMAT(' FS BY CONVENTIONAL METHOD =',F10.4/
     1          '              JANBU''S METHOD =',F10.4/
     2          ' NUMBER OF ITERATIONS      =',I5/)
580      FORMAT(' --[N.B. MINUS MEANS MOVING IN - X DIRECTION]--'/)
590      FORMAT(' SLIP SURFACE TOO HIGH ')
         END
         FUNCTION FOFIT(H4,GRAD4,GX,GY,SX,SY,COHES,PHI)
C
C----------------------THIS IS A FUNCTION FOR EVALUATING THE----------
C                       CORRECTION FACTOR FOR JANBU'S METHOD
         DIMENSION GX(4),GY(4),SX(4),SY(4)
         DIMENSION FO(9,2)
C
C----------------------JANBU'S CHART IS SUMMED UP IN THE--------------
C                       FOLLOWING TABLE
C
         DATA FO/1.000,1.018,1.029,1.038,1.044,1.048,1.051,1.054,1.054,
     1           1.000,1.036,1.064,1.084,1.100,1.110,1.119,1.124,1.127/
C
         ELL = SQRT((GX(4)-GX(1))**2 + (GY(4)-GY(1))**2)
         DH1 = H4+GRAD4*SX(2)-SY(2)
         DH2 = H4+GRAD4*SX(3)-SY(3)
         IF (DH2 .GT. DH1) DH1=DH2
         D   = DH1*COS(ATAN(GRAD4))
         DBYELL=D/ELL
C
C----------------------DBYELL IS THE d/L RATIO-----------------------
C
         X = DBYELL/0.05
         N = X+1
         IF (N .GT. 8) N=8
         PART= (X+1.0-N)
         FO1 = FO(N,1) + PART*(FO(N+1,1)-FO(N,1))
         FO2 = FO(N,2) + PART*(FO(N+1,2)-FO(N,2))
C
C----------------------WHICH CURVE WE FOLLOW DEPENDS ON WHETHER--------
C                      C OR PHI IS ZERO. WITH BOTH, THE CURVE
C                      IS JUST THE AVERAGE OF THE OTHER TWO
C
         IF (COHES .EQ. 0.0) THEN
                 FOFIT=FO1
                 ELSEIF (PHI .EQ. 0.0) THEN
                         FOFIT=FO2
                         ELSE
                                 FOFIT=0.5*(FO1+FO2)
                                 ENDIF
         WRITE(6,600) DBYELL,FOFIT
600      FORMAT(' d/L RATIO: ',F10.4,'   CORRECTION FACTOR ',F10.3)
         RETURN
         END
```

7 Water pressures in slopes

7.1 Introduction

In the analysis of the stability of slopes in terms of effective stresses, the porewater-pressure distribution is of fundamental importance and its evaluation is one of the prime objectives in the early stages of any stability study. This may involve extensive field measurements or modelling the seepage pattern.

A seepage pattern will develop if the necessary overall hydraulic gradients are present. In a slope, it is inconceivable that such a flow pattern would not develop, even solely under the effects of rainfall, because water infiltrating at the head or crest of a slope must have a higher potential energy than that infiltrating at the toe. Water-impounding embankments have an even more apparent hydraulic imbalance.

In addition to such flows, the water in the pores or void spaces in a soil may be under pressure because of stress changes experienced by the soil mass as a whole. Typical of such stress changes in a slope are those that accompany its formation, whether this is by the placing of fill or by excavation. Accordingly, the water pressures may be greater, or less, than those which would be in equilibrium with external hydraulic conditions. As a result of this, it may be found that more-or-less gradual changes in these non-equilibrium water pressures take place.

Other sources of non-equilibrium porewater pressures include the effects of vegetation (transpiration), although except locally, the effects of this are probably not significant in temperate climates.

Rapid changes in the external hydraulic constraints can cause an initially steady seepage pattern to become out of equilibrium, and need to adjust itself in response: drawdown of the water level retained by a dam, or rainfall on a parched soil slope, are two cases in point.

Hence, seepage, or the flow of water through soil or rock, can be treated in two separate parts: the steady flow of water in a pattern that is in equilibrium with the external hydraulic constraints (*steady seepage*), and the equilibration of non-equilibrium porewater pressures to the steady state (*unsteady seepage, consolidation* or *swelling*). In steady seepage, the flow pattern that results is dependent on the distribution of soil permeabilities in the slope, and the nature and extent of the hydraulic boundary conditions which control that seepage. Even so, the identification and quantification of these factors in order to make an entirely satisfactory prediction of porewater pressures for slope analysis and design is very difficult indeed. The situation for unsteady seepage is even more complex: the generation of non-equilibrium porewater pressures being the result of a sometimes long, but always involved, stress history, and the equilibration process may be partly complete, or proceeding concurrently with the stress changes in the slope. For many problems, the prediction of field porewater pressures may not be economic or practical, and it may be preferable or expedient to attempt to measure the porewater pressures *in situ* and then leave the predictive element to the likely future changes in those pressures.

The measurement of field porewater pressures is covered in Chapter 9, so only the modelling of seepage will be covered here. This modelling can be on the basis of physical (scaled-down) models, it may utilize analogies between seepage and other flow phenomena (for example the flow of electricity through a conducting medium) or it can be on a purely mathematical basis. The whole field of seepage analysis in detail is beyond the scope of this book, and indeed is covered adequately elsewhere (Cedergren, 1967; Polubarinova-Kochina, 1962). It is, however, necessary to recapitulate some of the principles of seepage theory in order to apply its results to the stability of slopes. This must start with *steady* seepage, with the theory extended later to the *unsteady* cases.

7.2 Steady seepage

Flow results from the movement of water with a high 'energy' to a location where water is present with lower energy levels. A measure of this is the sum of potential and kinetic energy per unit weight of water, termed the 'total head' in elementary hydraulics:

$$\text{Total head } H = z + \frac{u}{\gamma} + \frac{v^2}{2g} \tag{7.1}$$

In soil mechanics, the velocities of flow are sufficiently small (except perhaps in flow through coarse rockfill) so that the velocity or kinetic energy term of this can be safety neglected, leaving only potential energy terms. It is also more conventional to work in *pressure heads* than in pressures (p in the above), so that a new quantity, or *potential*, being the sum of the elevation z of a point in the water and its pressure head h is defined thus:

$$\phi = z + h \tag{7.2}$$

It is the gradient of the *potential* across a soil which causes flow to occur, the velocity in a particular direction (s in the following) being related to the *hydraulic* or *potential gradient* in that direction, by Darcy's law (Darcy, 1857):

$$v_s = -k_s \frac{\partial \phi}{\partial s} \tag{7.3}$$

A steady seepage pattern will eventually develop in any natural or artificial slope across which there is an overall hydraulic potential gradient.
There must therefore be a set of internal potentials and porewater pressures
that are in some way in equilibrium with the mean hydraulic boundary
conditions. The term 'mean' is used advisedly, since on the air face of the
slope there may well be seasonal moisture-content variations, and even deep
inside the slope, strata of high permeability can be connected to outside
sources of fluctuating pressure supply. Examples that spring readily to mind
are in coastal slopes, and those adjacent to rivers and ponds which have
seasonally varying water levels. It is assumed that any slope which contains
within it porewater pressures not equal to this steady seepage pattern, or
equilibrium distribution of porewater pressures, will gradually come closer
to it through the migration of soil moisture. The timescale of this migration,
or equilibration, depends on the permeability of the soil, its compressibility,
and on the distance water has to travel to escape from the soil: the
drainage path length. Broadly speaking, however, in respect of the timescale
of normal construction operations, the seepage patterns in gravels and
sands are to all intents and purposes the same as the steady seepage pattern.
On the other hand, in silts and clays, the steady seepage pattern may only be
reached after a great deal of time elapses.

To derive the governing equations, consider first a cube of soil in this
slope, orientated with respect to orthogonal coordinate axes as shown in
Figure 7.1. It is easy to show that with the considerations of continuity

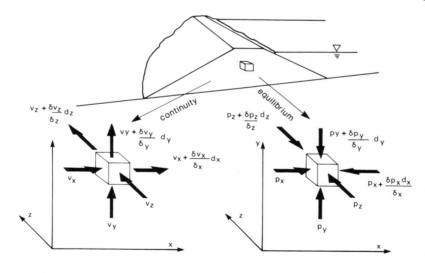

Continuity + Darcy's law ⟶ Laplace equation & finite difference solutions

Equilibrium + Darcy's law ⟶ Energy principle & finite element method

Figure 7.1 Basis of seepage analysis.

(what goes in must come out) that

$$\frac{dv_x}{dx} + \frac{dv_y}{dy} + \frac{dv_z}{dz} = 0 \tag{7.4}$$

With the insertion of Darcy's law, the following partial differential equation results:

$$\frac{d}{dx}\left[k\frac{d\phi}{dx}\right] + \frac{d}{dy}\left[k\frac{d\phi}{dy}\right] + \frac{d}{dz}\left[k\frac{d\phi}{dz}\right] = 0 \tag{7.5}$$

Taking the condition that the individual permeability coefficients do not vary in space, it is possible to take them out of the second differentiation, and if in addition, the permeability is identical in each direction, the conventional Laplace-type seepage equation results. For two-dimensional flow, this would give

$$\frac{d^2\phi}{dx^2} + \frac{d^2\phi}{dy^2} = 0 \tag{7.6}$$

(In principle, a solution of this equation within a particular *seepage domain*, subject to the *hydraulic boundary conditions* around its periphery, would give a complete set of potentials everywhere, from which the porewater pressures needed for stability analysis could be extracted. Solutions to this equation are, however, notoriously difficult to produce analytically (disbelievers should consult, for instance, Polubarinova-Kochina (1962), or Harr (1968)), and alternatives need to be sought. It is possible to represent the Laplace equation (7.6) approximately by a finite-difference mesh (Scott, 1967; Tomlin, 1970) but additional problems are introduced in dealing with arbitrarily shaped physical boundaries to the problem.

The conventional response to this difficulty is to use the property of the potential field arising from the Laplace equation that it is possible to define a function given the symbol ψ (termed the *stream function*) which is also governed by a Laplace-type equation, but the iso-lines (streamlines or flow-lines) of which are everywhere orthogonal to the iso-lines of the potential function (equipotentials). Indeed, through following simple rules, it is possible to sketch freehand an acceptable solution to the seepage problem. A part of the elementary training of every civil engineer and engineering geologist in geomechanics is likely to include the production of such *flow nets*.

Unfortunately, freehand sketching of the solutions to seepage problems becomes difficult where permeability coefficients are anisotropic: where streamlines and equipotentials are no longer orthogonal for instance, and in seepage domains which are zoned into quite different materials. In such cases, resort must be made to numerical methods. The author's preference in this respect is for the finite-element method. Since a treatment of that demands the direct inclusion of certain of the boundary conditions to the seepage domain, its presentation is deferred to section 7.4.

7.3 Boundary conditions

The general equation (7.6) for seepage cannot be solved without some knowledge of the hydraulic boundary conditions to the seepage regime. In simple cases these are of two types: either of *known potential* (usually constant in value, but not necessarily so) or of *known flow rate* (usually zero—as across an impermeable barrier). When seepage is defined solely in terms of these two types of boundary, the problem is termed one of *confined* seepage (Figure 7.2).

It should be noted, however, that at a boundary of the known potential type where the potentials are not constant, the flowlines *do not* cross the boundary at right angles. A typical example of this is where seepage emerges at an air face and runs down its slope. Pressures on such a boundary are at atmospheric pressure, and the potentials are thus defined solely by the elevation at which the water emerges. This is sometimes termed a *seepage surface* (Figures 7.2 and 7.3).

Unconfined seepage is intrinsically more complex. This is where water flows with a *free* or *phreatic* surface (Figure 7.3) and therefore occupies only a part of the available seepage domain. At such a surface pressures are atmospheric, and the potentials are definable in terms of the elevation of the free surface alone. In contrast to a seepage surface, there is no flow across the free surface which can therefore be seen to be a flowline or streamline. In most

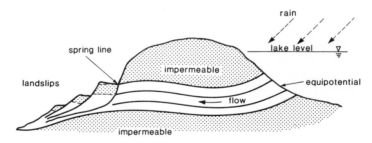

Figure 7.2 Hydraulic boundary conditions for confined seepage.

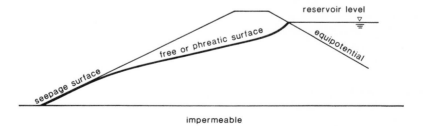

Figure 7.3 Hydraulic boundary conditions for unconfined seepage.

techniques of seepage analysis the position of such a surface cannot be predicted *a priori* and so must be found iteratively in the solution process. During the sketching of a flow net, for instance, several trial positions for the free surface may have to be investigated before the two conditions are both satisfied.

The free-surface concept is aesthetically satisfying, but has little reality. In most soils a free surface cannot be found because of a significant capillary rise in the soil. Indeed, in fine grained soils, the whole soil profile above the zero-pressure line may be fully saturated, but have less than atmospheric pressure in it because of the capillary effect. Soil-moisture movements on a seasonal basis in this capillary zone can cause significant changes in the effective stresses as suctions are created or destroyed, and these will obviously have a major influence on strength close to the surface. Even in a partly saturated zone similar, but yet more complex effects, will occur. At the present time, such effects lie outside the capabilities of routine engineering analysis except on a research basis. They should not be forgotten, however, as a mechanism for explaining the apparent stability of soil slopes in dry climates. It is, of course, unwise to rely on soil-water suctions as a slope-stabilization measure.

7.4 The finite-element method in seepage

The basis of the finite-element approach to seepage problems is not the continuity approach which yields the Laplace equation, but rather an energy approach. Considering the rate of dissipation of energy in the elementary soil cube of Figure 7.1, it can be shown that the rate of doing work, per unit volume of the element, to overcome the frictional drag resistance to flow is

$$\frac{\Delta W}{\Delta V} = k_x \left\{ \frac{d\phi}{dx} \right\}^2 + k_y \left\{ \frac{d\phi}{dy} \right\}^2 + k_z \left\{ \frac{d\phi}{dz} \right\}^2 \tag{7.7}$$

hence, integrating over the whole seepage domain, or volume V_r, the rate of doing work is:

$$W_1 = \int_{V_r} k_x \left\{ \frac{d\phi}{dx} \right\}^2 + k_y \left\{ \frac{d\phi}{dy} \right\}^2 + k_z \left\{ \frac{d\phi}{dz} \right\}^2 d_{vol} \tag{7.8}$$

Furthermore, along sections of the boundary at potential ϕ where fluid enters or leaves with discharge rate q, there is an overall contribution (or loss) of energy at a rate of

$$W_2 = \int_{S_r} q\phi d_{area} \tag{7.9}$$

Naturally, the total rate of energy dissipation $W_1 + W_2$ must be at a minimum, so the sum of these two integrals for the 'correct' distribution of potentials

G

in the seepage domain must be less than for any other, arbitrarily obtained, set of potentials.

If the seepage domain is subdivided approximately into a number of elements of finite size and simple shape, which connect at a number of discrete point (N), then the integration procedure, instead of being carried out over the whole seepage domain, could be performed in each element, and the results summed over all the elements. Naturally, the simpler the shape of the elements, the simpler the integration in each becomes.

Now suppose that the potentials within each element are entirely defined by the values of potential at the nodes to which that element connects. In a triangular element, for instance, the nodes could be at each apex, and the distribution of potentials could vary linearly from node to node. More complex elements could be imagined, and indeed, are in regular use in the field of engineering analysis (Zienkiewicz, 1972).

With this set of N nodes, we need N simultaneous equations in the unknown potentials in order to provide a solution to the potential distribution in the whole seepage domain. Such a set of equations can be extracted from the work equation (equation (7.8)) by differentiating it with respect to each of the unknown node potentials in turn. Since the subdivision into elements in effect isolates each node from all but the immediately connected elements, the resulting set of equations is quite simple, and contains coefficients which reflect only the size, shape, and permeability of elements immediately adjacent to each node.

The introduction of known potential boundary conditions is essential, as otherwise the coefficient matrix of this set of simultaneous equations is singular, and known flow-rate boundary conditions are introduced automatically through the term W_2 of equation (7.9). Free surfaces, however, do present some problems. Perhaps the simplest solution to the free-surface problem (Taylor and Brown, 1967) is to systematically adjust the position of the free surface between each of a series of re-analyses of the problem, in each case taking the free surface to be a no-flow boundary, and correcting its elevation by some percentage of the computed pressure-head error (it should have zero pressure-head once it has been located correctly). An alternative solution technique is to set the boundary potential along the free surface to the elevation at the first trial. Then, following a finite-element solution to the seepage pattern the in- or outflow rates on the free surface are examined. They should be zero if the correct position for the free surface has been found. Some iterative changes in position of the free surface nodes, and resolution of the problem, will be needed before the free surface has been located to the desired degree of accuracy. The movements of the nodes on the free surface can be computed from the computed in- or outflow rates.

The method can readily accommodate arbitrary variations in permeability, anisotropy, and virtually any shape of seepage domain, and at the time of writing, was usable almost as a 'black-box' technique even on desktop

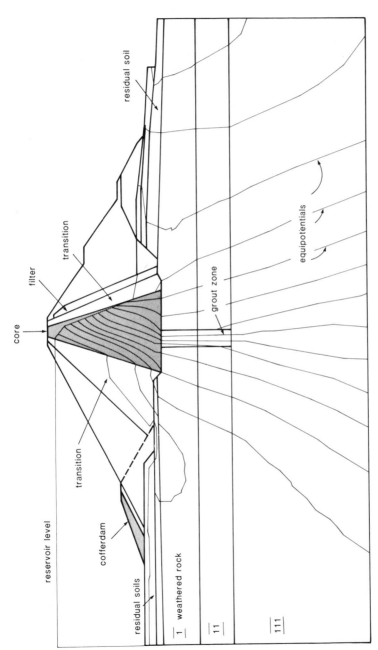

Figure 7.4 Steady-seepage regime in a dam predicted by the finite element method. This analysis was one of a series made during the design of the Kiteta Dam in Kenya (Pugh and Bromhead, 1985).

personal computers (Strucom, 1984; Bromhead *et al.*, 1985; Pugh and Bromhead, 1985). Figure 7.4 shows the seepage regime in a zoned-fill earth dam calculated using the finite-element method. One interesting feature of this particular analysis was that the seepage pressures were automatically abstracted from the finite-element seepage analysis for use in static and seismic stability analysis. Not only is the free surface of a complex shape, but so is the seepage pattern at the core–foundation contact. Only a numerical model could reveal this degree of complexity so clearly and easily.

7.5 Influence of variable permeability

It has been conventional for many years in engineering design and analysis to use flow-net techniques for solving problems of steady seepage. These techniques have been taught to generations of civil engineers, first of all as an aspect of classical hydrodynamics, and, since the 1950s, as a part of soil mechanics. Unfortunately, the flow-net sketching technique demands one thing above all others: an isotropic and homogeneous distribution of permeability. Unfortunately, soils are not like that. Intrinsically, they will be non-homogeneous, perhaps anisotropic to a greater or lesser degree, and almost certainly their permeability will change as they are subject to stress. Admittedly, there are techniques for dealing with a degree of anisotropy in the seepage domain, or of some non-homogeneity provided that this is in the form of a small set of quite homogeneous soil zones. However, the extensions to the flow-net technique rob it of its simplicity, and are quite difficult to apply.

The effect of anisotropy in the permeability is to prevent the equipotentials and flowlines from crossing at right angles (except in the special case of flow parallel to one of the axes of permeability anisotropy), and the effect of non-homogeneity is to cause the curvilinear quadrilateral figures of the flow net to depart from their 'square' shape. This departure can be abrupt, if the soil is zoned, or it can be gradual, if the soil varies in permeability across a zone. Both of these effects can be shown quite readily with a simple numerical model: finite-element techniques are particularly suited to this.

Bedding in soils can be the source of significant anisotropy in permeability. To a certain extent, the placing and compacting of fills can induce a similar preferentially orientated fabric in a soil which can cause some permeability anisotropy. Usually one of the axes of this is sub-horizontal, but in naturally occurring soils and sedimentary rocks the axes can be reorientated by folding (or for that matter, by landslide movement). The effect of anisotropy may be to lower or to raise the porewater pressures: whichever happens depends on whether the drainage path-lengths are increased or decreased relative to those in an equivalent, uniform permeability, case.

Non-homogeneity can occur for an even wider range of reasons—a change in deposition, for instance, bringing a change in grain size. Chemical effects, precipitating some mineral in the pores of a soil or rock, are another natural effect. In man-made slopes, however, gross changes in permeability are more often the result of a deliberate policy, with different zones of soil specifically chosen.

An important, and often neglected, source of non-homogeneity in permeability is the result of the nature of soil voids, or rock joint openness. When subject to increased effective stress, the volume of the joints decreases, and so does the permeability. This tends to cause a decrease in permeability along individual seepage paths, particularly when they have a downwards trend. Some solutions for this type of permeability variation are given by Morgenstern and Guther (1972) and by Bromhead and Vaughan (1980). Inevitably, the porewater pressures are *higher* in such a case than if the permeability was uniform. Furthermore, since the effect is scale-dependent, being much more pronounced in a large soil slope where the effects of stress change are commensurately larger, it may not be sufficient merely to extrapolate from previous experience in small slopes.

Some caution in the use of such simple seepage analysis tools as the flow net must therefore be exercised if great errors are not to be introduced. Note that a low estimate of porewater pressure causes an error in the calculation of stability which is on the *unsafe* side.

7.6 Undrained porewater pressures

Whenever loads are applied to the surface of a soil they set up stresses within it, in response to which the soil undergoes strain. If the pore fluid has a low compressibility, it will not easily change its volume, and as a result, pore-fluid pressures are set up. In soils of low permeability, these *excess* pore-fluid pressures cannot escape except after the passage of much time, and are therefore likely to be a major influence on the behaviour of the soil. Conversely, in solid of high permeability, the excess pore-fluid pressures escape so readily that to all intents and purposes they may be ignored. In soil mechanics, the term *drained* is used to denote the absence of excess or stress-change-induced porewater pressures, and the term *undrained* is used to denote their presence. It is possible to have partly drained conditions after the escape of some of the excess pore-fluid pressures.

It should be noted that these terms have nothing to do with the water content of the soil. 'Drained' does *not* for instance, mean 'with zero water content', nor does it even mean 'with zero porewater pressure'. The term should correctly denote the absence of stress-induced pore-pressure response in the soil.

Some terms are also used to describe the escape of pore fluid from the soil. The escape of elevated pore pressures, with its resulting decrease of ground volume, is termed *consolidation*. This term is loosely applied both to the ground strains (usually settlement), the decrease in moisture content, and to the porewater-pressure changes.

Where stresses are reduced, beneath a cut slope, or under an excavation, for instance, the pore-fluid pressures are reduced. The long-term increases in pore-fluid pressure, and the ground strains that accompany it, are termed *swelling*. Sometimes the expression *equilibration of porewater pressures* is used to describe situations where either swelling or consolidation or both may be going on.

In consolidation problems, it is often found that the porewater-pressure distribution re-equilibrates to that which existed before the load was placed. This is inevitable where the groundwater conditions are controlled externally by a lake, river or reservoir level. The excavation of a cut slope, however, is almost certain to change the external hydraulic boundary conditions to the steady seepage pattern, by changing their *shape* if not their *type*. It will be unusual therefore, to find an excavation reequilibrating to its original porewater-pressure state. Instead, it will equilibrate to a new seepage pattern.

Attempts can be made to model the stress and resulting porewater-pressure change in the soil. The total stress changes can be computed by a finite-element or other model and used via the porewater-pressure parameters A and B (Skempton, 1954; Bishop and Wesley, 1974) in the following equation which yields the pore-pressure change Δu:

$$\Delta u = B(\Delta\sigma_3 + A(\Delta\sigma_1 - \Delta\sigma_3)) \qquad (7.10)$$

A problem arising from this is that the neglect of the effect of intermediate principal stress changes means that A for plane strain conditions is not the same as that for triaxial conditions. This can cause serious errors if the value for A is measured in a triaxial test and applied to a plane strain field case without correction.

Henkel (1960) recognized the difficulties which surround the treatment of the intermediate principal stress, and proposed an alternative equation:

$$\Delta u = \beta\Delta\sigma_{oct} + \alpha\Delta\tau_{oct} \qquad (7.11)$$

in which the octahedral normal stress increment, $\Delta\sigma_{oct}$, is the *mean* normal stress increment:

$$\Delta\sigma_{oct} = (\Delta\sigma_1 + \Delta\sigma_2 + \Delta\sigma_3)/3 \qquad (7.11a)$$

and the octahedral shear stress increment, $\Delta\tau_{oct}$, is a sort of root-mean-square shear stress:

$$\Delta\tau_{oct} = \sqrt{\{(\Delta\sigma_1 - \Delta\sigma_2)^2 + (\Delta\sigma_3 - \Delta\sigma_1)^2 + (\Delta\sigma_2 - \Delta\sigma_3)^2\}} \qquad (7.11b)$$

The correlation between the two A values A_p for plane strain and A_t for triaxial test conditions is simple, however, since

$$\alpha = \frac{A_t - \frac{1}{3}}{\sqrt{2}} = \frac{A_p - \frac{1}{2}}{\sqrt{(\frac{3}{2})}} \tag{7.12}$$

Henkel's equation is more suited to general use, in computer models for instance, than is Skempton's equation, because it properly considers the effect of changes in the intermediate principal stress.

Naylor (1974) proposes an effective stress-based stress analysis in which the combined stiffnesses of the soil skeleton and pore fluid are used to calculate deformations. The pore-fluid pressure changes are then recovered from the volumetric strain in the pore fluid. This prevents the occurrence of pore-pressure changes resulting from shear strains in purely elastic soils. However, since it is precisely this effect that the A value is intended to predict, the usefulness of Naylor's procedure is open to some question.

The expression using A and B values can be rewritten in terms of a single pore-pressure parameter $\bar{B}$

$$\Delta u = B\left(\frac{\Delta\sigma_3}{\Delta\sigma_1} + A\left(1 - \frac{\Delta\sigma_3}{\Delta\sigma_1}\right)\Delta\sigma_1\right) \tag{7.13}$$

provided that the stress ratio $\Delta\sigma_3/\Delta\sigma_1$ remains constant; the pore-pressure response can then be calculated merely from the change in the major principal total stress. In the vast majority of cases, this will be approximately true, and the major principal stress can be represented by the vertical stress. Hence

$$\Delta u = \bar{B}\Delta\sigma_v \tag{7.14}$$

Changes in porewater-pressure can therefore be calculated approximately from the thickness of soil or rock placed or removed.

7.7 Changes in hydraulic boundary conditions

Drawdown of the water level in dams, or their filling, not only changes the hydraulic boundary condition on the upstream face, but causes stress changes in the soil under the weight of impounded water. This sort of effect is therefore considered separately in section 7.9. There are, however, situations in which the stress change is absent or negligible: the change from soil moisture deficit to surplus in a slope face throughout the annual seasonal cycle, for instance. In soils of low permeability and compressibility, the effect of such a cyclic variation may be damped out at depth, and the body of the slope may contain a steady seepage regime in equilibrium with the average conditions on the surface. It is this overall seepage pattern which controls overall stability.

On the other hand, the stability of surface layers of the slope may well depend on short-term changes in the infiltration pattern, and these readily reflect the changing nature of the hydraulic boundary condition on the surface.

By way of contrast, the degree of underdrainage in a slope could be influenced by a permeable bed at depth. Blocking of outflow from this, perhaps at some distance from the slope, or indeed, any change in the porewater-pressure conditions in the underdraining layer, will change the seepage pattern above. In the London Basin, for instance, pumping from the Chalk aquifer in the nineteenth and early twentieth centuries reduced the heads from artesian conditions to those of strong drawdown. Equally serious to the settlement of the London area due to underdrained consolidation of the London Clay is the possibility that as the various wells fall into disuse because of changed economic factors, or the penetration of the aquifer by saline water, or for whatever reason, the rising water levels change the underdrainage pattern in the reverse direction. The effects may well include the destabilization of some slopes.

Any change in the hydraulic boundary conditions which occurs more rapidly than can be followed by the water pressures in the body of the slope will leave some out-of-equilibrium porewater pressures. The analysis of the response of these, together with undrained stress-induced non-equilibrium porewater pressures is dealt with below.

7.8 Time-dependent seepage

It can be seen from the foregoing two sections that the porewater pressures in a soil or rock slope can be out of equilibrium with the hydraulic boundary conditions for two main reasons. Firstly, there may be undrained porewater pressures in the slope as a result of undrained loading or unloading, or there may have been a recent change in the hydraulic boundary conditions themselves which has not fully taken effect. An example where both of these situations apply simultaneously is dealt with in the next section, which reviews stability problems arising out of the rapid drawdown of reservoirs.

As the porewater pressures equilibrate, changes in stability occur. Underneath an embankment, for instance, porewater pressures will gradually reduce, and the embankment will become more stable with the passage of time. In contrast to this, porewater pressures in the side slopes of a cutting will tend to increase with time, and this leads to a progressive destabilization of the slope. This is graphically shown by Bishop and Bjerrum (1960) whose illustration is reproduced with amendments in Figure 7.6. On the whole, the problem of the generation and dissipation of porewater pressures under embankments as described above is widely understood. This contrasts with the situation in respect of cuttings, where the porewater pressures may be reduced so far as to cause them to become negative: this concept is not acceptable to many engineers. However, field measurements of the effect have been made (Vaughan and Walbancke, 1973; Chandler, 1984; Bromhead and Dixon, 1984) and predictions of the magnitude and rate of the effect are used,

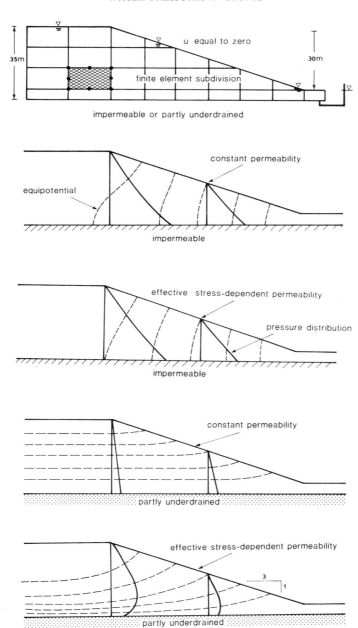

Figure 7.5 Effect of permeability variation. Even in a steady-seepage analysis, some simple alterations in the parameters which control seepage have a disproportionate effect. For instance, underdrainage is important, and where it is present, the influence of a decreasing permeability with depth is most marked.

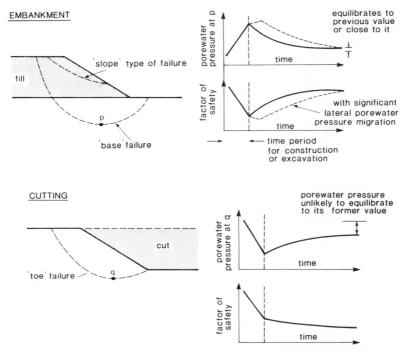

Figure 7.6 Undrained porewater pressures (Bishop and Bjerrum, 1960). Time-dependent changes in porewater pressure after the formation of a slope are important to its future behaviour.

for example by Eigenbrod (1975) as well as by Vaughan and Walbancke to explain the propensity for cutting slopes to fail long after construction. This finally provides a convincing explanation for a problem which has long puzzled engineers (section 1.7).

As a general rule, embankments are least stable at the end of construction, and cuttings progressively destabilize until the equilibration of porewater pressures is complete. There are some departures from this simple rule in the case of embankments. In the foundation of an embankment, it may be found that there is preferential drainage either vertically (through root holes or desiccation cracks) or horizontally (along the bedding of laminated soils). The closer to vertical that the majority of drainage occurs, the closer the simple rule is followed. If drainage takes place horizontally, there is a transfer of porewater pressures from the centre line of the embankment out to its toe, from a place where the porewater pressures matter little to a place where they have a significant effect. The effect will be particularly marked in interbedded silts and clays, such as recent alluvium often contains (Kenney, 1964), or in laminated clay (Rowe, 1968). This can lead to failures a day or so after the end of construction. However, the overall trend is for a decrease in porewater pressure, and the critical period is short-lived (Figure 7.9).

A second problem arises from the behaviour of compacted fills in the embankment itself. The effect of compaction is sometimes to induce porewater suctions. In addition to this, large lumps of clay dug from a deep cutting and used in a shallow fill may respond to the overall stress relief, rather than to the last in a series of stress changes. For both of these reasons the soil may have a suction in it. This will have then to behave more like a cutting side slope, taking in water and swelling, than the behaviour described above as typical of embankments.

A simple theory to account for the time-dependent changes in the porewater pressures in compressible soils can be obtained by considering the continuity equation (7.1), and instead of equating the net flow to zero, make it the volume of water released from storage in the soil pores by changes in effective stress. If the soil 'skeleton' has some mean compressibility represented by m_v, then this equation becomes

$$k_x \frac{d^2\phi}{dx^2} + k_y \frac{d^2\phi}{dy^2} = m_v \gamma_w \frac{d\phi}{dt} \tag{7.15}$$

This equation can be solved analytically, but with great difficulty, in seepage domains of simple shape, but recourse to numerical methods has to be made for realistic problems. Finite-difference, finite-element and boundary-integral methods are all available, and may be found in an extensive literature.

In principle, the porewater-pressure changes, and consequent effective stress readjustments, predicted with the use of this equation could be used to predict strains in the soil mass. It is the usual complaint of the authors of the many papers on the numerical solution of this problem that, except for the simple case where both drainage and strain is purely one-dimensional (it gives good results for that), the equation neglects the proper stress–strain behaviour of the soil skeleton and this invalidates the deformation prediction. However, the main usefulness of this equation and the simple theory that it represents is that it does give fairly realistic porewater-pressure predictions, and since it is these that we need for limit-equilibrium types of stability analysis, there is little additional merit at this point in exploring the many further complexities of the theory.

7.9 Rapid drawdown

When a reservoir is drawn down, the porewater pressures in previously submerged slopes can respond to this removal of water loading in a variety of ways, thus affecting their stability. For instance, in the upstream face of a rockfill dam where the permeability of the shoulder zone is extremely high, the water table in the rockfill will follow the external water level down, and water can freely leave the slope as drawdown occurs.

In the more general case of drawdown for slopes in very permeable soils, or rather soils with a high c_v value, for that is the real governing parameter, the phreatic surface in the slope may be influenced by other inputs to the seepage regime, for example from upslope, so that a horizontal water table may not result. Take for example the *time factor* T_v where

$$T_v = (c_v t / d^2) \qquad (7.16)$$

in which t is the time for drawdown and d is the maximum distance that water must travel to escape from the slope or *drainage path length*. The time factor must normally exceed 1 for this condition to apply. Except for very slow drawdown, this condition will only apply to soils of very high permeability, or those of moderately high permeability combined with negligible compressibility.

The other extreme case arises where the drawdown is so rapid that no porewater-pressure changes except those that occur in response to total stress change can take place. The total stress change is made up partly by the removal of reservoir water load, and partly by the lightening of riprap or other highly permeable surface layers. Hence the change in vertical total stress can be calculated from the difference in the depths of submergence before and after drawdown (multiplied by the unit weight of water) and the depth of drained riprap (multiplied by the unit weight of water and the porosity of the riprap). This extreme case will normally only arise where the c_v gives a time factor of less than about 0.1. Naturally, this effect can exist only in compressible soils.

There are two intermediate cases to consider. In incompressible soils, the lowering of the water table may follow the drawdown of the reservoir, but with a time lag caused by the permeability of the soil impeding the outflow. This case can be analysed graphically, or numerically. In the former case this involves the use of *gravity flow nets* (Cedergren, 1967), a series of which need to be drawn for various time stages in the drawdown. The phreatic surface falls from stage to stage at a rate which reflects the porosity of the soil and the downward hydraulic gradient in the vicinity of that surface. A small time interval may be needed if this procedure is to be at all accurate. Numerical

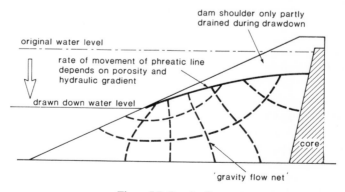

Figure 7.7 Gravity flow nets.

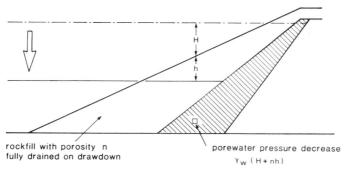

rockfill with porosity n
fully drained on drawdown

porewater pressure decrease
$\gamma_w (H + nh)$

Figure 7.8 Rapid drawdown. A method of computing the pore-pressure change when rapid drawdown occurs, and the slope is left unsupported following the removal or reservoir loading.

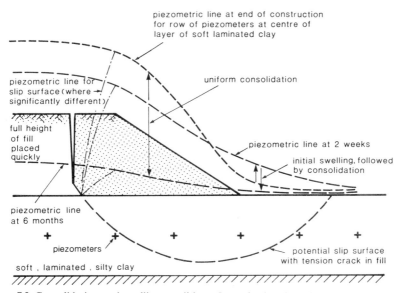

Figure 7.9 Consolidation and swelling conditions. Lateral migration of the excess porewater pressure under an embankment subject to undrained loading may cause swelling at the toe which is injurious to stability, and not offset by the effects of consolidation under the centreline.

methods for this problem, using the finite-element technique and variants of the procedure outlined above, have been described, for instance, by Zienkiewicz and Parekh (1970) and by Neuman and Witherspoon (1971). The method of dealing with a free surface outlined in section 7.4, due to Taylor and Brown (1967)), is less suitable than the alternative when extended to time-dependent problems in which the free surface is no longer a streamline. The *velocity* of the free surface at any instant can be computed from the hydraulic gradient and the porosity of the soil:

$$v = \frac{1}{n}\frac{d\phi}{dy} \tag{7.17}$$

In compressible soils, the analysis of drawdown seepage patterns must start with the *undrained* porewater-pressure change, and then treat the consolidation behaviour of the soil. The procedures of section 7.8 are applicable here.

7.10 Systematic analysis of seepage conditions

Systematic analyses of seepage conditions in slopes are often carried out in connection with embankment dam design, and to a lesser extent for studies of natural slopes.

An example of the former is given by Pugh and Bromhead (1985). As part of the design of the 34 m high Kiteta Dam on the lower Ngaa River in Kenya, systematic seepage studies were made. For each zone in the dam and foundation, mean soil permeabilities and the probable range of values for each zone were obtained. A parametric study of seepage patterns which would result if a particular zone had its maximum or minimum permeability was then undertaken. The first application of this technique showed some serious seepage problems with the original dam design, and a high sensitivity of porewater pressures in the downstream slopes to the permeability in some of the soil zones. Changes were therefore made to the design of the dam in a step-by-step sequence until a design was found which utilized the available materials in the most economic way while yet exhibiting the minimum sensitivity to probable variations in properties.

The short-term stability of the dam was analysed exclusively in terms of undrained strengths in critical zones, assuming no dissipation of construction-induced porewater pressures, so no 'generation–dissipation' analyses were carried out. However, the steady-seepage porewater pressures were used as the basis for drawdown and seismic analyses as well as for estimates of the long-term stability of the downstream slopes.

Analyses taking zoned permeability in tailings dams have been made by Kealy and Busch (1971). They also show that the distribution of permeabilities in the impounding embankment can make an enormous difference to the shape of the free surface and hence to the distribution of porewater pressures in it.

On a massive scale, Freeze and Witherspoon (1966–68) have examined regional groundwater flows and the influence of large-scale geological structures on this flow. This work is complemented by the slope studies of Hodge and Freeze (1977) and Lafleur and Lefebvre (1980) in small-scale slopes in clays in Canada.

Analyses of the undrained unloading effects in slopes are given by

Eigenbrod (1975), and of the effects of varying permeability arising from compression of pores and joints by Morgenstern and Guther (1972).

7.11 Field studies

Field studies in natural slopes to confirm seepage predictions are rarely carried out: many piezometers are needed to define a groundwater-pressure regime accurately, and if the costs of installing them are not prohibitive, the costs of reading them are. There are, however, a number of published case records which do demonstrate some of the important factors.

In the London Clay cliffs of Herne Bay and at Warden Point on the North Kent coast, large numbers of piezometers were installed for separate stability investigations. The slopes at Herne Bay are mentioned elsewhere in this book (Chapter 2) and at the time of the investigation were the subject of a controversial stabilization scheme. All in all, some 88 piezometers were installed at different depths on six separate cross-sections. These showed some quite extraordinary porewater-pressure distributions, and for want of a better explanation at the time, many of the readings were dismissed as a result of instrument malfunction. However, Bromhead and Dixon (1984) explain the observed porewater-pressure depth profiles as three superimposed effects:

an initially underdrained seepage pattern of flowdown through the clay into more permeable sands beneath (the Oldhaven Beds). Some elements of decreasing permeability with depth cause this porewater-pressure distribution to adopt a 'bowed' shape, similar in appearance to a consolidation isochron, despite the fact that this seepage pattern is for a steady flow

the superimposition of a stress-derived decrease in porewater pressure in the slope arising from marine erosion

partial equilibration of the resulting porewater-pressure distribution to a new steady seepage pattern.

In marked contrast to the Herne Bay slopes (which are formed of the lowermost beds of the London Clay), the cliffs at Warden Point are cut into the upper part of the deposit. Thus the natural downward seepage pattern does not demonstrate the bowed shape seen at Herne Bay since it is insulated from the effects of underdrainage by the joint factors of distance and permeability decrease with depth. Thus the effects of undrained unloading show more clearly. Indeed, considerable decreases in porewater pressure are evidenced well inland of the cliff edge, an effect which is demonstrated in stress analyses of the erosion process. Some of the Warden Point piezometers were installed in, under, and seawards of a large rotational landslide. All of these showed depressed porewater-pressure heads, or have indicated that suctions might be present (by their falling-head response after 'topping up' Casagrande-type piezometers with fresh water, or in the case of those piezometers which are situated between high- and low-tide marks, by repeatedly going 'dry'). Underneath the landslide, the decrease in (largely vertical) total stress has caused the change in porewater pressure: in the slip mass itself, the porewater pressures are the lateral stress-relieved set similar to those measured behind the rear scarp nowadays, but carried down the slope by the slide.

Porewater pressures set during dam or other embankment construction are regularly measured, and the results of such studies are sufficiently common-place in the literature not to merit special treatment here. A typical example of the sort of installation is shown in Figure 8.27.

7.12 Representation of information on porewater pressures

The end result of a field investigation of porewater pressures or a theoretical study of seepage is a set of porewater pressures in the slope. These values should ideally be fed into a stability analysis directly and not adulterated in order to simplify communication with a computer program.

Two concepts are commonly used for this communication: these are the *piezometric line* and the *pore-pressure ratio* r_u. The first of these is often confused with the free or phreatic surface (section 7.2). This is unfortunate, because even if a free surface actually exists, it cannot be used to define the porewater pressures in the soil beneath it without the accompaniment of the associated equipotentials. It is wrong, for instance, to compute the depth below the free surface and use it as a porewater-pressure head. Where there is downflow, this will give pressure heads that are too high, and where there is upflow, the prediction is too low (and is therefore an unsafe estimate).

If a slip surface is sketched on to a slope cross-section together with the porewater-pressure information, it is possible to construct a line, called the *piezometric line*, which has this desired characteristic of representing the porewater-pressure head at a point on the slip surface by the difference in level (the piezometric head). It will be readily appreciated that the position of a piezometric line is related to the initial choice of slip surface: take a different surface and a different piezometric line will be found. The more complex the porewater-pressure distribution, the greater the differences. A piezometric line is an excellent way of interpolating between a few piezometric heads measured on, or closely adjacent to, a restricted range of similar slip surfaces. It cannot be used to describe a porewater-pressure regime of any complexity.

Suction in the pore water could be represented by a piezometric line *below* the slip surface, although it is commonplace to represent zero pore pressure conditions with that convention.

The other main method of representing porewater-pressure data is through the use of a *pore-pressure ratio* r_u. Take any point in the slope section, and take the ratio of the porewater pressure at that point to the vertical total stress calculated approximately from the depth below ground level and the soil's unit

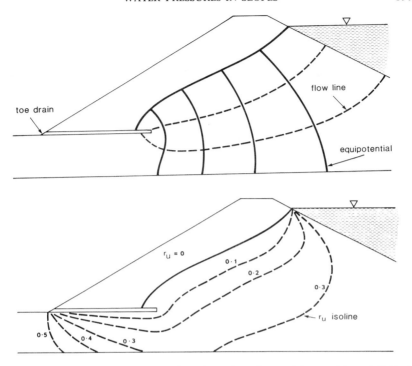

toe drain

flow line

equipotential

$r_u = 0$

0·1

0·2

0·3

r_u isoline

0·5

0·4

0·3

Figure 7.10 Zonation of a slope in terms of r_u. When using r_u factors, it is essential to subdivide the slope into these arbitrary zones if the correct distribution of porewater pressures is to be modelled. This subdivision may be quite different to that made on the basis of shear parameters, but is important if sensible and realistic failure mechanisms are to be predicted.

weight, i.e.

$$r_u = \frac{u}{\gamma_z} \qquad (7.18)$$

A value for this ratio could be found at any point in the soil, or could be expressed as an average along a slip surface, or could even be averaged for a particular soil zone. Such a parameter is a neat and compact way of representing porewater-pressure information: the steps in a computer program which yield the weight of each slice and select which soil strength parameters to use on that segment of the slip surface also produce the porewater pressure, and the total amount of data to store is small. The r_u parameter is also useful when stability charts (sections 6.8 and 6.9) are in use.

The negative side of the balance is more serious. First, the subdivision of the slope into soil zones of approximately equal r_u may differ radically from the subdivision on the basis of soil type. This in turn requires a fine subdivision of

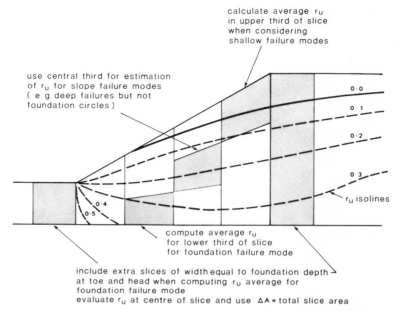

Figure 7.11 Calculation of an average r_u. Sometimes, however, particularly when using stability charts, an overall average r_u must be evaluated. The advice above is more extensive than that accompanying the majority of published stability charts.

the slope into sections, with a correspondingly high data-storage requirement. It is found that a great deal of averaging will be carried out, consciously or unconsciously, in order to reduce the amount of subdivision, and this leads to systematic but unintentional shifts in the porewater-pressure distribution: high porewater pressures are smoothed out and low ones are rounded up. Often this results in a change of emphasis in the computed factors of safety, with more serious deep-seated modes of failure computed as having higher factors of safety. Shallow modes of failure tend to be computed with lower factors of safety than they actually have.

It need hardly be stated that the appreciation of the stability of the slope as a whole may be seriously affected by such a systematic distortion of the data. This may be more serious than the 'errors' in the numerical results for factor of safety.

It is easy to confuse r_u with the other pore-pressure parameter, $\bar{B}$. The latter is an incremental response of porewater pressure to a change in vertical stress, under conditions where the ratio to vertical to horizontal stress stays approximately constant; the former is the cumulative effect of all influences on the porewater pressure expressed as a similar ratio. Compounding the problem is the tendency of early writers, including Bishop (1955), to use $\bar{B}$ where r_u would now be used. Of course, where the porewater pressures in a fill are solely the result of the application of total stress during placing, and no dissipation of porewater pressures has taken place, the two parameters may be numerically equal.

7.13 Porewater pressures for use in design

In dealing with natural slopes, the problem of selection of groundwater pressures is simple: use those that are measured on site. However, some caution must be exercised since it is by no means certain that the worst naturally occurring groundwater conditions will have been measured, especially if the period of observation has been short. At Hadleigh (Hutchinson and Gostelow, 1976), piezometers were read for a number of years before the worst porewater-pressure conditions occurred, and these seven-year maximum values are not necessarily the all-time worst water levels on that site. The problem is compounded near to the ground surface, or in soils of high permeability. In both of these cases, porewater pressures can change rapidly. A similar problem of estimation of *joint-water pressures* in jointed rocks can occur. If the joints can become blocked, by freezing for instance, then quite large pressures can build up. These are quite distinct from ice expansion forces.

If in doubt as to the maximum water pressures to use, assume that the groundwater table rises to ground level.

Where cuttings are formed in the ground, an estimate of the long-term steady seepage pattern should be made. This will give rise to the worst porewater-pressure conditions, and to the lowest predicted factor of safety. If failure is indicated in the long term, even if there appears to be an adequate reserve of safety in the short term (whether calculated using undrained shear strengths of undrained porewater-pressure response with effective-stress-based shear strength), *use great caution*. Such a case can only be allowed if the consequences of failure have been assessed. A case where it might be allowable is where a policy decision to allow some earthwork failures in the future has been taken to defer expenditure to a later date, and where the behaviour of the slope will not put people at risk. It is most certainly not acceptable where, for instance, construction work has to be carried out at the toe of the slope and workers are threatened by slope movements. The equalization of porewater pressures can be surprisingly fast in some soils.

Undrained porewater pressures must be considered where these are set up under fills, regardless of whether the fill is placed on a flat foundation, or on sidelong ground. If data is not available as to the porewater-pressure response in the affected ground, it should be calculated using the height of fill that has been placed and a $\bar{B}$ value of 1. The subsequent behaviour of the porewater pressures must be carefully considered, especially if there is the possibility of a lateral spread or migration of porewater pressures to the toe of the slope.

Inside a fill, the effect of compaction is normally to set up some slight suctions initially. Further fill placement sets up positive porewater pressures so that the suctions are short-lived as the soil is buried deeper. Again, a $\bar{B}$ value of 1 should be used for design if better information is wanting, as this will usually lead to a conservative result.

Porewater pressures arising from reservoir filling need some consideration. The stress changes in the upstream face of a dam, and their effect on stability through porewater-pressure change, are normally more than offset by the additional support provided by the water load, although checks should be made, with conservatively assessed porewater pressures. This is not the case on the downstream side, where the increased porewater pressures can affect stability. A choice as to whether to monitor the pore-pressure response during impounding, and to slow down reservoir filling if undesirable water pressures occur, is an alternative to attempting to predict them.

Impounding may also destroy suctions that are present in the soil or rocks present in a dam fill or forming the sides of a reservoir. Other damaging effects can include wave action, or small failures during drawdown, and the effect of small slides can be compounded if they act in such a way that toe loading is removed from larger potential or pre-existing landslides.

As a steady seepage regime through a dam becomes established, porewater pressures may gradually increase. However, if impounding is started while there are still undrained porewater pressures, in the core for instance, significantly higher than the long-term seepage values, then consolidation will continue.

The case of rapid drawdown must be investigated (section 7.9 above), even if the reservoir is not to be operated in that way. This is so that the performance of the dam is fully understood, and so that it will not behave unexpectedly if it is necessary to draw the reservoir down suddenly. In a large river, one may have to draw a reservoir down in order to accommodate a flood from upstream: breaching of a landslide-dammed lake in a tributary being one such instance.

7.14 Design charts for estimating r_u

Where a preliminary estimate of stability is to be made using effective stress-based stability charts, an estimate of the average pore-pressure ratio must also be made. There are two main methods for this. If a seepage analysis has been made, in however rudimentary a form, the average pore-pressure ratio can be found. A method for this is given by Bishop and Morgenstern (1960), and the following is a variant of their approach.

At any point in the slope, the ratio between porewater pressure and vertical total stress (evaluated as an 'overburden pressure') can be found by the use of equation (7.18). If the slope is divided into strips, or areas, r_u can be found in each strip, and an area-weighted average r_u found:

$$r_u \text{ (average)} = \frac{\Sigma \Delta A r_u}{\Sigma \Delta A} \tag{7.19}$$

A suitable subdivision is from five to eight vertical strips. For non-cohesive soils, the critical area to be sampled need only occupy the triangular space under the slope face itself, as deeper slips are likely to be less critical, but when the soil has appreciable cohesion, somewhat deeper strips must be taken. It is probable that the pore-pressure ratio at the base of each strip will be higher than at its centroid, and to introduce some conservatism into the procedure, pore pressures should be sampled at this position. The procedure is outlined in Figure 7.11.

As an alternative, one could resort to the use of a systematic parametric study of seepage in slopes, with the results summarized in the form of a chart or charts, analogous to those used in stability analysis. Two such charts are presented in Figure 7.12. The two diagrams relate to slopes formed in soil strata which overlie respectively an impermeable bedrock, and a permeable soil horizon which permits the soil stratum to be underdrained.

It need hardly be stated that the hydraulic boundary conditions for seepage in an earth slope are of vital importance to the resulting porewater-pressure distribution in the seepage domain. These boundary conditions are at least as important as the distribution of permeabilities in the slope, and its shape. At the upper surface of the slope in a temperate climate, zero pore-pressure conditions are likely to be found at, or just below, ground level. This surface should therefore be treated as a 'wetted' boundary, and not as a 'free surface', in the manner shown by, for example, Skempton and Brown (1961) or Hoek and Bray (1978). If necessary, the boundary potentials here can indicate the presence of soil suction, which in effect pushes a 'wetted' zero-pressure boundary slightly lower into the slope.

Considering the capillary suctions in an estimate of the pore-pressure ratio will lower its value, and since they may be transients, and act only close to the surface, they should be neglected. For the present, zero pressure will be taken above the zero-pressure line, and the seepage regime will be assumed to occupy only part of the available seepage domain or complete slope section.

At the base of the section, the *shape* of the boundary is of less significance than its *nature*—whether it is permeable or impermeable. An absolutely impermeable basal boundary appears unlikely to occur in practice, but it may be a convenient fiction where a large decrease in permeability occurs rapidly with depth. In contrast, the majority of cases to which this approach would be applied would be slopes in clay. It is not probable that the necessary one to two orders of magnitude permeability decrease at the base of this stratum will occur. It is more likely that at the base of the clay stratum there would be an *increase* in permeability: clays often tend to be underlain by sandy facies.

This brings us to the permeable-base case. The potentials in an underlying

(a)

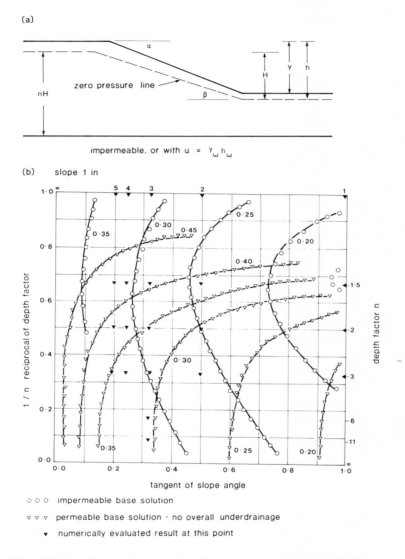

impermeable, or with u = $\gamma_\omega h_\omega$

(b)   slope 1 in

Figure 7.12 Key diagram for evaluation of average porewater pressures in slopes.

aquifer are variable within a wide band, from complete underdrainage to artesian conditions. Natural artesian conditions are rare, partly because of a lack of the requisite geological structures, and partly due to the exploitation of these aquifers with wells.

In order to use the chart, it is necessary to establish a numerical parameter to represent the degree of underdrainage. This we will term the *underdrainage ratio*, D_u, defined as follows:

$$D_u = \frac{\text{potential drop across the thickness of the stratum}}{\text{thickness of stratum}} \qquad (7.20)$$

We will refer this to the crest of the slope, rather than to the toe, for the purposes of the chart, but it would be possible to talk of a local underdrainage ratio with respect to the toe of the slope. (This would be lower than the overall ratio.) The underdrainage ratio may not exceed 1, but might be zero, or negative, the latter corresponding to artesian conditions.

The high degrees of underdrainage will invariably reflect a degree of artificial groundwater abstraction. Caution should be exercised, however, because the effect can be obscured by a reduction in permeability with depth which has the effect of insulating the slope itself from the basal underdrain. The effect of decreasing permeability in this way can be represented by distancing the slope from bedrock with a transformation of the thickness of clay soil under the slope.

The results presented in the two charts come from a finite-element seepage analysis in which different slope geometries have been represented, and the method of use is to approximate the slope to the one shown in the key diagram, Figure 7.12*a*. The thickness of the stratum may need to be altered from its actual value to a larger or smaller one, to cope with a difference in permeability in the soil below the slope toe. From the charts, Figure 7.12*b*, it is possible to read off *R*, which for the impermeable-base situation corresponds to the ratio

$$R = \frac{\text{average porewater-pressure head in zone } A}{H} \qquad (7.21)$$

If a permeable-base problem is to be solved, the underdrainage ratio D_u must be evaluated. In the setting of the key diagram, Figure 7.12*a*, this is given by:

$$D_u = \frac{nH - h_w}{nH} \qquad (7.22)$$

The value read off from the chart must be multiplied by $(1 - D_u)$ in order to obtain *R* for this case. Since the average vertical stress is of the order of

$$\sigma_{av} = \frac{\gamma H}{3} \qquad (7.23)$$

the average r_u ratio in the critical section of the slope can then be obtained from

$$r_u = \frac{3R\gamma_w H}{\gamma h} \qquad (7.24)$$

Subsequently, r_u can be used with effective stress stability charts to determine an approximate value for the factor of safety as outlined in sections 6.9 and 6.8.

7.15 Case records of seepage in slopes

Some case records will be used to illustrate the principles outlined in the earlier sections. First of these are the slopes at Herne Bay, mentioned several times in this book for their landslides in the London Clay. Just as striking as the landslides, but in some ways more important, are the porewater-pressure conditions in the unslipped slopes. At first, the apparently anomalous behaviour of the piezometers was taken to be malfunction, but it later became clear that a self-consistent pattern was indicated, although the explanation was complex. Subsequently, a number of the measurements were repeated with new instruments and these broadly confirmed the original measurements in terms of the magnitude and distribution of porewater pressure in the slopes.

The first factor is the underdrainage of the London Clay into the underlying Oldhaven Beds. In combination with a decrease in the permeability of the clay with depth, the effect gives rise to a 'bowed' porewater-pressure v. depth distribution, noted particularly at clifftop locations (Figures 7.13 and 7.14),

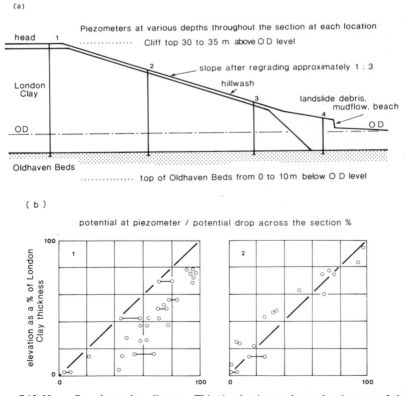

Figure 7.13 Herne Bay slopes: key diagram. This simple picture shows the elements of slope geometry within which the Herne Bay porewater pressure data should be viewed.

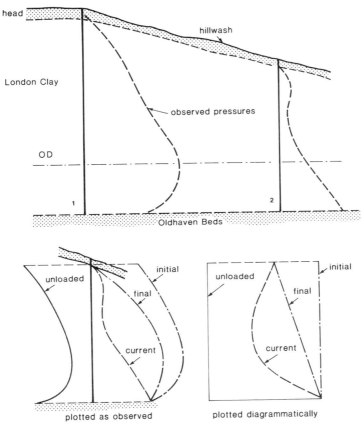

Figure 7.14 Herne Bay slopes: porewater pressure observations. The porewater pressure data at Herne Bay shows the effects not only of depressed porewater pressures (partly equilibrated) but also the effects of underdrainage in a soil with decreased permeability with depth.

but also seen in the landslide debris at the foot of the slopes, where present. In the latter case, there is a large permeability discontinuity between the landslide debris and the small thickness of *in-situ* London Clay beneath the debris, and this makes the porewater pressure nearly hydrostatic; the majority of the head loss is in the *in-situ* material. The bowed porewater-pressure *v.* depth distributions are also seen in other underdrained slopes, such as the slope of till at Cow Green Dam (Figure 7.15). Such bowed porewater-pressure distributions look like isochrons, as though the slope were in the process of changing from one set of porewater-pressure conditions to a more under-drained case, but they are in fact steady seepage patterns. The impact of this distribution of porewater pressure on a stability analysis, compared to what

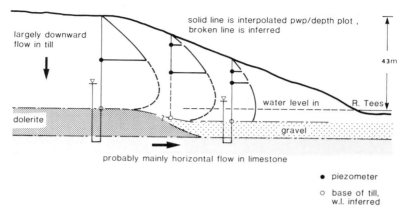

Figure 7.15 Porewater pressures in a till slope at Cow Green. These illustrate the combined effect of underdrainage, and a decreasing permeability with depth. (Vaughan and Walbancke, 1976.)

might be predicted from an analysis assuming constant permeability, can readily be appreciated.

To complicate matters, the porewater pressures in the body of the slope have been reduced by stress relief, and the depression in piezometric levels has partly equilibrated. It would seem that the porewater-pressure reductions are almost entirely in response to vertical stress relief; the argument for this is that the cliff crest piezometers show very little evidence of depressed porewater pressures, as they surely would if lateral stress relief was included. Bromhead and Dixon (1984) examine this in some detail, and conclude that the observed porewater-pressure patterns are well accounted for by taking first a period of relatively rapid erosion, and then a period of quiescence during which the porewater pressures equilibrated to a significant degree. This is indicated (although in an oblique fashion) by the first issue of the large-scale Ordnance Survey map (1869) which shows at least part of the slope to be covered in vegetation. Subsequently, coast erosion restarted, and the majority of the stress relief to which the porewater pressures are presently responding has been vertical stress relief.

In contrast, the slopes at Warden Point on the Isle of Sheppey, west from Herne Bay up the Thames estuary, show the effects of both lateral and vertical stress relief. Although again in London Clay, these slopes are at the *top* rather than the *bottom* of the deposit. They are therefore completely insulated from the underdrainage effect by the large thickness of clay beneath sea level. What is more, they have a better-documented history of erosion, on a far greater scale than the slopes at Herne Bay. In an investigation, porewater pressures in the slopes were found to be substantially depressed (Figure 7.16) and complicated relative to the Herne Bay case by the carrying downslope of

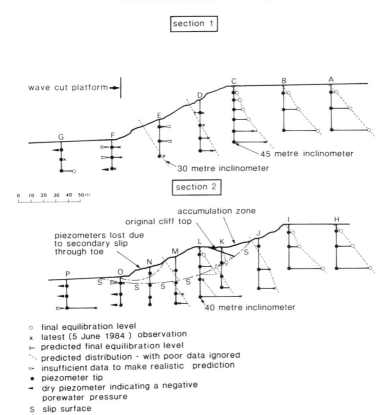

section 1

wave cut platform ➡

45 metre inclinometer

30 metre inclinometer

section 2

0 10 20 30 40 50m

accumulation zone

original cliff top

piezometers lost due
to secondary slip
through toe

40 metre inclinometer

o   final equilibration level
x   latest (5 June 1984) observation
⊢   predicted final equilibration level
ˋˋ predicted distribution - with poor data ignored
=   insufficient data to make realistic prediction
•   piezometer tip
←   dry piezometer indicating a negative
     porewater pressure
S   slip surface

Figure 7.16 Field conditions in the Sheppey slopes. Quite considerably depressed porewater pressures have been observed in actively eroding slopes. These play an important role in the failure mechanism, and in post-failure behaviour (Bromhead and Dixon, 1984).

porewater-pressure patterns within slide blocks (lower part of figure). There is therefore a major contrast between the fresh slide block in the slope examined and the degraded slide remnants of the Herne Bay slopes. The Sheppey slopes were able to reveal quite significant reductions in the porewater pressures adjacent to the cliff crest, evidently the result of lateral stress relief.

Needless to say, the porewater pressures in these slopes could not be reproduced with a simple flow-net analysis, nor indeed by any analysis using steady seepage and a uniform homogeneous permeability. A similar conclusion can be arrived at in respect of the seepage through some dam cores. In Figures 7.17 and 7.18 are two sections, the first a vertical central core, the second an inclined upstream core. Both cases reveal porewater pressures significantly different to those indicated by simple seepage analysis. Nor was it that these porewater pressures were the result of incomplete equilibration: in

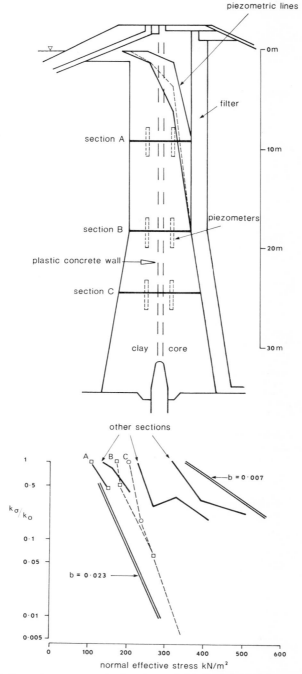

Figure 7.17 Porewater pressures in the core at Balderhead Dam (after Vaughan *et al.*). The observations are best accounted for by making the assumption that the permeability of the core material is dependent on effective stress.

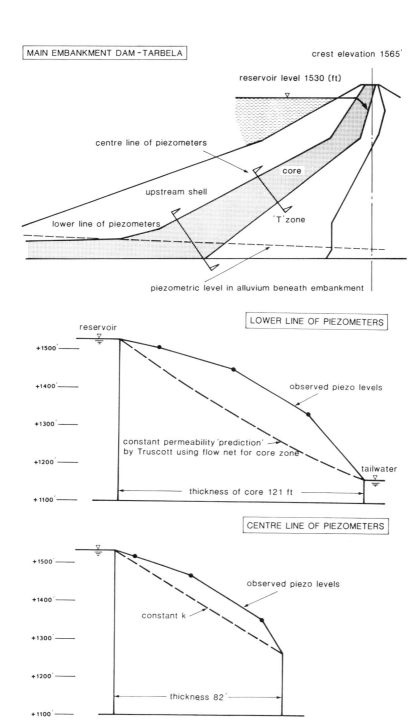

Figure 7.18 Porewater pressures in the inclined core of Tarbela Dam (after Truscott, 1976). These too show the effect of effective stress-dependent permeability changes.

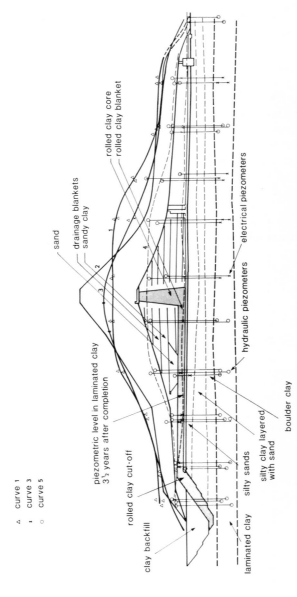

△ curve 1
• curve 3
○ curve 5

rolled clay core
rolled clay blanket

sand

drainage blankets
sandy clay

piezometric level in laminated clay
3½ years after completion

rolled clay cut-off

clay backfill

laminated clay   silty sands

silty clay layered
with sand

boulder clay

electrical piezometers

hydraulic piezometers

Figure 7.19 Undrained porewater pressures in the foundation of Derwent Dam (after Rowe, 1972). Rowe attempted to model porewater pressure generation in the foundation of this dam using a number of simplified theories. These include the simple use of the height of fill, and various elastic theories.

the Tarbela Dam, problems with spillways led to the drawing down and reimpounding of the dam repeatedly, and the core porewater pressures were observed to re-equilibrate to those shown in the figure on more than one occasion. Again, care is needed in the interpretation of simple seepage analyses if erroneous conclusions regarding the stability of the slope are not to be drawn.

One last example will, however, restore a little confidence in the applicability of simple methods to the calculation of porewater pressures in soils. This is the record of foundation porewater pressures in the Derwent Dam. In a series of analyses, Rowe (1972) shows how quite trivial refinements to the basic principle as taking $\bar{B}$ to be 1 and calculating the change in porewater pressure as $\bar{B}$ times the vertical stress increment can give an extremely close fit to the observed porewater-pressure response. In the case of a large embankment on a soft foundation, the rate of gain of strength of the foundation is crucial to the overall performance of the embankment. It may be the dominant factor in controlling the rate of construction, but at least one must have confidence that these high undrained porewater pressures will eventually dissipate to lower, more stable values. Regrettably in the unloaded slopes the converse is true. The prognosis there must be a continued rise in groundwater levels, and a future of decreasing stability.

8 Remedial and corrective measures for slope stabilization

8.1 Introduction

In the design of side slopes for earthworks it is often the case that stability checks show undesirably low factors of safety. The design must then be altered to make the slope acceptably stable. It is possible to make these design changes systematically and rationally, if the original stability analysis has been performed and interpreted correctly. Furthermore, the principles behind this rational approach to redesign are identical to those which govern the design of measures to correct instability in both natural and artificial slopes. In these latter situations, it is not a systematic stability computation that is called for in the first instance, but rather detailed on-site (surface and subsurface) investigation into the forms and causes of the instability that has developed.

Whether it is a slope in the design stage, or an actual example of a real slope which has shown instability, it is essential to understand the causes of the instability which is indicated in the analysis or which has developed in practice. These causes may be summarized simply as follows:

the slope is too high or too steep for the materials of which it is composed

the materials are too weak to sustain the slope at its present profile

the porewater pressures are too high, and thus adversely affect the soil strength

the slope is affected adversely by some external influence, for example seismic loading, or applied loads from structures.

Of course, all of these factors are interrelated. If a slope is too high or steep for the materials of which it is composed, it is possible to correct the situation in two ways: lower or flatten it while retaining the same material, or replace the material with something stronger. This replacement may mean changing a design section, the constraints on which are what materials are available, and what they cost; or it may mean a physical replacement of material in a slope. Alternatively, it may mean an alteration in the physical properties of the soils or rocks making up the slope while leaving them *in situ*.

The way in which the factors are viewed also depends on the viewpoint of the engineer. A mine waste spoil tip, for instance, becomes too high or steep to the mining engineer because he is presented with a large quantity of material of relatively uniform and predefined properties; whereas, to the highway engineer, who must produce an embankment of certain proportions to suit a road alignment, the problem is inverted, and it is the materials which are too weak for the slope design. The problem is the same: the viewpoint, and indeed the solutions, may well be different.

Once the problem has been identified in this way, the preferred method of solution immediately presents itself: correct the main problem. If a slope is too high or steep then lower or flatten it; if the materials are too weak then strengthen or replace them; if the porewater pressures are too high then lower them; and finally, if the slope is subject to undesirable external influences, then insulate it from them. Such solutions will always be the cheapest, and technically will be the easiest, if the slope can be considered in isolation from its surroundings.

8.2 Cut and fill solutions

Take, for example, regrading a slope in order to improve its stability. Broadly speaking, there are three available alternative procedures (Figure 8.1):

grade to a uniform flatter angle

concentrate the filling at the toe of the slope, creating a step or berm in the section

reduce the overall slope height while keeping the profile unchanged.

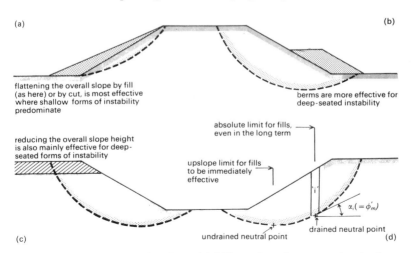

Figure 8.1 Types of regrading. Although a trivial illustration, there are quite clearly some situations in which some of the alternatives are impractical, or where some are more valid than the others.

H

The first procedure is most effective for correcting shallow forms of instability in which movement is confined to soil layers close to ground surface. It does relatively little for deep-seated modes of sliding, where the same amount of fill, placed in a concentrated form at the toe of the slope, can have a much greater stabilizing effect. The same considerations apply to making cuttings instead of placing fill: slope flattening is relatively better for treating shallow modes of instability than forming berms, which are more effective for deeper-seated slides. Choosing between cut and fill solution is rarely a problem, since the section is usually constrained by a requirement that either the toe or the slope crest is located in a specific position: in stabilizing (usually natural) slopes without such a constraint a combination of cut and fill is adopted to balance the quantities of each. It will be found that fill slopes have normally to be a little flatter than cut slopes in the same material, but that fill slopes made of selected imported material can often be formed steeper.

A slope with a stepped or bermed profile may continue to be subject to shallow, near-surface forms of instability, but in forming the berms, the extent of such a failure is controlled, and the seriousness of any movements is reduced.

The third procedure, lowering the overall slope height, is of much less practical application. The design height of an earth embankment (a bridge approach, for instance) is not normally open to change; remedial measures must retain the original heights as a primary design objective. In addition, reduction of slope height influences deep-seated modes of failure most of all and does little for shallow slides. For instance, reducing the overall height of an embankment will give a substantial increase in stability if the most critical mode of failure is a 'bearing capacity' failure into a soft foundation stratum; it will do little to prevent shallow slips of the embankment fill itself. For the same amount of muckshifting, the creation of a bermed profile will ordinarily be found to yield a better improvement in stability.

With natural slopes, the choice of regrading scheme is less clear cut. Often, the slope profile is highly irregular, and with larger natural landslips of some antiquity, development on the surface may make some areas unusable as sites for cut or fill. Where the subsurface form of such an old landslip is complex, it may be that the placement of fill in one area will destabilize another. Some tool is required to enable the location of suitable areas for both cut and fill at an early stage in the design process which does not require too great an input of computation. Such a technique does exist: it is termed the neutral-line theory (Hutchinson, 1976).

8.3 The neutral-line theory

Suppose a cross-section through a landslip, for example that in Figure 8.2, is considered. For simplicity, a circular-arc-shaped slip surface will be taken; the

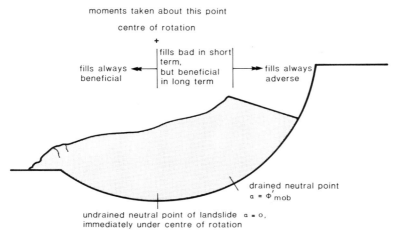

moments taken about this point

centre of rotation

fills always beneficial ← | fills bad in short term, but beneficial in long term | → fills always adverse

drained neutral point
$a = \Phi'_{mob}$

undrained neutral point of landslide $a = o$,
immediately under centre of rotation

Figure 8.2 The neutral line theory. (After Hutchinson, 1978.)

result is the same for all shapes of slip surface. This slip is then divided into slices in the conventional manner. Now a small load is placed on one of the slices. It can have two independent effects: it can alter the net destabilizing moment, and it can change the effective stresses acting on the slip surface, thus influencing the shear strength and hence the moments which resist movement. As this small load is moved from slice to slice from the toe of the slip to its head, the influence on the destabilizing force goes from a large positive effect, through a position where it has no effect at all, to an entirely adverse effect when it is located in the upper part of the slice. However, the load always increases the effective stresses, wherever it is placed (subject of course to the porewater-pressure response in the soil under the load), and thus even at the position of no net change in the destabilizing moment the overall effect is an increase in stability. The load must move still higher up the slope to a position where it causes an increase in the overturning or destabilizing moment equivalent to the increase in resisting moment before the position of net effect on stability, or *neutral point*, is reached.

If a series of neutral points for different sections through a landslip can be found, then the trace of these in plan is termed the *neutral line*.

The position of the neutral point on a given slope section is easy to find. On the slip shown in Figure 8.2, the initial factor of safety F_0 can be defined as

$$F_0 = \frac{M_r}{M_o} \tag{8.1}$$

and the effect of a small load dW placed on slice i is to change the factor of safety to F_1 where

$$F_1 = \frac{M_r + (dW \cos \alpha_i - dP_b) \tan \phi_i'}{M_0 + dW \sin \alpha_i} \tag{8.2}$$

in which dP_b is the change in the water force on the base of the slice arising from the application of load dW at the surface. These two could be related through the pore-pressure parameter $\bar{B}$, such that

$$dP_b = \bar{B} dW \sec \alpha_i \tag{8.3}$$

as a result of which it can be shown that at the neutral point

$$\tan \alpha_i = (1 - \bar{B} \sec^2 \alpha_i) \frac{\tan \phi_i'}{F_0} \tag{8.4}$$

Thus a position on the slip surface can be found, immediately above which an applied load will cause no net change in the factor of safety. Further up the slope, there is an adverse effect on stability; further down the slope, stability is improved. The position of this neutral point depends on the porewater-pressure response in the soil under the applied load via the $\bar{B}$ parameter, and, as the undrained porewater pressures set up in the slope dissipate, the neutral point migrates upslope. It cannot move further, however, than the position for $\bar{B} = 0$. The position of the migrated neutral point may be found as follows.

Suppose the initial porewater-pressure response was 80% of the applied stress (i.e., $\bar{B} = 0.80$), which subsequently dissipated by 25%. This would leave 75% of the original response, equal to an initial generation of porewater pressure of 60% (75% of 80%) of the applied loading. The neutral point position could easily be computed for this using an equivalent $\bar{B}$ of 0.60.

In a slope with complex stratigraphy and slide forms, there may be many of these neutral points, some related to the whole mass of landslide debris, others related to its constituent parts (Figure 8.3). Where a landslide follows a sequence of bedding planes, there may even be a number of neutral points for that one sliding surface alone. Finding these, and drawing them on a section, rapidly reveals possible areas for cut or fill, and delineates practical corridors for the construction of road or rail earthworks to cross the landslide complex.

It is noteworthy that some areas in which the placement of fill would initially be harmful may well prove suited to the placement of fills to take effect later. This is because of the migration upslope of the neutral point which accompanies drainage and consolidation. Care must be taken to adopt other measures in concert with this filling to cater for the short-term worsening of stability which can occur.

Changes in the slope profile are the simplest, most reliable and effective remedial actions. In comparison, all the other techniques are very much second best. That is not to say that they do not have their place in a corrective scheme, merely that to be used alone there has to be very good reason indeed why regrading cannot be undertaken.

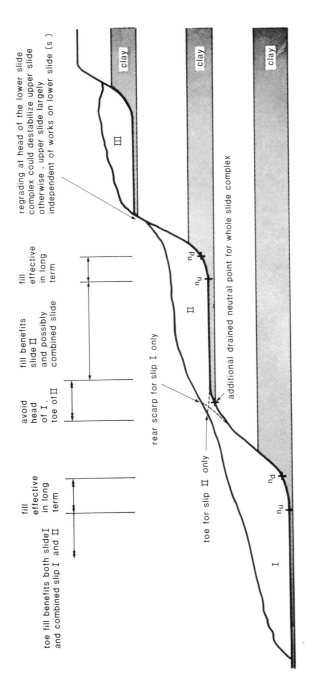

Figure 8.3 Complex slides. Applying the neutral line theory is more useful, if a little more involved, where complex slides are concerned.

In the equations presented, the *initial* factor of safety is used to locate the drained neutral point. Of course, as the factor of safety increases, the position of the drained neutral point migrates downslope; in cases where significant increases in factor of safety are expected, it is advisable to replace F_0 in the equation with the increased factor of safety. This downslope migration of the drained neutral point is entirely separate from the movement of the effective neutral point from the undrained to the drained position with consolidation under the applied loads.

8.4 Case records of the use of slope reprofiling

Examples of the use of cut and fill in stabilizing landslides abound. Large-scale examples include the slide at Folkestone Warren, where extensive filling was placed on the foreshore as a toe load (see section 10.6 and Figure 10.19). This fill had no other purpose than to load the toe of the slope. Not so the analogous fill place on the foreshore at Llandulas, in front of a coastal landslide developed in tills overlying a limestone bedrock. This fill was provided to stabilize the landslide (already crossed by an earlier road and a railway line) but was also the location of the new A55 coast road in that vicinity (Figure 8.4). The fill is protected from wave action by armouring on its seaward face (Wilson and Smith, 1983).

When analysing the influence of rockfill toes placed as a stabilization measure, it is important not to extend slip surfaces which tend to rise up at the toe into and through the rockfill. Inevitably, when this is done, the calculated factors of safety rise considerably because of the high ϕ' value of the fill material. Testing the analysis with just the dead-load effect of the fill invariably shows much reduced increases in stability. The latter are seen to be much more realistic when the (often small) lengths of slip surface through the toe fill are considered. This finding was first made in the early analyses for this particular slide. Neglecting this additional component of added resistance makes it largely unnecessary to explore the effect of alternative toe positions, or the influence of some loss of toe weight by marine erosion during storms, whereas to include it makes the 'improved' factor of safety very sensitive to those effects.

Inland slopes may also be effectively stabilized by the construction of toe loading. In some cases, it may not be practical, because of the presence of a river, or some development at the toe of the slope. They tend to be most effective in the case of deep-seated slides, or at the toe of an accumulation zone, since 'overriding' slip surfaces are likely to form in a shallow slide. The fill would not in such a case be a really effective solution (Figure 8.5). An example of the successful use of toe loading on a slide of the 'accumulation zone' type in an inland slope is given by Early and Skempton (1972) for the landslide at Walton's Wood in Staffordshire (Figure 8.6). In contrast to toe loads on coastal slopes, erosion protection is not necessary, and weaker materials may be used

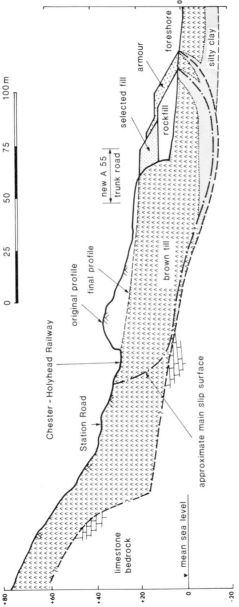

Figure 8.4 Toe loading fill on landslide at Llandulas, North Wales. The rockfill toe, with its wave armour, has an extremely high ϕ' value, and where analyses with slip surfaces rising through this toe fill are made, excessive factors of safety are computed. An underriding slip surface is a more appropriate model.

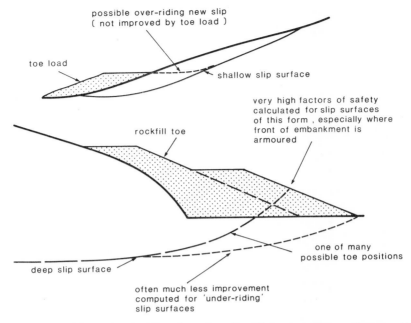

Figure 8.5 Overriding and underriding slip surfaces. Toe fills have no effect at all on overriding slip surfaces, and may not be quite so effective as appears at first sight on underriding slip surfaces.

as fills. This then removes the difficulties in stability analysis associated with strong shear-strength properties in the toe fill.

A toe fill is generally to be preferred to an excavation at the head of a slide, because the former is frequently found to be more cost-effective than the latter, particularly where it can be situated over a rising part of the slip surface in a base-failure-mode slide. Furthermore, in compound and complex landslides, toe fills at the extreme toe of the slide are less likely to interfere with the interactions between individual slide elements except in a positive way. It *may* prove possible, by careful application of the neutral-line method, to find locations within a slide complex where a fill can be placed to stabilize one or more slide elements, without adversely affecting others, but in every case, this must demand detailed exploration of the subsurface conditions.

Reprofiling slopes to a flatter angle by placing fills against them is relatively free from problems, providing that the fill has adequate strength to stand at the angle to which it is placed. Reaching the same profile by excavation may be more risky, if the excavation unloads the toe of an existing or potential deep-seated failure. Examples of this are not common, but there is evidence that a failure in 1962 in Lyme Regis, Dorset, was the reactivation of a pre-existing deep-seated failure in Lias Clays by unloading due to slope reprofiling at its

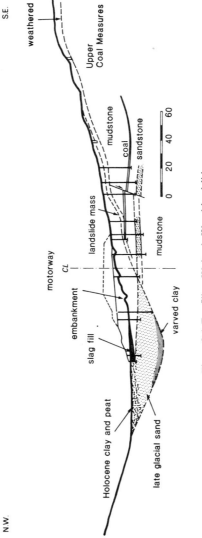

Figure 8.6 Toe fill at Walton's Wood landslide.

Figure 8.7 Placing a counterweight fill, and regrading a slide.

toe (Pitts, 1979). A similar, but smaller-scale slide in London Clay at Highlands Court, near Crystal Palace in South London, occurred where a building site redevelopment at the toe of a slope in London Clay close to the junction with the overlying Claygate Beds was excavated. Slope reprofiling, probably to improve light levels reaching the lowermost windows of this block, stimulated slides along pre-existing slip surfaces in the London Clay (Figure 8.8). A model experiment illustrating this effect is given by Hutchinson (1984).

Some major earthmoving to stabilize coastal cliffs has also been undertaken successfully. This involves regrading the whole slope profile, rather than just adding toe fills. The Herne Bay cases are described elsewhere in this book under the treatments of the different landslipped sections, but here, as elsewhere in the coastal slopes on both sides of the Thames estuary, slopes have been stabilized by cutting back to 1:3 or flatter. Ordinarily, a flatter apron, formed in landslipped debris and poor-quality fill, is left at the toe of the slope, and the whole is protected from marine erosion by a sea wall. Examples

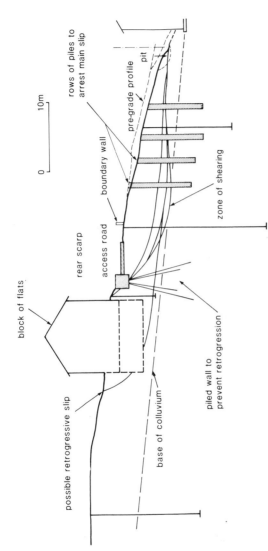

Figure 8.8 Excavation and slip in South London. The excavation of only a metre or so of fill to allow more light to the ground floor of the lower block of flats triggered this slide. Construction of the 600 mm dia. piles through the slide mass effectively arrested the main movements, although construction was difficult due to restricted site access. A Pali Radice wall protected the upper flats from further retrogression. Shallow drainage works were also required.

of such schemes are described, for instance, by Duvivier (1940), Allan (1968), and Barratt (1985).

8.5 Rock and soil anchors

Anchors in soil and rock slopes can be of two types: they can be unstressed, and rely purely on a dowelling action to increase the resistance to sliding; or they may be stressed. In this latter type, the axial load in the anchor increases the effective stresses at depth in the soil or rock, improving the strength. A vector component of the anchor force may also act to reduce destabilizing forces and moments. It is these stressed soil and rock anchors that the following section describes.

The design problems are:

to evaluate the total anchor force required to stabilize the slope

to establish the number, size and location of the individual anchors

detailing the anchor head at ground level, so that the forces are suitably distributed into the ground

ensuring that the lower end of the anchor is located in stable ground, and that the anchorage at depth can sustain the applied loads

providing a suitable corrosion proofing system for the individual anchor components

monitoring the performance of the installation, both with respect to individual anchors, and their overall effect on the slope.

Including the anchor forces in a stability analysis is more difficult than it at first seems. In most variations on the method of slices it is easy to incorporate vertical forces; these are merely added to the weight of the slice in which the anchor head is located. It is the horizontal force component which is difficult to apply.

Anchor loads not only affect the forces which cause instability, but also, by increasing the stresses on slip surfaces, can increase the resistance to sliding. While it is scarcely conceivable that anchors installed to stabilize a slope would decrease normal stresses on a slip surface, other external loads could do so. In order to produce a general treatment of such loads here, saving the necessity to duplicate much of the mathematics later, such external loads on the slope will be lumped in with anchor loads in the following.

These external loads might, for instance, be applied by guy ropes staying a mast or tower, or the lateral thrusts from a water-filled tension crack, in addition to the wide range of structural loads that the engineer may require to be carried. Some of the forces that may have to be considered are shown in Figure 8.9.

Take, for example, a single block of soil or rock set to slide down a slip

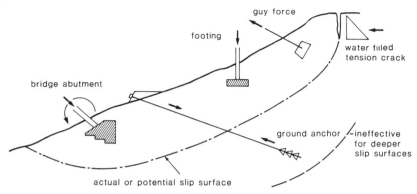

Figure 8.9 Anchor loads. Horizontal forces can arise from a variety of sources, not just anchor loads. They are all equally difficult to accommodate in most methods of slope stability analysis.

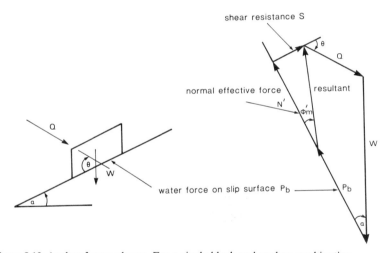

Figure 8.10 Anchor force polygon. For a single block and anchor combination.

surface inclined at an angle α to the horizontal. An anchor load Q acts to restrain the block, and this force acts at an angle θ to the slip surface (Figure 8.10). Whereas the factor of safety against sliding initially takes the value

$$F_{o} = \frac{(W \cos \alpha - P_{b}) \tan \phi'}{W \sin \alpha} \tag{8.5}$$

after placing and stressing the anchor, the factor of safety becomes

$$F_{1} = \frac{(W \cos \alpha + Q \sin \theta - P_{b}) \tan \phi'}{(W \sin \alpha - Q \cos \theta)} \tag{8.6}$$

It can be shown, by differentiating this equation with respect to θ, that the optimum inclination for the anchor is at an angle θ_c to the slip surface in which

$$\theta_c = \tan \phi'/F_1 \qquad (8.7)$$

or an angle equal to the mobilized angle of shearing resistance.

Just as in the neutral-line theory, this mobilized angle of shearing resistance is the key parameter.

The position in which the anchors should be placed is not predicted by this simple formula, but, logically, they should be placed at or near to the toe of the slope. This prevents the tension which could develop if the head or upper part of the slide was to be treated in isolation.

8.6 Additional factors in ground anchor design

Consolidation of soils under anchor loads

An anchor load applied to the surface of a soil mass will increase the porewater pressures in it. Eventually, water will escape, and the porewater pressures will return to their original state: in equilibrium with the hydraulic boundary conditions to the seepage pattern in the slope. As this takes place, the ground reduces in volume, or consolidates, resulting in a loss of stress in the anchors. The amount and rate of this consolidation can be approximately predicted from elementary theory, and allowed for in the initial load applied to the anchors. However, since even a small error in this computation has a major influence on the loss of anchor load, and because there will be some loss of stress in the anchors from slip at the anchorages or creep in the strand used in the anchor, it is preferable to allow for periodic restressing of the anchors to take up these losses as they occur.

Increasing the load on the anchor to cater for expected losses will at best only worsen the conditions which gave rise to losses in the first place, and at worst can cause failure in one or more of the components.

To programme this restressing correctly, it is necessary to monitor anchor loads. A variety of load cells with the necessary long-term stability, and durability, are marketed for this purpose. Photoelastic glass types seem more likely to perform consistently in the long term than those operating by electrical means. It is sometimes the practice to trim the ends of an anchor short, and to fit a protective steel cap. Where restressing is envisaged, however, such an arrangement is not possible, since it is then difficult to grip the end of the strand to apply more load. Care must be taken not to do anything that prejudices the proper treatment of the anchor installation in this way. It may also be necessary to keep vehicular access to the heads of the anchors available for several years after installation, to facilitate these adjustments.

Corrosion problems in ground anchors

Corrosion problems with ground anchors arise in two places: the anchor bar or tendon, and at the anchor head at ground surface. The bar or tendon may be protected to a certain degree by grouting the anchor drill-hole after installation and tensioning. This must only be undertaken when it is known that further retensioning will not be required, because, when the drill-hole is full of hardened grout, further tensioning of the anchor will only serve to put this grout into compression. Wire strands are more susceptible to corrosion than bars, in view of their greater surface area. Accordingly, the individual wires are usually greased, and the strand is covered in a PVC sheath. The wires protrude from this only at the ends.

It is difficult to ensure that the individual wires are surrounded by mortar or grout at the end of the drill-hole where bond is required. One scheme is therefore to strip, splay and degrease the end of the strand, and then to cast the wire into a corrugated plastic container with epoxy resin. This container is then concreted at the bottom of the drill-hole to provide the anchorage at depth. The crucial elements of the anchor can then be worked on at the surface, and be fully inspected before installation.

Corrosion protection at the head of the anchor is normally confined to the application of tar or bitumastic paint to exposed surfaces; a steel cover bolted down over the exposed wires, to prevent them flying out should one break, provides a little weather protection.

Forming the anchorage at depth

The normal method of forming an anchorage at depth is by grouting the end of the anchor at the bottom of its drill-hole. This may involve the use of cement mortar, concrete, epoxy, or other chemical grouts. The use of an expanding agent is advised if cement products are used. In soft rocks and soils, the end of the drill-hole may be expanded using a reaming tool to provide a greater pull-out resistance than could be obtained with a straight-shafted anchorage.

Where rock-bolts of lesser capacity are used to provide support to surface elements of a rock slope, it is common to find expanding details at the end of the rock-bolt which provide a friction grip on the sides of the drill-hole. These rock-bolts are extensively used in underground excavations, and by and large are proprietary products usually protected by patent. A large number of the available types are illustrated by Hoek and Brown (1982), and the subject is discussed in detail by Littlejohn (1970, 1980).

8.7 Case records of ground anchoring in soil slopes

Two case records show the use of anchors in the stabilization of slips in soil, as distinct from failures in rock slopes.

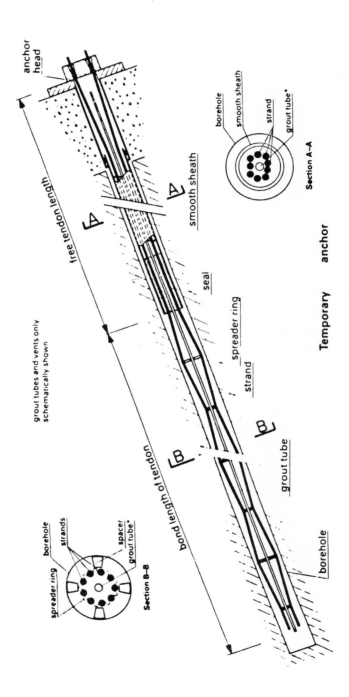

Temporary anchor

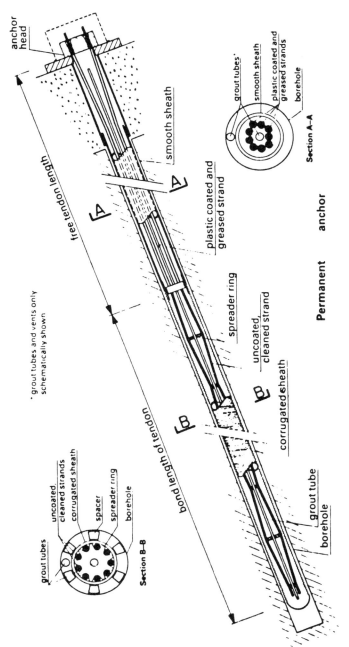

Figure 8.11 Anchor details for a typical stressed tendon anchorage. Much of the anchor head detail is identical to prestressed concrete anchorages.

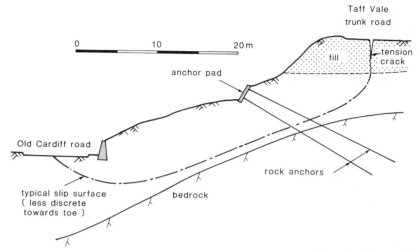

Figure 8.12 Ground anchors used for slope stabilization at Nantgarw, S. Wales. Although in principle anchors at the toe of the slope would be most effective, here the works had only to retain the new road. The anchors are positioned in the optimum position for that.

A landslide in glacial and glacio-fluvial deposits affecting the Taff Vale Trunk Road at Nantgarw, South Wales, is reported by Hutchinson (1984). Filling for the new dual-carriageway road at the head of this slide appears to have reactivated movements on a pre-existing slip surface. These movements were up to 15 mm per year at the head of the slide, and lane closures in the cracked carriageways were necessary. Movements at the toe of the slope were less, at about 2 to 5 mm per year.

The engineers for this project were severely constrained by lack of space, so that slope reprofiling was not practical, and an anchor scheme was adopted. The location of the anchors (Figure 8.12) about halfway up the slope, was also not ideal from the point of view of stabilizing the whole slope, but they are located in a convenient position for restraining the important new road at the top of the slope, and are positioned so that works could be carried out without disruption to either the old or the new road. A total of 128 anchors, each of 500 kN working load, were installed at inclinations of between 25° and 30° to the horizontal. They all penetrated the underlying Coal Measures bedrock. At ground level, the anchor loads were transferred to the soil through reinforced concrete anchor pads each taking four anchors. These pads were 3 m high by 4.8 m long.

A second case, again in south Wales, is the landslide on the Risca–Rogerstone bypass. Although following the line of an abandoned railway, the road alignment needed further cut both in level and back into a hillside to

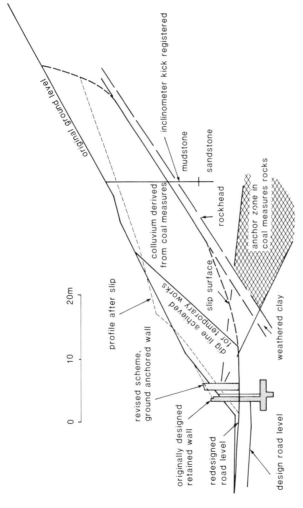

Figure 8.13 Ground anchored stabilization scheme on the Risca–Rogerstone bypass. An early design demonstrating principle is shown. In the final design, anchors higher up the slope were used to counter potential overriding slips.

reach its design section, and a little extra for the temporary condition prior to the construction of a permanent reinforced concrete retaining wall. At this stage a slip occurred in the colluvium into which the cuts had been made (Figure 8.13). The slide occurred after a period of wet weather, although it had stood for some three weeks previous to that, which showed porewater-pressure transients to have had at least some bearing on the failure.

At this point the earlier, pre-construction, site investigation was supplemented by further boreholes, with both inclinometer and piezometer instruments. These showed that the groundwater levels were low, at or below rockhead, which was 6 to 8 m below ground level, and close to which lay the slip surface situated in a weathered clay layer mantling the bedrock surface. Inclinometers situated uphill of the slide showed a tendency for movements to retrogress upslope, and those in the slide picked out the level at which movements still continued. Inclinometers in slide masses (section 9.5) can be useful adjunct to examination of cores for the location of slip surfaces, and can allow the engineer to decide which of the, possibly many, shears he discovers is the currently active one. They also allow the effectiveness of remedial measures to be monitored immediately. However, at relatively small displacements, where these are concentrated on a discrete shear surface, the access tube distorts so that the passage of the instrument is no longer possible, and the tube must be abandoned. When planning the dual-purpose use of inclinometers, it is essential to undertake the remedial measures without delay.

At this slide, the road level was raised, so that no further excavation in level was required. A remedial scheme based on ground anchors installed near the toe of the slope was prepared. The designed anchors were to be drilled into the Coal Measures bedrock, at a variety of inclinations. The dominant factor was to obtain the 900 kN per metre horizontal thrust from the anchors needed to obtain stability not only for the slide mass itself, but for the filling needed *at the head of the slide* to control the further retrogression of slip movements upslope.

Figure 8.14 shows a ground-anchored wall at a site near Sandgate, Kent, during the tensioning phase. Slides here had affected a small number of dwellings, and were initially controlled by placing a counterweight fill at the toe of the slide (foreground). The anchored wall was needed as a permanent measure to allow the removal of the fill and reinstatement of the road surface on which it had been placed.

8.8 Drainage

Drainage as a method of slope stabilization can be very effective, but as a long-term solution it suffers greatly because the drains must be maintained if they are to continue to function. Often, the design is such that maintenance is impossible; whether this is so or not, proper maintenance is rarely planned, and even more rarely practised. With this in mind, it is possible to see the role of drainage more clearly in the wider picture.

Figure 8.14 Ground anchor scheme at Sandgate, Kent.

Drains can do three things. They can control the movement of surface water, and through their influence on the hydraulic boundary conditions to the seepage regime in a slope, bring about the desired reductions in porewater pressures at depth. Such drains are shallow, or surface, drains. Of all the types, they are the easiest to maintain, but the ones most likely to fall rapidly into disrepair. Deep drains act to modify the seepage pattern within the soil or rock mass. Often much more costly than shallow drains, they are usually more effective because they remove water, and decrease the porewater pressures, directly at the seat of the problem. In the final category are those drains which are installed to dissipate undrained porewater pressures arising from total stress change. These drains are normally short-lived. They may be of the shallow or deep category, but are most likely to be the latter. It will be the case in many situations that more than one type of drain is installed, or that drains are multipurpose and perform more than just one of the three roles.

It is often difficult to quantify the effects of drainage *a priori*, so it is usual to monitor the effectiveness of the drain installation with piezometers. This is particularly so where drains are incorporated in fills to eliminate the porewater pressures which arise as a result of increased total stresses as construction proceeds. In such cases, the piezometer readings may be the controlling factor in selecting a rate of construction, regardless of any calculations of the allowable rate of construction based on laboratory test data.

8.9 Shallow drains

Shallow drains normally take the form of lined or unlined ditches. They may alternatively be shallow gravel-filled trenches. Ditches are perhaps the most difficult form of drain to keep operational: if the mean discharge is too high, they scour out; if it is too low, they become blocked by weeds. Sometimes they are lined with concrete, or with slabs, the better to resist scour. This is expensive, and can easily be disrupted should small movements take place. They do have the advantage that short-duration high discharges can usually be accommodated, because of the large section of channel available. At the outfall end, they may need substantial works to take the discharge together with materials swept down by the flow.

A use for open ditches in natural slopes is to carry away discharge from springs, or to lower the maximum level of landslip ponds, or to re-route streams.

Gravel-filled trench drains are used to intercept run-off as well as forming the shallowest of the categories of 'deep' drains. They must then be open at the top (not covered with earth, as is sometimes done), and the upper layers of drainage material need to be protected by a filter from the ingress of fines from the catchment, at least until a vegetation cover has become established. These drains are normally used as access paths by pedestrians, since their granular infill is not slippery and does not stick to footwear. It is impossible to prevent their use in such a manner: they may be protected during construction operations, but are vulnerable afterwards. It is probably best to protect their upper surface with large-sized stones that will not easily be displaced or carried off.

A chevron or herringbone pattern (Figure 8.17) on the face of the slope can be effective in catching runoff, although other patterns have been used with equally satisfactory results. Unless discharges are expected to be high, the expense of providing loose-laid open-jointed or perforated pipes at the bottom of these drains is probably not worth the additional benefit in capacity. The same argument may be applied against the provision of slabbed or concreted inverts to protect the base of the trench from erosion. If, however, the soil is readily erodible, or if experience shows it to be necessary, then both of these expedients may be adopted.

Figure 8.15 shows a coastal slope on the north side of the Isle of Sheppey, Kent, stabilized by regrading but with a pattern of shallow drains installed over the slopes to control runoff and infiltration. The prime purpose of these drains is not to achieve a substantial lowering of the piezometric levels, unlike other installations where trench drains of one sort or another have been used.

Figure 8.15 Shallow drains. Various trench and cut-off drains are under construction here. This unstable railway cutting was stabilized by the construction of contiguous bored-pile walls.

8.10 Deep drains

The purpose of deep drainage is to modify the shape of the seepage regime, rather than to control its hydraulic boundary conditions. Trench drains can do this to a certain extent, as well as collecting runoff, and they must therefore be included in this category. A selection of types of deep drain is shown in Figures 8.17 and 8.25–8.26.

Vertical bored sand- or gravel-filled drains can be effective where there is a perched water table and a convenient permeable stratum beneath which can carry the discharge. However, one should beware of overloading the capacity of the draining layer, possibly causing seepage erosion at the outfall. A proprietary system of sand-filled geofabric tubes lowered into small-diameter drill-holes is very effective: the strength of the fabric prevents the drains from being severed if any small movements take place before the installation is fully effective. Larger-diameter holes can be filled with sand or gravel to provide the necessary permeability characteristics. An example of the use of 'sand wicks' at Herne Bay is shown in Figure 8.19.

Horizontal, or near-horizontal, bored drains are rarely filled with a drainage medium, but instead are lined with perforated pipes, or alternatively with no-fines concrete pipes or earthenware pipes laid open-jointed. These

Figure 8.16 Drainage under construction: trench drains.

keep the hole open while allowing water to escape. On a larger scale, drainage headings excavated in from the toe or face of the slope can fulfil the same purpose, but are easier to keep on line than drill holes unless the latter are quite short. In soils and weak rocks, drainage headings have to be supported with a tunnel lining but in sound rock may be left open. Inside the slide, the influence of the heading may be increased with extra drives, or with arrays of bored drains.

Naturally, there is a degree of hazard associated with working below ground in an active landslip, especially if the heading goes through a major slip surface:

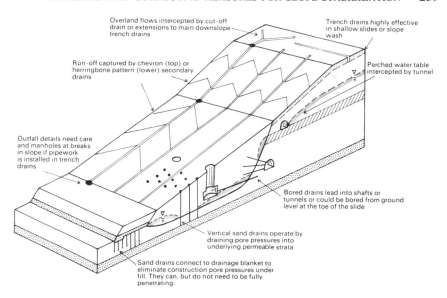

Figure 8.17 Deep drains. A number of different drainage measures are illustrated in this composite diagram.

Figure 8.18 Deep drains. Shaft sinking is in progress. Bored drain arrays were drilled to stabilize this relict landslide. (See also Figure 8.25.)

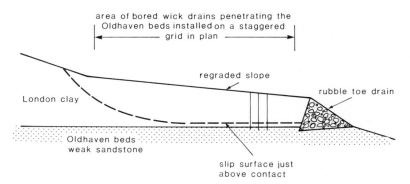

Figure 8.19 Vertical, sand-filled 'wick' drain scheme used at Herne Bay.

movements could trap the workforce. If this is a likely peril, then it may be preferable to construct the drainage galleries under or behind the landslip and, if the slip surface must be penetrated, to do it with bored drain arrays drilled from chambers in the comparative safety of the main drainage gallery. Drainage galleries under the slip have the advantage that the maximum drawdown on the water table in the slip can be achieved using gravity alone, although the gallery itself may need to be pumped if it does not have a convenient outfall. Constructing tunnels behind a slide can also be effective, particularly when they intercept flows from an aquifer. As well as stabilizing landslides, such a drainage gallery can be effective in improving the stability of fills placed over springlines on sidelong ground.

Sometimes there is no reasonable access point for drainage galleries or headings, and so shafts can be sunk from the surface with fans or arrays of bored drains drilled out from the sides. Inevitably, the discharge from these needs to be pumped from a sump, incurring a continual maintenance charge and running cost.

For maximum effectiveness, deep drains should extend well beyond the point at which the reduction in groundwater pressure is desired, probably as much as two or three times the drain spacing being about the minimum further projection.

Deep drainage measures are now increasingly being used in the United Kingdom. In Chapter 10, specific reference is made to the drainage galleries in the slipped masses at Folkestone Warren. These are in lined tunnels excavated from the top of the sea wall back into the slopes. Most of them do not penetrate as far as the slip surface at the rear of the slide complex, but some do. An early system of these tunnels, under construction in the period 1914–15, was disrupted by a major slide, despite their already having effected noticeable lowering of groundwater levels, by the draining of a large pond at the surface, for instance. The later series of drainage headings appears to have been more effective, although they operate in conjunction with massive toe loading. Toms (1953) observed that the tunnels had lowered the groundwater levels in observation wells by nearly 20 m. The Warren drainage headings are similar to, but on a much larger scale than, the sub-horizontal bored-drain arrays sometimes used for smaller slides. They discharge under gravity.

Bored-drain arrays can be more effective when installed at a lower level. An example of this at Herne Bay, in the Beacon Hill landslide (illustrated both in Figure 2.8, and in respect of the remedial measures again in Figure 8.25), involved bored-drain arrays radiating from four 4 m diameter 14 m deep shafts. Each of the bored drains was 100 mm in diameter, and was lined with porous land drains. Water discharging from drains at each of several levels in three of the shafts collects through connecting drains to the sump in the lowermost shaft (D in Figure 8.25) from where it is pumped to the surface by an intermittently operating pump. Lower-level drain arrays, not all of which penetrate the slip surface, are provided at the base of shaft D. With a calculated improvement in factor of safety of some 30%, the slide movements were reduced from some 13 mm per annum (although they had been higher some years earlier) in the year before drainage to zero in two years. The scheme is described by Berkeley–Thorne and Roberts (1981).

Hutchinson (1984) describes a number of schemes where vertical drains are intercepted by horizontal drains or by drainage tunnels in order to provide a means of discharging the water. In one case, submersible pumps were provided to reduce groundwater levels before the headings were constructed to intercept the vertical drains.

Such deep drainage measures can also be used to effect by reducing the groundwater pressures in intact soil and rock under a landslide, and thus prevent it from being fed with water from behind or underneath. This is implicit in the diagram of Figure 8.17, and the principle has been used effectively in stabilizing sliding coal waste tips in South Wales. These formerly unstable tips have been built on sidelong ground, over the springline where the perched water table (in a sandstone, overlying a coal and seatearth) discharged on the slope. This supply of water fed the groundwater body in the tip, and directly contributed to its instability. The main element of the remedial scheme was a tunnel passing behind the tip to intercept these groundwater flows and channel them safely away (Siddle, 1984).

8.11 Trench drains

Shallow rubble- or gravel-filled trench drains are used extremely widely in slope stabilization works. In terms of the definitions of drains as being shallow or deep, these trench drains can be either or both. This section is devoted to simple design methods for trench drains assuming that they behave as 'deep' drains.

The term 'counterfort drain' is sometimes used to describe trench drains which penetrate into solid ground below the soil which is being drained, and may therefore provide some mechanical buttressing effect as well as their effect on the porewater pressure and hence shear strength in the drained soil. In modern practice such trench drains are used mainly for their drainage action, and counterforts of rammed chalk such as were formerly used extensively on the railways are now rarely seen. These may, of course, have had considerable success, not only through their (slight) drainage action, but also through the inadvertent changes in soil strength resulting from the alterations in soil chemistry which the presence of chalk in large quantities inevitably caused.

The major use of these trench drains is to stabilize shallow slides and slides of a principally translational character. In these types of slide it is often totally impractical to attempt extensive regrading. Often, the landslide may involve highly disturbed soil, mudslide debris or a solifluction sheet, which will have a much larger permeability than underlying undisturbed soil. The seepage pattern in the landslip debris, in between the drains, can then be treated as having an impermeable base.

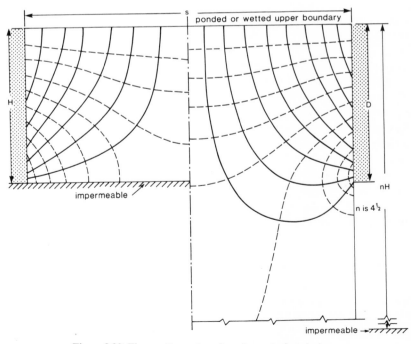

Figure 8.20 Flow patterns: trench and counterfort drains.

The basis for a theory of the use of the drainage effect of trench drains is shown by Henkel (1957) with a flow net, although the appreciation of this may indeed be much older (see for example, Richardson (1908)). Solutions to the flow pattern to trench drains were obtained by means of flow nets (Figure 8.20), analogue experiments, or occasionally by means of finite-difference relaxation methods. These all show that while for maximum effectiveness the drains should penetrate to the base of the permeable stratum (usually to the slip surface at the base of landslide debris), some beneficial effect can be obtained even with partly penetrating drains. With the availability of finite-element methods, a set of solutions was obtained, so that the theory could then be put into a systematic form by Hutchinson (1977), and compared to the available data from case records.

The various key parameters in the design of a counterfort or trench drain installation are shown in Figure 8.21. There is an interdrain spacing s, and an effective depth H: the effectiveness of the drain is obviously inversely related to s/H. However, the effect of permeability anisotropy must also be taken into account, so a composite parameter taking into consideration the ratio of horizontal and vertical permeabilities, k_h and k_v is

$$R_s = \sqrt{(k_v/k_h)}s/H \qquad (8.8)$$

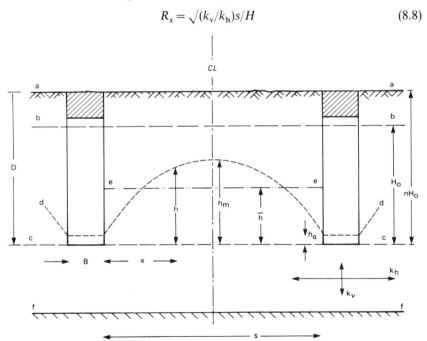

Figure 8.21 Key diagram: trench and counterfort drains, a–a, ground surface; b–b, original piezometric elevations for drain invert level c–c; d–d, e–e, actual and average piezometric levels respectively for drain invert level following the establishment of steady-seepage conditions after drain installation; ff, base of soil layer being drained: k_h, k_v, horizontal and vertical permeability coefficients.

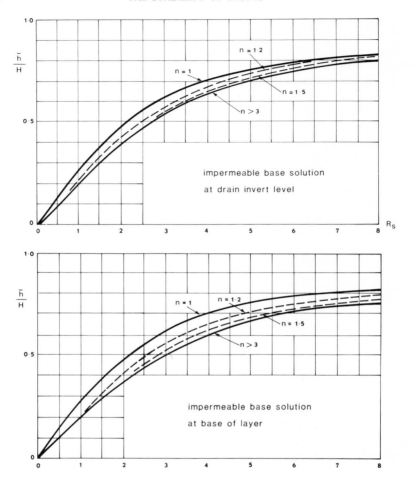

Effectively, this reduces the drain spacing where the horizontal permeability is larger than the vertical permeability.

Drains may penetrate to the base of the stratum in which they are installed, i.e. they are counterforts, or they may be partly penetrating. The degree of penetration is denoted by a depth factor, n, which is the ratio of effective stratum depth to effective drain depth.

Now a set of these drains may be installed in a layer where the underlying material is of such lower permeability that 'bedrock' may be considered to be impermeable. Alternatively, the underlying soil or rock may be more permeable, and thus constitute a 'drainage' boundary. In the latter case, the potentials effective at that drainage boundary will have an effect on the seepage pattern in the vicinity of the drains.

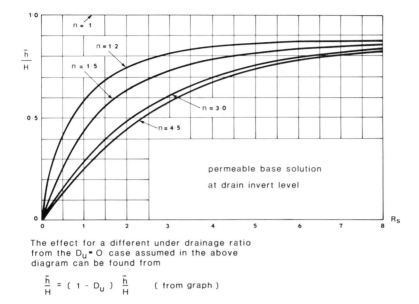

The effect for a different under drainage ratio from the $D_u = 0$ case assumed in the above diagram can be found from

$$\frac{\bar{h}}{H} = (1 - D_u) \frac{\bar{h}}{H} \quad \text{(from graph)}$$

Figure 8.22 Design charts for trench drains.

Impermeable-base solution

Finite-element calculations have been carried out for a variety of depth factors, and the average change in piezometric head relative to that originally present in the soil calculated for two positions: at drain invert level and at the base of the stratum. These changes are shown in the upper two boxes in Figure 8.22. At other positions, located between the drain invert and the base of the stratum, linear interpolation between the two values is admissible. Between drain invert level and the effective original water surface, linear interpolation between the result from the chart for drain invert level and zero may be made. Thus, for a given drain installation, a fresh solution need not be made; these charts can be used instead. They can give the piezometric head reduction at any level, for any drain spacing likely to be encountered in practice.

Permeable-base solution

As in the impermeable-base solution, partly penetrating drains in an underdrained soil layer have no effect on piezometric pressures at the level of the original water surface. Their effect then increases until the level of the drain invert, and then decreases again, broadly in a linear fashion, until the base of the stratum is reached. In contrast to the impermeable-base case, however, there is no effect at this lowest level, where piezometric levels are controlled from below. So the chart for this case, the lowermost box in Figure 8.22, shows the effect of the drains at drain invert level only: at other levels, interpolation must be used.

Also, the effect of the drains is related to the amount of *underdrainage* (or its inverse, where artesian conditions are present). An underdrainage ratio D_u must be defined:

$$D_u = \frac{\text{potential drop across the thickness of the stratum}}{nH} \tag{8.9}$$

and the value provided by the chart modified to allow for the effect of this underdrainage as follows:

$$\frac{\bar{h}}{H} = (1 - D_u)\frac{\bar{h}}{H} \tag{8.10}$$

(to use in design)   (from the chart)

In every case, the drain width has been assumed to be negligible, and the effect of finite drain widths is to indicate a somewhat larger reduction in mean piezometric head at the drain invert level. Furthermore, of all the cases, only one has proved amenable to solution by means of pure mathematics; that is the counterfort drain case. A solution for this is given by Bromhead (1984).

The mean porewater pressure remaining at any level above the base of counterfort drains solved by Bromhead (1984) is reproduced in Figure 8.23. This is used by first identifying the required mean porewater-pressure head reduction on the slip surface (the infinite slope method, equation (5.56), can be used for this with great success in the majority of translational slides. If the

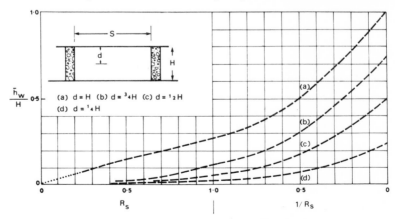

Figure 8.23 Design chart for counterfort drains.

drains are of appreciable width, then this required head reduction is then modified to take account of the zero porewater pressures in each drain. The R_s parameter is then read straight off the chart, and this is turned into a drain spacing.

An example may help. Take fully penetrating drains of 5 m depth at 20 m centres in a layer of colluvium. Assume that they are of negligible width. Since they are counterfort drains, Figure 8.23 could be used. The soil is isotropic, and groundwater level is at ground level.

R_s is given by the ratio 20/5, i.e. it is 4. For this, the mean piezometric level to drain depth ratio is 0.75, and thus the mean piezometric elevation is reduced from 5.0 m to 3.75.

If the drains were 1 m wide, what spacing would be required to obtain a reduction to 3.0 m average piezometric level? A trial and error approach is needed. Take drains at a closer spacing of 16 m centres. The distance between them is 15m, thus R_s is 3. From the figure, the ratio h/H is 0.66, hence h is 3.33 m. But, the drains, which run empty, have a width of 1.0 m, so that the true average piezometric height is $(1 \times 0 + 15 \times 3.33)/16$, or 3.12 m. A little extra manipulation will yield the desired result. Some further examples are given by Bromhead (1984).

Case records on this subject are reviewed by Hutchinson (1977), and some examples are shown in Figure 8.24. Broadly, the piezometric observations are in agreement with the theory outlined above, but the piezometric curves between drains are flatter than the theory predicts. This may be the result of smear on the drain sides, or other secondary effects not taken into account in this sample analysis. Hutchinson also shows that the terrain slope has no effect on the results of the theory, being based on proportions rather than on absolute values.

8.12 Drains to eliminate undrained construction porewater pressures

Drains installed specifically to eliminate construction-induced porewater pressures normally fall into two types: drainage blankets installed in between lifts of fill, or between a fill and its foundation; or vertical drain wells. In the general context of 'deep drains' (section 8.10) the use of sand-filled geofabric sandwicks has been discussed. These are of course equally useful in shortening a drainage path-length to accelerate consolidation in an embankment foundation. Sand drains or drain wells of this sort have been constructed utilizing a variety of materials including geofabrics used alone, cardboard and sand-filled boreholes. They are most effective when they can tap some preferentially horizontal flow. The theory of consolidation is used to evaluate drain spacing, and attempts to model the process are littered throughout the history of soil mechanics. Notable contributions to it have come from Barron (1947), Moran *et al.* (1955), and Rowe (1968).

J

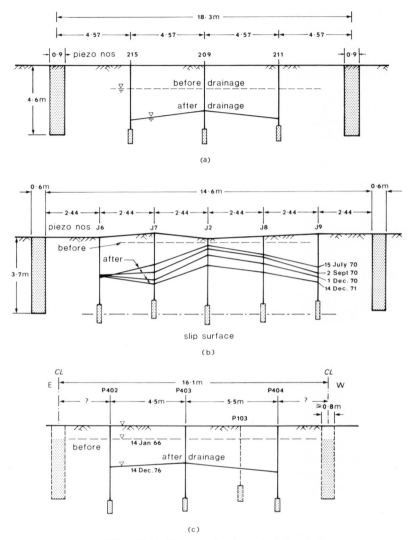

Figure 8.24 Case records of trench drains.

It is essential in the construction of these drain wells to minimize the smearing around their periphery if horizontal flow is not to be seriously impeded. However, unlike the permanent vertical drain wells mentioned previously, it is not necessary for them to penetrate an underlying bed of higher permeability to take advantage of gravity drainage. Excess porewater pressures can escape quite happily at ground level if the tops of the drains are capped with a drainage blanket with suitable outlet details.

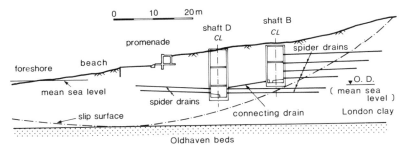

Figure 8.25 Deep drainage scheme used in the Beacon Hill landslide, Herne Bay. (See also 8.18.)

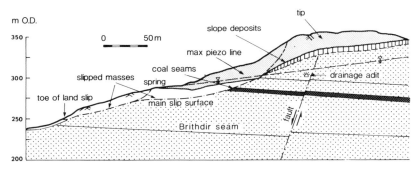

Figure 8.26 Drainage adit at the National Colliery spoil heap, South Wales. An example of tunnelling to intercept groundwater flows towards an unstable colliery spoil heap in South Wales, given by Siddle (1984).

Drain wells can be pumped out if necessary. This is rarely done, except as a means of performing field trials on the rate of consolidation: here pumping is cheaper and simpler to organize than the construction of test fills.

Drainage blankets in fills can allow low-permeability materials to be used which would otherwise be rejected as unsuitable due to their build-up of pore pressures during placement. Gibson and Shefford (1967) review this sort of application in depth. Many dams have such blankets in both upstream and downstream shoulders. Those in the upstream side have the additional role of allowing water to escape from the dam under rapid drawdown conditions (section 7.9).

8.13 Drains in dams

It is vital to control seepage in earth dams. This is always done at the cost of extra throughflow: a drain shortens the drainage path-length and increases

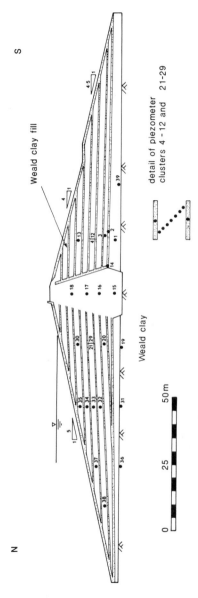

Figure 8.27 Drainage blankets in dams. A typical clay fill with frequent drainage blankets: the presence of these differentiates between the 'core' and the 'shoulders'. A piezometer installation is shown; this would be used for monitoring the effectiveness of the drainage blankets.

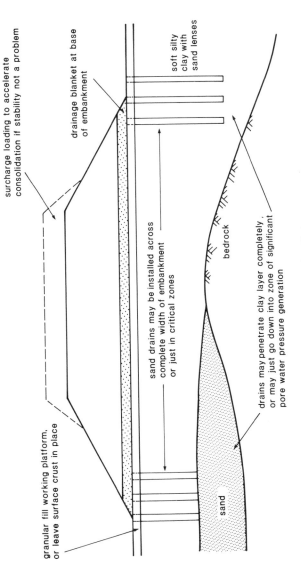

surcharge loading to accelerate
consolidation if stability not a problem

drainage blanket at base
of embankment

soft silty
clay with
sand lenses

sand drains may be installed across
complete width of embankment
or just in critical zones

bedrock

drains may penetrate clay layer completely,
or may just go down into zone of significant
pore water pressure generation

granular fill working platform,
or leave surface crust in place

sand

Figure 8.28 Sand drain installation.

flow velocities as well as reducing porewater pressures. Since any increase in the velocity of seepage brings with it an increase in the stream power of the flowing water, and an increased potential for internal erosion, the design of filters for these drains is of paramount importance. Experiments in the 1940s gave rise to design rules for this. Essentially, the pore size for the filter should not allow the passage of fines from the soil which is to be 'protected'. This is usually done by specifying the shape of the particle-size distribution curve for the filter as some function of the distribution for the protected soil. Examples of such an approach are the filter rules from the US Army Corps of Engineers and Terzaghi.

Where it proves impossible to obtain the desired permeability in the drain when it is constructed according to these rules, then *graded* filters may be used. The drain is then made up of a series of layers, each of which has a filter grading with respect to the next outermost layer. At the centre of this system lies the drain itself.

Such rules may not be entirely satisfactory for clays. (It takes a clay to filter out clay particles!) Vaughan and Soares (1982) propose a more involved treatment based on the average flocculated particle size, which is to a great extent dependent on the porewater chemistry.

A clogged filter makes a drain less effective because its overall size is reduced, and not because the filter becomes 'impermeable'. It is unlikely that a clogged filter will be less permeable, for instance, than the soil it protects. However, a clogged drain ceases to act as a drain, since it adopts the same permeability as the surrounding soil. Some filters in dam drains are specifically intended to become clogged in certain circumstances. A major problem, for instance, arises if the core of an earth dam cracks due to tension induced by differential settlement, or stresses arising from the filling of the reservoir. If the downstream side of the core is protected by a filter which clogs up and prevents the carrying of fines by the water which escapes through these cracks, then they will 'heal' of their own accord. Once water is no longer travelling quickly, the sides of a crack will soften and swell until it closes.

The design thicknesses both of drains and of their attendant filters must be sufficient to cope with settlement and other deflections: the effects of earthquake shocks for instance (section 6.6). This militates against the use of geofabric filters which may be torn by ground strains, or punctured by sharp particles. Advances in this technology, however, make the use of geofabrics as filters, and indeed, as soil reinforcement, an increasingly attractive proposition.

8.14 Retaining walls in slopes

The forces and moments which act on walls or piles constructed as a slope stabilization measure can be very large in magnitude. A way of estimating

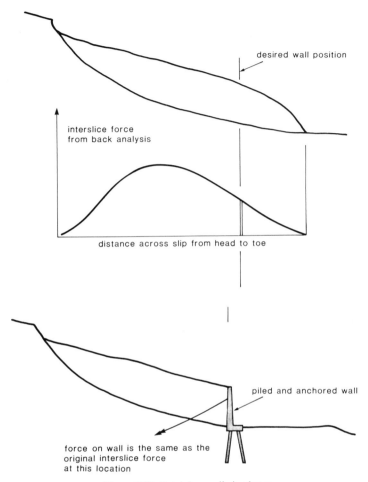

Figure 8.29 Retaining walls in slopes.

these is to use the interslice force resultants from stability analysis. Stability analysis derived forces are those for equilibrium of the soil mass. If at the proposed position of the wall there is a thrust E coming from upslope, there must be an equivalent reaction from downslope. The wall serves only to provide additional resistance, which can only be mobilized in the case of a cantilever wall, or one without stressed anchors, by further deformation. This transfers load from its line of action through the soil mass into the intact soil or rock beneath.

Steel-sheet pile walls may be used for very small earth movements, or as a very temporary expedient, but in the majority of cases they will be found to have little flexural resistance and low stiffness. They do have the advantage that their resistance is mobilized immediately upon insertion, in contrast to concrete walls which take time to acquire strength. Additional capacity can be gained by the use of high-modulus section piles.

Access for plant is normally so bad on actively moving landslides that such walls can only realistically be considered when valuable fixed installations are threatened, or to make especially certain of the stability of old landslides which are not moving at present but which might inadvertently be reactivated.

The forces on a wall which crosses a landslide can therefore be estimated from the output of a stability analysis which yields the interslice forces. Where the landslide changes in sectional shape across its width, the loads on the wall, and possibly their point of action, will also change. This effect may be estimated by analysing different sections through the slide, bearing in mind that conventional stability analysis assumes unit thickness for the sliding mass, so that the interslice forces are expressed as forces per unit width.

Walls which cross only part of a landslide, or other constructions, such as bridge piers within a landslide and penetrating the slip surface, can be subject to additional loads. This can be visualized by taking the case of a single pier in a large sheet slide. Suppose the pier is immovably fixed in *in-situ* soil below the slip surface. It will not act as a satisfactory stabilization method for the slide as a whole, and movement of this may continue, sweeping landslide debris past the pier. Just as the slide may be seen as sweeping past the pier, the pier might be imagined to be moving up through the slide mass, and in order for the pier to move in this way it would have to push out a passive wedge in front of itself. In the limit, the loading on the pier would therefore be this passive force. Calculation of the force should take account of all the normal earth pressure factors, including the upward slope of the ground surface, and the side shear on

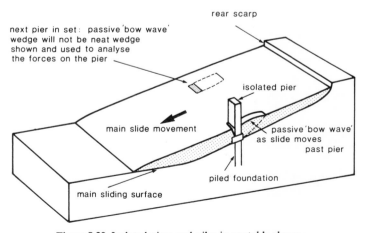

Figure 8.30 Isolated piers and piles in unstable slopes.

the passive wedge (although this will be of lesser proportional importance where a wall of any length is considered).

A passive force on an isolated pier may be many times the local intensity of the interslice force calculated in the normal way at that location. Care must be taken not to produce an absurd result where a passive force is computed which exceeds the total force needed to stabilize the landslide, since in such a case it is preferable to enlarge the structure so that the landslide is in fact stabilized.

Recent examples of the use of pile solutions to slope instability are given by Ellis (1985) and Snedker (1985) in the context of soil nailing: a contiguous bored-pile wall solution is discussed by Leadbeater (1985).

8.15 Use of geogrids in embankments and stabilization generally

Geogrids are synthetic soil-reinforcement materials. Typical of these is Tensar (manufactured by Netlon Ltd), which is manufactured by punching a regular pattern of holes in a sheet of plastics material, and then stretching it to orientate the long-chain molecules in the plastics material. The resulting grid can then be incorporated into an embankment fill, for example, to make it considerably less extensible. Two possible modes of action are then available.

The first of these is to consider the reinforcing grid as an anchor. With a slip-circle mode of failure, the additional resisting moment could be calculated, and a geogrid of sufficient section chosen to provide the additional reaction. Such a mode of operation is indicated in the literature disseminated by the manufacturers of geogrids (Figure 8.32). In contrast, a potentially more useful feature of geogrid reinforcement may be in the control of ground *strains*, reducing the likelihood of progressive failure in brittle soils, or of excessive plastic deformation in weak foundation materials.

Geogrids have also been used effectively to repair small slides in engineering earthworks.

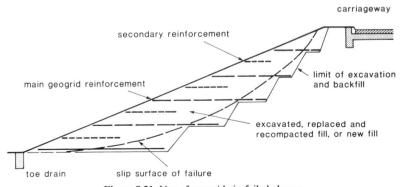

Figure 8.31 Use of geogrids in failed slopes.

method 1 :   geogrids used to control lateral
              strains and reduce possibility
              of progressive failure

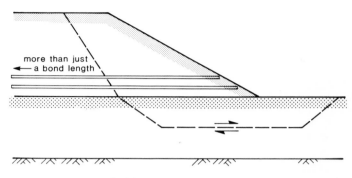

method 2 :   geogrids used as an untensioned
              ground anchor

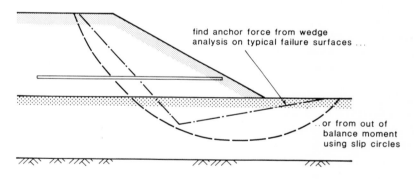

method 3   geogrids used to retain embankment
              fill at steeper slope

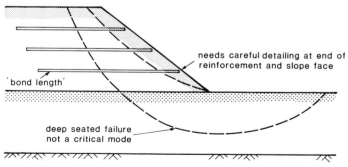

Figure 8.32 Use of geogrids in embankment construction.

8.16   Temperature treatments and grouting for slope stabilization

The effect of elevated temperatures on clays, first driving off excess moisture and then baking the material, suggests that temperature treatment may be a possible means of slope stabilization. At low temperatures, too, freezing of the pore water can produce a strength improvement. Ground freezing is most likely to be a successful expedient in those soils such as silts or fine sands where temporary control of slope stability is all that is required. On the other hand, high-temperature treatments are of a rather more permanent nature.

Early railway practice for treating unstable slopes in clay soils included lighting wood or coal fires in pits dug into the slopes to bake the surrounding soil. It can be imagined that this would only be of use where the slips were shallow, such as in embankment or cutting slopes, or where a solifluction sheet was moving. Elsewhere, the depths of pit, or quantity of fuel required, would rule out the technique or render the effort ineffective.

Fuel costs are a major disincentive to the consideration of the technique, but in special cases heat may be available as a waste product. Beles and Stanculescu (1968) describe the use of natural gas which would otherwise have been flared off, to stabilize mudslides on the Black Sea coast of Bulgaria. Burners were introduced into large-diameter boreholes to form baked clay cylinders. These then act in a similar way to piles, and give extra shearing resistance to the sliding surface.

In contrast to the beneficial effect on clay soils of high temperatures, in some natural soils and rocks uncontrolled combustion is undesirable. Ayres (1985b) reports fires in fill used in railway embankments formed of poorly compacted combustible materials. Similar fires occur in colliery shale waste tips, and are not unknown in natural slopes of oil-bearing rocks (the 'Burning Cliff' at Kimmeridge in Dorset being one such example). Treatment of such fires in thin fills may involve excavation of fire-break cut-off trenches quickly backfilled with material down to silt size to control air movement. Grouting with cement, pulverized fuel ash or limestone dust (the latter giving off carbon dioxide on heating) can also control the effect, and will be useful in thicker fills. Deep compaction (e.g. dynamic consolidation) is also useful in controlling air movement. Water injection is worse than useless, as water gas is generated.

Some beneficial effects have been reported (Ayres, 1961) where chemical or cement grouting has been used in unstable slopes, principally in connection with railways, in the United Kingdom. Particularly where an embankment has been formed of relatively loose material, by end tipping for instance, the introduction of grout to fill the voids will make for a much less compressible fill, and one which is less subject to distortion and which may therefore 'bridge over' foundation movements. The grout may then have lowered the permeability sufficiently that water is subsequently excluded from that area, giving further improvements in strength, and in stability.

It is less easy to see the effect on clay slips. Here, as a result of ion exchange in the pore water, clays may change their mineralogy. This can cause an improvement in shear strength by altering fundamental effective stress properties, or by causing more of the moisture present to become bonded to the clay particles. It is conceivable, too, that penetration along slip surfaces can take place, giving a cementing action to the soil masses on each side of the slip surface. If this is the desired mechanism, however, it is essential to keep grout pressures low, and to limit the grout injected from any single hole, so that the liquid grout does not act as a high-pressure pore fluid with the obvious and opposite result to that intended!

Detailed accounts of the technique and its application in the field are given by Ayres (1985).

8.17 The choice of a stabilization method

The first response to any slope stabilization problem must always be to consider the use of regrading or cuts and fills as the main element of the scheme. This is because the effects of fills at the toe of the slope are immediate, if anything improving further with the passage of time. What is more, once

Figure 8.33 Other stabilizing techniques. If the slope cannot be stabilized, as in the case of the hazard from boulders on this slope in Hong Kong, then a barrier structure can be built. 'Dams', sheds and shutes of various forms are used. This is a gabion wall built of 'Tensar' geogrids (courtesy Netlon Ltd.).

emplaced a fill is unlikely to be removed or to become ineffective. It is only in the comparatively uncommon cases where regrading is impossible that another stabilization technique has to be considered as the principal element of the scheme.

Drainage alone can be considered as the sole stabilization method where stabilization by regrading is impractical. Low-angled natural slopes are a typical example: regrading would require massive muckshifts just to effect a slight change in overall slope angle. It becomes immediately effective where large flows in permeable strata can be intercepted. In fine-grained soils, it takes some time for the drainage to become effective; there, drainage must be considered a longer-term expedient, and not as a method of bringing slides to rest. On the other hand, treatment of surface water flows is essential in controlling many mudslides and mudflows in fine-grained soils, and in stiffer deposits the excavation of trenches for shallow drains can decrease *in-situ* porewater pressures by an undrained unloading effect, so some immediate improvement is realizable in theory at least.

The long-term susceptibility of drainage measures to blockage makes the use of this technique for stabilization subject to caution and the employment of a regular maintenance programme.

Structures constructed in an unstable slope are the third means of primary stabilization. These have been extensively discussed in the foregoing, but fall into two main classes: active (stressed ground anchors, rock-bolts) and passive (walls, unstressed dowels, piling through the slip surface, etc). Both active and passive schemes can be used successfully, but to become effective, passive schemes rely on further movement of the slide, which can be undesirable.

The digging out of slide materials and their replacement can be considered in certain cases, notably when small slide masses are involved. Landslide debris can be hauled to tip, or replaced after suitable treatment; or the cavity can be refilled with imported materials with intrinsically better properties. Suitable treatment for the replacement of landslide debris in clay soils can include recompaction to destroy slip surfaces, lime stabilization, compaction in thin layers between drainage blankets and the use of reinforced earth techniques with geofabrics and geogrids. The method does rely on the excavations remaining stable for long enough to allow the works to be completed.

After the selection of a primary stabilization technique, secondary treatments may be required to preserve the long-term effectiveness of the primary treatment, or to prevent deterioration of untreated areas of the slope. Drainage is the key element of this, both in its shallow and deep forms.

Analogous to the treatment of unstable natural slopes is the redesign process when inspection reveals that a preliminary design is unsuitable. The first reaction must be changing the slope profile; the second, drainage. Subsequently, redistribution of the available materials in the slope, reinforcement techniques, and similar procedures can be considered.

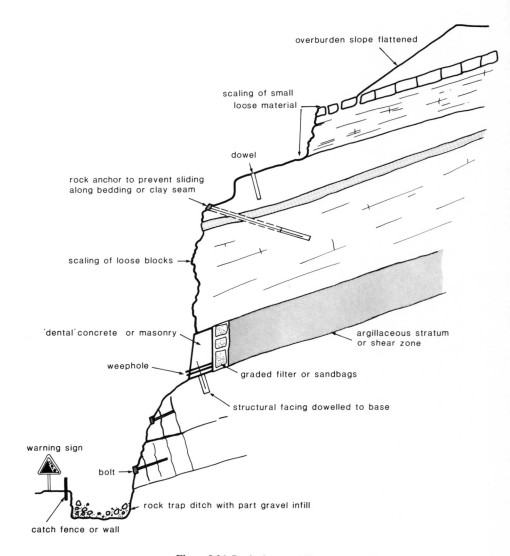

Figure 8.34 Rock slope stabilization.

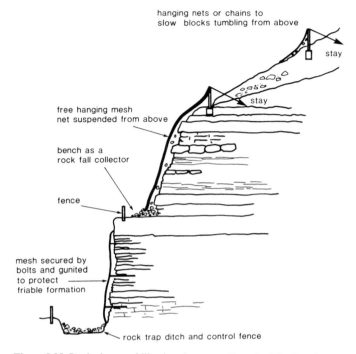

Figure 8.35 Rock slope stabilization, by controlling the falls that do happen.

8.18 Rock slopes

Rock slope stabilization is often undertaken on a fairly *ad hoc* basis as the condition of the rock mass is exposed. Some possibilities are shown in Figures 8.34 and 8.35, which draw heavily on the review of the subject by Fookes and Sweeney (1976). The monitoring of rock slopes is described by Franklin and Denton (1973). A comprehensive treatise on the subject is given by Hoek and Bray (1974).

9 Investigation of landslides

9.1 Land surface features: methods of mapping

Surface mapping is an integral part of landslide studies. For preliminary or route design purposes on a very large scale, satellite photograph interpretation may be used. No doubt resolution of available satellite photographs will soon be much better than the currently available Landsat imagery. This resolves features of the order of a kilometre, and is therefore of little use for routine engineering work.

Ordinary air photographs, taken from civil or military missions, are much more useful. When commissioning photographs, the flying height should be about 10 times the maximum vertical relief—more, and the scale of a contact print from the negative will be too small; less, and the distortions induced by parallax effects will make it difficult to use the photograph for quantitative work. Most air-photography cameras have a focal length of about 150 mm, and this then limits the maximum scale for a contact print from the negative. Maps can be prepared photogrammetrically from these air photographs provided that the position and level of a few points are known: this is a routine, and (compared to detailed terrestrial survey) a relatively cheap procedure. It is also, in principle, simple and straightforward, although without the use of the appropriate apparatus together with the assistance of a skilled technician, it is difficult to employ the techniques to do more than just survey a cross-section or two.

Appropriate overlap between photographs enables them to be viewed stereographically. This gives the illusion of exaggerated vertical relief, and can show quite subtle landforms. However, features which have only small relief are better revealed by the shadows thrown by low-angle lighting than by the illusion arising from stereoscopic viewing. Stereoscopes vary in complexity, but all demand that the viewer is capable of seeing stereoscopically in the first instance. Some people find it difficult to fuse the images because of poor eyesight. I have found some success in digitizing points on a pair of air photographs, and using a computer to operate on the coordinates thus obtained to solve the equations relating plan position and elevation to photograph coordinates. The accuracy of the digitizers can be quite low, but still make this an acceptable technique, with accuracies of perhaps plus or minus a metre or so in plan position and in level obtainable within a 500 m-square area with about 40 m of elevation difference.

Although conventional photogrammetric techniques are described in a wide variety of standard texts, the use of a simple digitizer tablet is not: a brief review of the steps necessary in the application of the method for those interested in pursuing it further is therefore in order here. First of all, the fiducial points on each photograph are digitized. The program can then relate digitizer tablet coordinates to photograph coordinates. Secondly, three points of known position and elevation which appear in the overlap between the photographs are digitized. Since the angles subtended at the point on the ground representing the centre of each photograph are the same as the angles subtended by the images of the points at the centre of the photograph (at least for photographs of small tilt), then the relationship between photograph and ground-plan coordinates can be established. Finally, other points digitized on both photographs can be positioned on the ground by the use of the two transformations already obtained.

Level information is obtained as follows. From the three points initially digitized, the scale of the photograph is calculated. This is then related to the approximate flying height via the focal length of the camera, remembering to add the elevation of the ground in the vicinity of the control stations. Elevations of other points are obtained by computing their parallax (which is a simple function of the relative position of the point on both photographs) and applying the conventional parallax equation for heighting. Such height data is poor, however, because the measured parallax is greatly influenced by inevitable inaccuracies of measurement on the digitizer, and the effects on the photograph scales of the aeroplane not flying perfectly level during the photographic mission. The errors that arise are distributed in the form of a hyperbolic paraboloid over the surveyed area. The constants of this *error surface* can be obtained from the known and calculated levels of five points (the three original control stations and two others), and corrections computed relative to the position of any surveyed point can then be applied to the height computed on the basis of the parallax equation. Even when the flight path was level it is worth applying the correction, so that systematic digitizing errors can be eliminated.

Orthophotographs are photographic images corrected for ordinary camera distortions to make true maps. Level information (e.g. contours) is generated as part of the rectification process, and can be added to the orthophotograph at no extra real cost. The orthophotograph will be much more detailed than a line drawing from an aerial survey, and of course, is not therefore an *interpretation* of the features present in the way that a drawing is. An example of a contoured orthophotograph of a landslipped slope is given by Hutchinson and Gostelow (1976).

The features which can be seen from an aerial photograph will depend to a large extent on the vegetation cover, and the lighting at the time the photographs were taken. Naturally, a dense tree cover obscures the ground, so that elevation information can be related only to the canopy rather than to the ground itself. It is therefore more desirable to photograph during the winter months when trees are bare. Sometimes, too, the interpretation of the features seen is difficult, and ground verification is necessary to decipher what is seen indistinctly on the photographs, or to establish details which are obscured by vegetation cover.

The end result of an air survey for engineering purposes will usually take the form of a contoured plan, with the contours superimposed on a *base map* showing largely man-made features such as fences, power lines, buildings and so on. *Physical features*, where shown at all, will be gross features such as the edge of a cliff, rather than subtle ones such as landslip toes, tension cracks, etc. These latter features need a greater degree of interpretation on the part of the surveyor, but when included on a map they change its character. It then becomes an interpretive map, rather than a simple record of observations.

The contoured map is by no means the only base medium for recording such observations, and an often satisfactory alternative is the technique of *geomorphological mapping* (Savigear, 1967) in which the slope is subdivided into facets of relatively uniform slope. The map consists of the plan positions of the boundaries of these slope facets, represented by symbols which show whether the transition is a sharp break of slope or a gentler transition, and whether it is concave or convex in section. Further data, showing measured slope angles and the directions of dips, can also be recorded. The technique is thus a replacement of the contour information, and may need the information of a base map to be related back to the field situation at a later date. Strictly speaking, such a morphometric map should not be interpretive; at least, not in the first instance.

Geomorphological mapping is often presented (in the geomorphological literature!) as the answer to all landslide problems. However, there are some fundamental difficulties. It is, for instance, impossible to decide what scale of surface features are significant, and should therefore be mapped, until one has an idea of the subsurface structures. These structures cannot be elucidated without expensive and time-consuming subsurface work: this could best be planned if one had an idea from surface features of the nature of the deeper structures. In practice, an *iterative* technique has to be adopted, involving alternating surface and subsurface phases, each chosen to supplement the information arising out of the previous stages.

An excellent example of this is given by Hutchinson *et al.* (1985) for the slopes at Lympne described in section 10.4.

At some stage, the features present must be interpreted in terms of the structure of the site, and it is essential that the interpretation is quite clearly separated from the factual observations on which it is based. This enables the interpretation to be questioned more readily at a later date when more information is available. This interpretation will be based on more than just the slope geometry; it will also take into account a range of other factors, best found on a walk-over survey of the site. To my mind, the principal benefit of the preparation of a geomorphological map is that it requires somebody to examine the site in detail: it is therefore just as well that the engineer or geologist prepares the map as that the services of a trained geomorphologist are engaged. The latter, although more proficient at the business of simply mapping, may not be sufficiently involved later in the investigative process to reassess the information gained from the walk-over.

The features to look out for in a walk-over survey are listed in Table 9.1.

Table 9.1

Landform	Water course or feature
Slip scarps	Streams
Counterscarps or graben	Landslip ponds
Soil waves	Springs and sinks
Successive soil ridges	Marshy ground
Backtilted blocks	Hydrophilic vegetation
Toe bulges	(marshy grasses, rushes,
Lateral ridges	teasels, horsetails, etc.)

Disrupted artificial feature
Breaks in walls
Distorted openings in buildings
Damaged services
Cracked roads
Dislocated fence lines

9.2 Locating slip surfaces by ground surface measurement and inference

Regardless of the utility or otherwise of surface-mapping techniques, a complete investigation of a landslide requires subsurface data. This may include finding the depth of slip debris, or the positions of blocks of soil and rock (perhaps to some degree preserving their original stratification although now displaced), or locating the position of sliding surfaces. The latter is an art in itself, whereas the other two fall into the general field of ground investigation generally, and therefore need little amplification here.

The techniques for finding slip surfaces in landslides have been reviewed exhaustively by Hutchinson (1983). It is necessary to locate the position of these slip surfaces in order to correctly position piezometers and other instruments, to carry out reliable analyses of the causes of instability and to correct them, and to permit the design and safe execution of remedial measures.

There are techniques available both for use in moving landslides and some applicable to stationary ones. In general, the latter methods are equally useful in the former case, but not vice versa, although some of the methods for

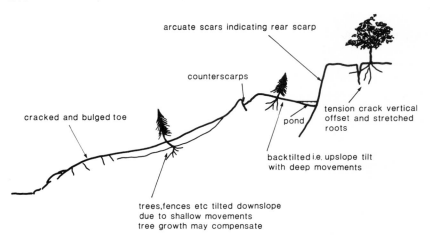

arcuate scars indicating rear scarp

counterscarps

cracked and bulged toe

tension crack vertical
offset and stretched
roots

pond

backtilted i.e. upslope tilt
with deep movements

trees,fences etc tilted downslope
due to shallow movements
tree growth may compensate

Figure 9.1 Surface features indicating landslide form.

stationary slides may not be usable in moving slides for reasons of practicability or safety.

An important first step is always the examination of surface manifestations of instability. These may be found at the head, sides and toe of a slide. At the head, or crown, of a fresh slide, the outcrop of the slip surface is often clearly visible, and not only its position, but also its slope, can be determined from a fresh slip scar. When degraded, however, the inclination can be deceptive, and the original position of the rear scarp of the slide lost. Shallow slides and flows often leave lateral ridges at their sides. This happens as a result of the spillage of material onto stable ground, or thrusting of material outwards. At the toe, however, bulged, heaved or broken ground is a poor indicator of the original position of emergence or breakout of the toe. Such indications should always be measured and recorded properly on an engineering geological or geomorphological map: it may be difficult to survey them directly, in which case photogrammetric methods may be applied.

The occurrence of intermediate scarps (Figure 9.1) can be a pointer to a multiple character for a landslide: graben structures or other surface disturbances reveal the presence of sharp bends in the slip surface. A number of such features will be seen in the illustrations to the foregoing chapters of this book, notably Figures 2.9, 2.10, 2.12 and 2.15, and in the case records, most notably in Figures 10.6, 10.26 and 10.27.

Sometimes surveys before and after a landslide will have been made, or can be made retrospectively, from air photographs. Features which are present on the surface in both surveys, but which have become displaced, show the magnitude and direction of movement, and such *movement* vectors can be used in an inferential procedure to find the approximate location of the slip surface.

The method was originated by Jakobson (1952), and is shown diagrammatically in Figure 9.2.

The principle is to take the vertical section underneath the location of a typical point as remaining vertical during the slide. Any heave of ground at the toe must equal the volume swept out by the section, and hence an approximate depth to the slip surface can be found. An appropriate allowance for bulking of the slipped material, or even for toe erosion after the slide, can easily be made. If estimates of slip surface depth are made working both from the toe to the head of the slide and vice versa taking loss of section rather than heave, then a smoothed average curve will give a better result. Ideally, the slip surface inclination should be parallel to the associated vectors of surface movement.

In cases where the slide has a clear rotational character, a similar procedure, but making use of radial sections, may be considered.

Other data on the depth and shape of slip surfaces can come from the

Figure 9.2 Principles of Jakobson's (1952) method.

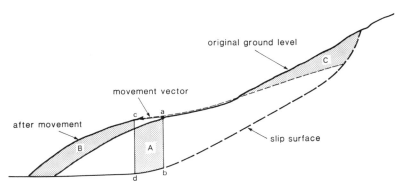

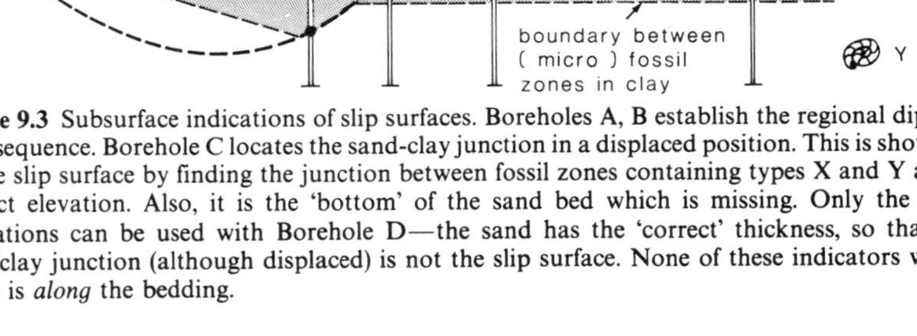

Figure 9.3 Subsurface indications of slip surfaces. Boreholes A, B establish the regional dip and rock sequence. Borehole C locates the sand-clay junction in a displaced position. This is shown to be the slip surface by finding the junction between fossil zones containing types X and Y at the correct elevation. Also, it is the 'bottom' of the sand bed which is missing. Only the fossil indications can be used with Borehole D—the sand has the 'correct' thickness, so that the sand/clay junction (although displaced) is not the slip surface. None of these indicators works if slip is *along* the bedding.

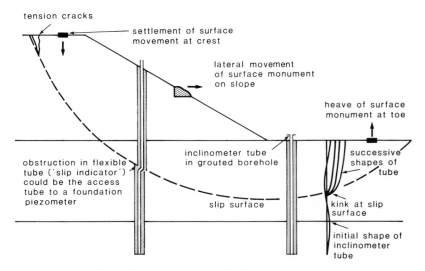

Figure 9.4 Instrumentation for finding slip surfaces.

movement of trees, telegraph poles, fences and structures at the surface. If after movement these show a backward tilt (*into* the slope), then the slip surface is relatively deep. Forward tilts, especially of trees, where the tap root anchors the tree into intact material below the slip surface, are usually an indicator of shallow depth to the slip surface. Deep-rooted species of tree die when their tap roots are severed, but shallow-rooted species may survive a landslide, and continue to grow in their displaced position.

Geophysical surveys made from the surface can also be used in landslide investigations, but they do require calibration against boreholes. Three methods have been used: seismic reflection and refraction, and electrical resistivity measurement. In all methods, lateral variations in the nature of the landslide debris and the depth to the slip surface can make the results difficult to interpret, and specialist interpretation is essential. Complex groundwater conditions can also affect resistivity studies.

9.3 Direct logging of faces and cores

Direct observation of slip surfaces in exploratory excavations including boreholes, or in samples recovered from such excavations, is the most satisfactory of all the techniques. Shallow slip surfaces can be found in the sidewalls of trenches excavated with backhoe-type excavators: depths of 3–4 m being attainable with wheeled excavators and 6 m or more with larger, track-mounted vehicles. The latter are preferable on steep or slippy terrain. It is comparatively difficult to find slip surfaces in the floors of trenches unless by accident the excavator rips out the material above the slip surface. One cannot be perfectly sure that such a slip surface has not been formed by the excavator, until it has been followed carefully into ground undisturbed by the digging.

Any programme of trench or pit investigations needs to be planned systematically: it is usually best to start from the toe of the slope to investigate the break-out of slip surfaces where they can be reached easily. This allows the engineer or geologist to learn to recognize the particular features of slip surfaces in the materials on that site, before moving upslope to investigate other geomorphological features such as small scarps where it cannot be certain at the outset that the full depth of slipped material can be penetrated.

Logging such pits is best done on a scale drawing of the sidewall elevation. Roughly evenly-spaced vertical sections are then picked clean and logged from ground-level downwards—a horizontal string line or tape pinned to the side of the pit is essential in correlating these, as it will not usually prove possible to level the tops of all the sections before starting the recording. Important features can be traced across from section to section later, perhaps by withdrawing critical sections of the trench strutting.

Slip surfaces in clay soils may be found by locating the striated, or fluted, and polished surface (Figures 3.4, 3.5) on which movement has taken place. The directions of the individual striations show the sense of movement on the slip surface, information which can sometimes be supplemented by noting the displacement of geological features such as marker beds. Roots can sometimes grow along slip surfaces, but may be dragged into a dog-leg shape by movement, or severed. Fresh slip surfaces in clay soils are almost invariably 'polished' or shiny. This is partly as a result of the alignment of platy clay mineral particles in the direction of shear, so that the slip surface is formed from the flat faces of myriads of clay particles, but is mainly a porewater phenomenon. This can be seen by virtue of the fact that exposed slip surfaces lose their shine as they desiccate, although they remain smooth.

Old slip surfaces sometimes take the form of a 1–2 mm thick zone filled with a soft gouge which differs in colour from the surrounding soil. This is due to the percolation of water taking place more readily along the slip surface; where this happens, the groundwater brings with it free oxygen to oxidize the soil, or weak organic acids to reduce it. It is sometimes found in highly disturbed landslide debris, particularly solifluction debris, where the landslipped soil consists largely of clay materials, but with occasional small fragments of harder material, that these fragments tend to collect just above the slip surface. In such cases the slip surface is often found to take the form of a thin lamina of soft gouge described above.

Exploratory shafts and adits are less commonly used than pits because of their cost. They come into their own, however, when they can be incorporated into the permanent remedial works, or are used in effect to confirm design assumptions when the permanent works are in progress. The need for permanent support means that they have to be logged concurrently with their construction. Large-diameter boreholes can perform the same function as a trial shaft at much lesser cost if appropriate plant comes readily to hand.

Pits, trenches and shafts are very dangerous and should be entered with caution. A major source of hazard is the casual bystander or idler who may kick loose materials down on to the workers below. Helmets must be worn: the chinstrap variety is preferred as it will not fall off when the head is bent. Support for the sides of a trench or shaft must be provided in the form of timbering or shored steel trench-lining sheets. These obscure much of the face, which is unfortunate but necessary. Proprietary hydraulic ram-type temporary shores are useful for protecting the workmen who place the main shores, and have an important role even when the main shores are to be removed. It may also be necessary to use pumps to dewater pits, particularly in the winter months. A source of hazard not often taken into account is the collection of fumes or noxious gases at the base of the excavation. These include vehicle exhaust, hydrogen sulphide, methane, carbon monoxide and carbon dioxide arising from reactions between groundwater and the soil. Some gases, present dissolved in the ground water, may be released into a pit by a change in the barometric pressure, and a pit which is safe at the start of a day may become hazardous with the passage of time.

Downhole cameras can be used in small-diameter boreholes which will stand without casing, enabling discontinuities which can include slip surfaces to be located and logged. Opacity of water infilling the hole, and smear on its sides, are major problems.

When using samples taken from a borehole, in the place of directly logged excavation faces which cannot be used for reasons of practicability or cost, great care must be taken to obtain sufficient numbers of samples, each of adequate quality, for examination. Sample or core spacing down boreholes should be as close as possible, possibly with the use of an adjacent borehole with a staggered sampling pattern with respect to the first hole. In very soft soils, piston-type samplers should be used. The special Swedish foil sampler or the Delft sampler can improve sample recovery in these soils. In slightly stiffer soils push or drive samples can be taken, and in very stiff or hard soils rotary drilling techniques are as applicable as they are in rock. Many proprietary samplers intermediate in character between drive and rotary type—for example the Denison core barrel—can be employed. The samples should be split longitudinally and logged when fresh, taking specimens for index and residual strength testing from one half only. The other half should then be air-dried and re-examined. This can reveal features missed in the first examination. The author's experience is that rotary core sampling is practical in firm clays provided that some elementary precautions are observed. These include the use of a double- or triple-tube core barrel with a face discharge bit, and a synthetic polymer additive in the flush water to increase its sediment-carrying capacity so that the absolute minimum of flush return can be utilized. Drilling penetration needs to be very slow.

Since core or sample loss is most likely to occur in the disturbed zones, several runs of core or samples should be laid out and logged concurrently: different lithologies or soil textures in adjacent runs are important indicators that such a loss has taken place. Spurious slip surfaces may also occur in the samples as a result of the sampling procedure: overdriving open-drive samplers in stiff clays, for instance, gives rise to neat 45° slip surfaces of negligible throw in the top of each sample.

9.4 Compilation of the results of direct logging and boreholes

Observations from direct logging of exploratory excavations and samples from boreholes should be plotted on a cross-section at a natural scale. It is difficult to be precise about the scales to be used—these depend on the size of problem that is being investigated. As a rule of thumb, however, it is useful to plot the slope on an A3 size drawing. Features which do not show clearly at that scale are probably not significant to the overall slope behaviour. If small features are significant, it means that some slope facet, rather than the whole slope, should be the subject of investigation. This facet should then be redrawn at the A3 size, and so on.

Boreholes outside the landslipped area show the undisturbed succession and its regional dip. This data can be supplemented by surveyed outcrops of marker horizons. Where there is a shortage of outcrop, geomorphological indicators can be used to infer the positions of these markers (Figure 9.1). Comparison of the lithology, colour, texture, micro- or macro-fossil content, chemical composition (including salinity of pore water) or mineralogy between the undisturbed and intact successions reveals the position of slip surfaces very clearly. Lithological and palaeontological indicators are of most use because they are readily observed and are not susceptible to change with time in quite the same way as are colour and porewater chemistry, for instance. It should, however, be borne in mind that where slip is along a bedding plane, all the characteristics of the soil related to its original deposition are useless as indicators of the location of any slip surfaces. Several boreholes in the same locality which penetrate the slip surface can provide a dip for the slip surface as well as its level.

The material above a slip surface may have a very different texture from that beneath, as a result of shearing distortions in the slipped mass. This is even more marked where the slip is itself the accumulation of very many smaller movements. Examples of this come from mudslides, movements affecting rockslide and rockfall debris, or soliflucted soil. The degree of weathering of the slipped mass, and its water content (expressed, say, as the ratio of water

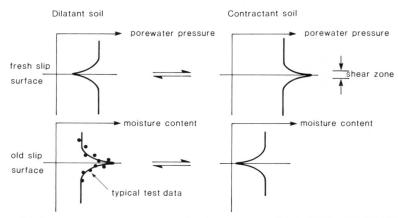

Figure 9.5 Cusps of porewater pressure and moisture content. It is common to use moisture-content cusps to confirm a visual identification of a slip surface.

content to liquid limit) will often be found to be much larger than for the parent material. Naturally, when a large movement occurs, the landslide overrides a variety of materials at its toe. Where these are strikingly different to the slip material itself, for example beach, river terrace or other readily identifiable alluvial materials, then the position of the slip surface may be inferred to lie at or close to the contact.

A check on the location of a slip surface may be made if cusps in the distribution of porewater pressure or of moisture content can be detected. In a dilatant soil, intense shearing causes a locally depressed porewater pressure. This attracts soil moisture and may therefore be a short-lived feature. It is replaced by a soil zone with higher moisture content, which is therefore softer. The reverse effect is implied for contractant soils. In these, the porewater migration ought to act to 'heal' slip surfaces. Measurements of water content cusps have been made quite frequently (Henkel, 1957; Skempton, 1964; D'Appolonia *et al.*, 1967) but the measurement of the equivalent porewater-pressure effect, requiring rapid mobilization on site and sensitive, rapid-response piezometers, is a much rarer event (Hutchinson, 1961). The presence of soft zones, not of a depositional origin, is an indirect demonstration of this effect. These may sometimes be detected through the use of penetrometers, vane-boring techniques, or use of a resistivity sounding device.

9.5 Instrumentation

Instrumentation in slopes has a number of uses, including the provision of specific data in known unstable slopes, the verification of design assumptions upon which the stability of earthworks may depend, or providing a basis for

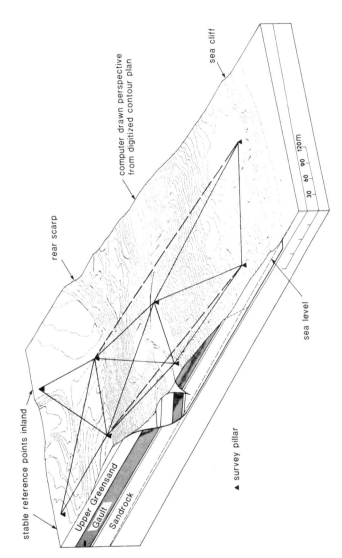

Figure 9.6 Surface movement measurement. Three braced quadrilaterals (two shown, the third encompassing the upper and lower pair of geodetic pillars) were surveyed by triangulateration. The upper pair of pillars were located with reference to further fixed points inland.

the control of a construction process. In extreme cases, the monitoring may be used to give advance warning of impending movements when the only possible response to that particular hazard is evasion or escape.

In landslides it may be desirable to install instruments to measure the surface and subsurface deformations of the landslide, or to monitor groundwater pressures.

Boreholes which penetrate the shear surface of an active landslide will be sheared off at the level of the slip surface. When this is complete, the borehole may be plumbed to find the depth at which shearing has taken place. Slow movements may take unacceptably long to do this, so a smaller-diameter tube can be grouted into the hole, which will kink at smaller total movements. This can be tested by running a short length of rod attached to a cord down the tube. Wherever there is a constriction it will be impossible to pass the rod. A similar rod, left at the bottom of the tube, can periodically be hoisted to the surface. The formation of a thick shear zone will be detected if the eventual blockage for the 'pull up' rod is found to be deeper than for the 'drop down' one. Such a simple and cheap *slip indicator* can be made especially attractive if combined with a piezometer standpipe. Variants on this theme include wells large enough to be entered, in which the deformations can be measured directly, or the use of markers buried in the side wall of a trial pit or put into a

Figure 9.7 Inclinometer (courtesy of Geotechnical Instruments Ltd.). The data logging read-out unit is on the right, and the operator is about to lower the sensor into the special casing, engaging the spring-loaded wheels into tracks in the casing.

borehole and subsequently, perhaps years later, uncovered and the deflections measured.

More sophistication, and the chance to measure much smaller deformations. is provided through the use of inclinometers. Small-diameter casings are grouted into a borehole. These casings may be of aluminium, which is stiff and liable to corrosion in alkaline grouts, or plastic. Having tried a variety of types of plastic inclinometer tube, it is my opinion that there is much room for improvement still. Typical faults include inadequate connections (designed to make the alignment of keyways between consecutive lengths of the tube simple), poor detailing of the grooves (so that inclinometer probes jump out of them), twist, and poor detailing at the transition between consecutive lengths of tube. In contrast, the extruded aluminium tubes tend to perform better, but have a limited lifespan, and this must always be a concern for a long-term reading programme, regardless of the corrosion-protection system employed.

Normally, four keyways are formed down the inside of the tubing: it is difficult to make plastic tube without a degree of twist in these keyways, and this disorientates the sensor. A wide range of sensor probes are now marketed. These operate by sensing the inclination of the hole to the vertical, and the deflected shape of the access tube can be estimated by summing the offsets along the tube. Gross and systematic errors can be detected or minimized by measuring with the instrument run down opposite keyways in a manner analogous to 'changing face' in surveying. Comparison of 'before' and 'after' profiles enables the displacement pattern of the ground to be found.

Traditionally, inclinometers have measured the deflection in one plane only, and have been designed with a low cross-axis sensitivity. Modern instruments can measure both deflections simultaneously, in the simplest way by having two independent measuring systems mounted at right angles. These can even automatically correct for the cross-axis sensitivity of each measuring system on the basis of the other system's output.

Deep inclinometer access tubes are expensive, and so is the probe. The access tube can become kinked at even small shear deformations, and so careful consideration should be given before the system is used in actively moving landslides. It is therefore sensible to try out the tubing and its access ways with a dummy probe, before inserting the genuine instrument. An inclinometer can be useful, even if very short-lived, in an active landslip if it enables the main slip surface to be discriminated from others seen in borehole samples but not currently active.

In Hutchinson's account (1983) of methods of locating slip surfaces he reviews a large number of methods that have had only a minor usage, or are not widely applicable. The reader is referred to that detailed paper, or to Hanna's *Foundation Instrumentation* or to Penman and Charles (1974) for more information on these.

Wilson and Mikkelsen (1978) describe a number of devices used in the USA for sensing movements of landslides. Rather than run the risk of rapid

Figure 9.8 Extensometer. This extensometer for measuring both vertical and lateral movements was designed by Dr M. P. Chandler (illustrated).

technological progress outdating this section, I would advise consulting geotechnical instrument manufacturer's catalogues for up-to-date techniques and devices.

Instruments of the above types are especially useful where movement is known to be occurring, but at a slow rate, and the object is to quantify this movement, and to pinpoint its location. A more immediate need can sometimes be to have advance warning of the onset of movements which will take place rapidly. Large rockfalls can be in this class. Automatic monitoring is essential for this, and extensometers and tiltmeters have been used with some success. Occasionally, one needs, not advance warning of a slide, but to prevent access to the slide area when movements have already taken place. If a secondary road crosses an area of slide danger, and the costs of reconstruction are not immediately merited, then the rupture of an electric circuit to switch on a warning sign may be all that is warranted or necessary.

9.6 Piezometers: types and response

The simplest of all types of piezometer is the Casagrande *standpipe* type. This comprises a plastic or metal tube with a porous tip. These can be driven into the ground, if the soil is soft, and the piezometer tubing and tip robust enough. In the Cambridge drive-in piezometer, for example, the tip is protected during driving by a sheath which is driven further by means of a rod down the centre of the tube when the tip is in the correct location. For many years the porous tip was made of a ceramic, but nowadays cheaper tips made from sintered plastic grains are available.

Borehole piezometers of the standpipe sort are placed down the hole, and a sand filter is poured around them. Ideally, the remainder of the hole should be filled with a cement and bentonite grout of a syrupy consistency, placed via a tremie pipe to the bottom of the hole to displace any water which has collected in the borehole or which has been poured down the inside of the piezometer tube, to ensure that the porous tip and the sand filter are saturated. A layer of compressed bentonite tablets placed on top of the sand filter is desirable if the swirling of the grout is not to disturb the sand. Putty-consistency bentonite balls rammed into position can give the same protection.

Where the use of a grout mixer, pump and tremie-pipe arrangement is not practical, grout can be mixed at the surface and just poured down dry boreholes. This cannot be done with water-filled holes: the use of large quantities of compressed bentonite pellets or bentonite balls is recommended. Small quantities of the arisings from each hole can be added to the grout without adverse effect, if the individual lumps are small. The spoil derived from continuous-flight augering in stiff clays is good; the large plugs of clay from shell and auger boring in the same materials have to be broken up or they jam in the hole and lead to undesirable voids in the backfill.

Major advantages of this type of piezometer are simplicity and robustness. Water rises up the tubing to find its equilibrium level, and an electrical indicator which buzzes or bleeps when the water level is reached can be used to discover where this is. Readings can be taken by unskilled staff. An additional advantage is that the device is self-de-airing, and any air bubbles which do collect in the sand filter or 'piezometer cavity' escape of their own accord up the tube. However, the response of such a piezometer to changes in the ground-water level is slow, and with the simple reading probe ('dipper') the top of the tube must be permanently accessible. This often rules out this type of instrument for use in or under fills, for instance, or where the top might become submerged. In the latter case, an air- and watertight cap must be provided, and the response of a closed system like this is further impeded. Where, however, such a system is to measure extremely slow groundwater responses, and the caps can be released periodically, it may nevertheless prove satisfactory. Finally, such a piezometer can only register positive pressures.

A bubbler device has been used to take remote readings of these piezometers. The pressure required to discharge air through a fine-bore tube inserted into a standpipe-type piezometer tube reflects the head of water present above the opening of the air tube. This enables remote readout of the required air pressure to be made. It is little dependent on the relative elevations of the readout and the piezometer tip.

Other piezometer types include hydraulic, electrical and pneumatic types. A hydraulic piezometer usually has two fine-bore plastic leads through which de-aired water is passed after installation. When this flow is cut off, the piezometer

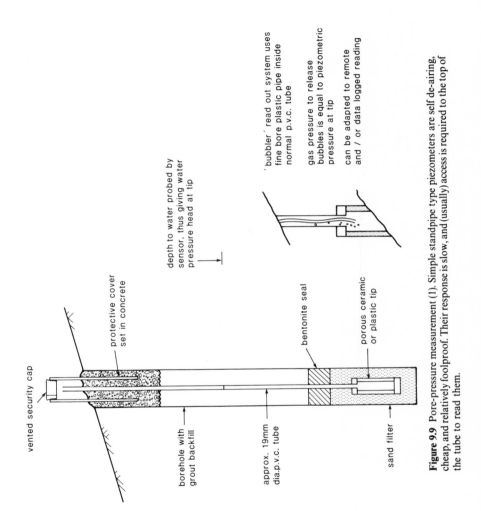

Figure 9.9 Pore-pressure measurement (1). Simple standpipe type piezometers are self de-airing, cheap, and relatively foolproof. Their response is slow, and (usually) access is required to the top of the tube to read them.

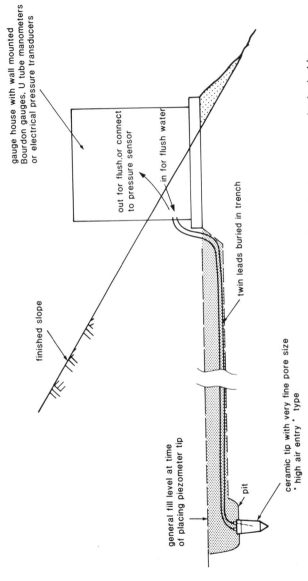

9.10. Pore-pressure measurement (2). Closed system hydraulic piezometers can be de-aired by flushing, and are often used in and under embankments. A permanent gauge house is usually provided. Response is rapid when the system is free from air.

K

rapidly equilibrates to the groundwater pressures. The readout system is a manometer, bourdon gauge, or electrical pressure-transducer system in a remote gauge house, the elevation of which relative to the piezometer tip and to any high spots in the connecting leads is crucial if cavitation is to be avoided. Air-bubble formation in the leads, tip, or measuring system can give rise to substantial errors in the indicated porewater pressure at the tip. This is inhibited by frequent de-airing, the use of air- and water-penetration-resistant tubing and a special ceramic for the tip which has a high resistance to the entry of microscopic air bubbles. This is termed a *high-air-entry* material because it takes a high air pressure to force bubbles to pass the small pores in this ceramic. These tips are often cast in plaster of Paris to ensure a good contact with the soil.

Electrical piezometers have the pressure transducer located in the porous tip. It is difficult to get a long-term reliability out of them, but response is rapid. High-air-entry tips are normally used, but unless separate flushing facilities are provided, any air entry will ultimately lead to inaccurate readings.

Pneumatic piezometers operate by an air pressure from the surface actuating a small-diaphragm valve. This allows return of the gas (dry nitrogen) to the surface via a second tube. The actuating pressure should be the same as the piezometer cavity pressure, but diaphragm flexibility can cause an error in this. Another bound to the pressure at depth can be found on venting the supply line. This allows the diaphragm to shut, and the pressure trapped in the return line gives the range of reading caused by diaphragm stiffness. Pneumatic piezometers cannot indicate suctions.

It is essential in planning and carrying out an instrumentation programme that attention is given to two easily neglected factors. These are the costs of protection of the instruments from damage, including vandalism, and the costs of actually reading the instruments. The latter is particularly important where instruments with a slow equilibration time are used (or those where a long-term record is required for other purposes). Standpipe piezometers need vented caps, and these must project well above ground level if the site is liable to flooding during heavy rain. This in turn makes them much more obvious, and prone to vandalism or accidental damage, than those set flush with ground level. Padlocked protective caps or gauge houses are an open invitation to vandals, and should not be used.

The costs of reading the instruments, including travel and staff time, can easily exceed the costs of installation in many cases, and it is easy to overlook this. When instructing staff, or writing a contract, provision should be made to read piezometers daily at first, particularly while site work is an progress. This enables the early, and most rapidly changing, part of the piezometer response curve to be obtained, and at least real cost. Although this initial equilibration curve may be of little use in estimating the permeability of the ground, since factors such as the setting of the grout interact with the simple groundwater equilibration, it is important since it enables a close approximation to the final water level to be made.

9.7 Historical research

An important element in landslide studies is the historical investigation. There are many instances of this in the literature, but some examples of the technique will be given here. In the United Kingdom, we are fortunate in having an unparalleled collection of large-scale topographical maps dating back to the middle of the last century. These show features such as the coastline, and record changes in these features from edition to edition. In urban areas cover exists at 1:1250 scale, and elsewhere at 1:2500. Relatively small changes in position can therefore be found, and dated to some time between two editions, a period of usually about 30 years.

Changing patterns of development can show signs of instability. On the coastal cliffs east of the pier in Ventnor, Isle of Wight, for example, early editions of the large-scale Ordnance Survey maps show a series of large houses on what would be a prime site overlooking the sea. This development was short-lived, being replaced by a greater number of smaller houses by the turn of the century. Nowhere else in Ventnor is this rapid redevelopment seen: the exploitation of the Ventnor area is a mid-Victorian phenomenon. Even these houses were demolished in the period following the Second World War, as a result, it is known, of ground movements. One can therefore ascribe the earlier redevelopment to slope movements too, although this was perhaps not understood at that time.

On the older maps, features on the steeper slopes are shown by hachuring. Brunsden and Jones (1976) regard these indications as highly inaccurate, and for illustration only, but at several sites I have found them to be accurate and useful in reconstructing former slope profiles. On the North Kent coast, for instance, the hachuring faithfully reflects features known to have existed from old photographs, and on the more exposed and rapidly eroding coastal sections of the Isle of Wight (Hutchinson et al., 1981; Bromhead et al., in prep.) features existing in the present-day slopes are shown in earlier maps in the hachuring. These latter features have migrated inland, and changed in relative position with the passage of time, but these changes are readily accounted by the slope degrading processes demonstrably acting at the present.

In addition to the Ordnance Survey map coverage there are many older maps, often commissioned in connection with engineering works. A selection of these, recording the interruption of littoral drift upstream of Folkestone Harbour Pier, and the consequent denudation of beaches downdrift and destabilization of slopes, is dealt with below in the case history of the Folkstone Warren landslides. These may be supplemented by other records, including nineteenth-century and earlier romantic etchings (Hutchinson and Gostelow, 1976; Pitts, 1979) or by (often fragmentary) earlier non-cartographic records.

In the United Kingdom, the technique of strip farming was abandoned relatively recently with the passing of the Enclosure Acts. Chandler, (1973) cites disruption of the relict strips at Gretton, Northamptonshire, as evidence of recent movements affecting slopes in Lias Clay. Elsewhere, the presence of unaffected strips shows the absence of slope movements since their abandonment, although the late date of abandonment of many of these features can sometimes lead to the evidence being somewhat inconclusive.

9.8 Premonitory signs of slope instability

In an earth or rock fill the important signs of impeding slope instability are increasing rates of lateral movement at the toe of the slope and formation of cracks at the crest. Initially the cracks will merely open in width, but when the failure has progressed there will be a vertical separation. Ordinarily, the lateral movements and any toe heaves will be imperceptible at first, and can only be picked up by appropriate subsurface instrumentation and surface surveying. These are therefore an essential and integral part of any construction control programme.

In natural slopes the premonitory signs may be more varied. Tension-crack formation and lateral movements are just a small part of the repertoire of signs that a slide is imminent or just under way in a natural slope. Take first natural slopes containing pre-existing landslides, where the problem is one of reactivated movement. If the slip surfaces are shallow, then movement along them may sever tree roots and cause the affected trees to die. Shallow-rooted species (e.g. *Cupressus macrocarpa*) may appear unaffected by this. Fencelines, pipelines and buried cables may be alternately stretched and compressed along their length, and small ripples or waves may form in levelled surfaces, for example tennis courts, lawns, or parking areas. At the margins of the worst-affected areas, roads or other tarmacadam paved surfaces will have Riedel shears like wrench faults running irregularly through the surface.

Small changes of level will occur. Without an appropriate frame of reference these may not be obvious, but will be picked out by surface-water drainage patterns: the movement of streams from their normal paths, or changes in the pattern of springs and well levels when compared to their usual behaviour, can all be evidence of these slight movements. Ponds may fill or empty as a result of these changes.

Large-scale movements may give rise to the 'intervisibility problem'. This is when two points formerly intervisible become no longer so, or vice versa. This is a difficult problem, as the recollections of the observers may be at fault: if there is not a long intervening period, then it will usually be obvious that an earth movement is in progress! One such case occurred at St Catherine's Lighthouse on the south coast of the Isle of Wight. Here, a lighthouse-keeper recollected that on his first tour of duty he was able to see the windows of a cattleshed of an adjoining farm about 100 m distant. On his second tour of duty at that lighthouse some 18 years later he could barely see the roof slates of the shed; but whether one or both of the locations had sunk, or the intervening land had risen, could not be conclusively stated in the absence of any reliable level data (Hutchinson *et al.*, in preparation).

Animals are frequently sensitive to small ground movements and micro-tremors which occur with increasing frequency as an earth movement (or seismic event) approaches. The fringe literature is full of instances of cattle refusing to enter a field, or moles leaving a patch of ground, that was soon to be affected by a slide. Such evidence is useful in retrospect, but is not reliable enough as an early warning system.

9.9 Landslide hazard and risk assessment

A natural hazard is one defined by Varnes *et al.* (in press) as 'the probability of occurrence within a specified period of time and within a given area, of a potentially damaging phenomenon'. In the context of slope movements, the 'potentially damaging phenomenon' is a landslide, fall or flow. Varnes defines other terms in detail, which have otherwise been used synonymously, as follows:

elements at risk include the population, properties, public utilities and economic activities within the considered area

vulnerability means the degree of loss to a given element or sets of elements at risk resulting from a landslide of a given scale, and is expressed as an index in the range 0 (no damage) to 1 (total loss)

specific risk is the expected degree of loss due to a particular landslide: it is the product of natural hazard and vulnerability. It may be compared to the *total risk* which encompasses the sum total of the elements at risk as well as the specific risk from any given incident.

It should of course be noted that without 'elements at risk', there is no 'hazard', because the damaging aspect of a particular landslide event is related to its impact on those at risk—people and property. Hazard must therefore be assessed in a timescale of several human lifetimes, and must take into account all types of slope movement, and all persons (however fleeting their exposure to the hazard) and property potentially at risk.

Landslide hazard can often be expressed on a qualitative basis, and in some cases may even be quantified. The principles involved are:

precedent: that is, past landslide activity in that particular locality, or similar localities, is likely to continue in the future, and to follow broadly the established patterns

susceptibility: that is, certain geological structures, soil and rock types, etc. are known to be prone to landsliding (either by experience, or through a knowledge of their engineering properties)

predictability: that an engineering or geological assessment of the conditions and processes that promote instability can be made.

Landslide hazard assessment at a particular site therefore has several facets. Firstly there is the identification of existing landslide activity on the site, adjacent to it, or in geologically similar localities. Following this, there is the evaluation of susceptibility to landslides in that general area due to factors such as geological structure, lithology, climate and so on. The site-specific factors must be taken into account. These include slope angles, aspect (microclimate may be influenced by this) and local erosive conditions at the toe of the slope.

Where there has been earlier slope movement, the major risk is for re-activation of that movement. Is the earlier movement the response to stimuli now no longer present, or is it a perhaps temporarily quiescent response to stimuli still in action? The answer to this question must obviously colour the perception of future hazard.

Slope instability of a minor nature in green-field sites is often of negligible significance, but the development of such a site for dwellings renders these movements of considerable importance. Any assessment of hazard must therefore include an appreciation of the susceptibility of structures on the site to movement. In this context we must consider services, roads, engineering works, etc. to be structures.

Finally, account must be taken of future changes on or off site which could affect stability. Examples of this could include the cessation of pumping from an underlying aquifer allowing regional groundwater levels to rise. Pollution of a potable water supply, or the abandonment of mineworkings, could be the cause of such a cessation in pumping.

9.10 Courses of action

Once an assessment of landslide hazard has been made, there are basically four courses of action. These courses are much the same whether one is considering landslides or any other hazards arising out of the earth's surface processes, but humankind's response may reflect not only the scale of the problem, but also the perceived level of risk, and quite simply the funds available.

These four courses are as follows:

avoid the problem by going somewhere else

solve the problem by treating its causes

adopt a form of construction that is less susceptible to the problem

accept it fatalistically (an ostrich-like head-in-the-sand attitude of merely ignoring the problem is an unthinking variant of this approach).

Taking these one at a time, the first approach really does demand a great deal of foresight. Preliminary geomorphological assessments are of enormous value at this stage. They most certainly give a rational basis for initial site selection if there is a choice, and while obviously at their best when considering virgin sites in areas of currently active natural processes, when skilfully done can be equally effective in quite highly developed, and thus partly obscured, areas in which the main hazards arise from processes long past. At a late stage, however, the freedom of choice to move elsewhere may no longer exist. A town may have too much investment in its infrastructure to move elsewhere: a road, railway or canal may be partly constructed, and it may be preferable then to try to correct the problem which has become evident.

Treating the causes of the problem is much more to the liking of the civil engineer (and I am one). Civil engineers are paid for doing works, not for suggesting that they are not done. What is more, it is usually far more cost-effective to prevent a problem arising than to try to rectify it later. Hence slopes can be investigated and stabilized for a fraction of the cost of reinstating major failures. The only exception to this is where long lengths of cut and fill (as in a highway scheme) are undertaken: a few failures here and there probably indicate an overall economic design, and their rectification may only be a fraction of the cost of uniformly flatter side slopes with their consequential costs in land-take and muckshift. Even in that case, a blind approach will lead to an imbalance one way or the other, and appropriate investigations and design work must be carried out to achieve the requisite cost-effectiveness ratio.

Minimizing susceptibility to earth movement is not always practical. In the case of seismically-induced instability, earth slopes can be designed to be less susceptible (although never entirely proof against it). It is not normally possible to design structures which sit on an untreated but active earth movement and which remain unaffected. There are cases where some structures have been built on large, slowly moving or nearly quiescent, landslides in such a way that they are not subject to *differential* movement — being built on large blocks for instance. Most of these are not the result of planning, but have been shown up when surrounding buildings have been much more severely affected. A similar effect can be obtained by making provision for foundation movements.

Finally, one comes to the area of fatalistic acceptance of landslide hazards. Not only the magnitude of movement, but also the frequency with which it recurs, can affect both the perception of landslide hazard, but also human response to what is perceived. In Table 9.2, an attempt is made to classify the scales and frequency of movements.

A single rare or infrequent movement in the small to moderate classes, particularly if widespread, will cause severe damage to property, roads, services and the like. It will cause homes to be abandoned. At the bottom end of the size range, the majority of these will be reoccupied, with isolated demolitions of less repairable property, but at the upper end there will be total

Table 9.2

Amount	Classification	Return period	Classification
>10 m	catastrophic	>1000 y	rare
1–10 m	large	100–1000 y	infrequent
0.1–1.0 m	moderate	10–100 y	frequent
0.01–0.1 m	small	1–10 y	common
0.001–0.01	very small	0.1–1 y	almost continuous
<0.001	imperceptible	<0.1 y	continuous

The time for an individual movement to occur is taken to be less than 0.1 y.

(Based on a concept by Squire *et al.* 1978)

abandonment of permanent dwellings. A long period of very small to small movements, on the other hand, even though they have been continuous or almost continuous, will give rise to very little real concern; the occupants of that area will reconcile themselves to repeatedly repairing their homes. This is notwithstanding the fact that the total displacements may soon exceed those discussed previously.

10   Case histories

10.1 Flow slides in tips and spoil heaps: the Aberfan tip failure of 1966

The 1960s were marked by a series of devastating mass movements in tips and lagoons of waste from mining and quarrying, and in waste fly ash from coal-burning power stations. In Bishop's classic paper (1973) on the stability of tips he covers many of these: particularly the disastrous flow slide in colliery spoil from Tip No. 7 at Aberfan. Coincidentally, on the same day that this disaster happened, there was also a failure of the enclosing lagoon wall at Fiddler's Ferry power station which released large volumes of waterlogged fly ash (Al-Dahir *et al.*, 1970). This series of failures focused attention on the problems of the disposal of waste from mining and quarrying operations and, in the United Kingdom, led to legislation to control tips and spoil heaps of such materials, and also lagoons in which fine materials are stored in suspension in water.

Flow slides in colliery discard were not unknown in South Wales, even before the Aberfan disaster. In December 1939, for instance, a few miles south of Aberfan at Cilfinnydd Common, a slide in a tip developed into a flowslide which would surely have been more destructive than the Tip No. 7 slide had there been anything substantial in its path. In Bishop's report of this slide, he estimates that about 180 000 tonnes of waste were involved, which travelled some 435 m from the toe of a 46 m-high tip (Figure 10.1).

The main movements took place in only two to three minutes, in which time the flow slide had cut a power line, blocked 180 m of a road to a depth variously estimated as being between 2.5 and 7.5 m, blocked a canal, blocked and diverted the River Taff, blocked a railway and cut a large-diameter compressed-air main. During 1971, road widening operations cut into the debris from this flow slide, exposing the decomposed grass beneath about a metre of the black colliery waste. There was no sign of the contact being sheared, so that the mechanics of the latter stages of the movement at least cannot have involved sliding. High pore-fluid pressures in the debris are invoked by Bishop (1973) to account for the relative mobility of the debris, and its rapid solidification when they escaped. The thin layer of debris would have facilitated this escape of pore-fluid pressure, compared to the situation applying during the early stages of the movement when the debris was much thicker.

Not only was there evidence of such movements outside Aberfan; in the tip complex itself, a number of flow slides had already taken place. Fortunately, none of these had reached the town. These flow slides took place in 1944, 1963

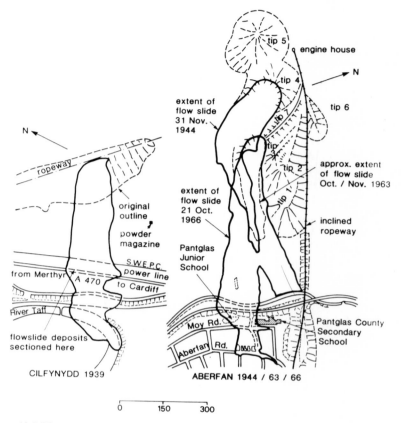

Figure 10.1 Plan comparing Cilfinnydd Common and various Aberfan tip slides. (After Bishop, 1973.)

and 1966; the last, on the morning of 21 October, caused the disaster. Some 3 m of displacement had already occurred at the crest of the tip by the time the slip was first reported, by the day shift arriving for work, and this doubled in the following hour. A little later, visible movement occurred in the toe of the slip, developing into the flow slide, which travelled down the 12.5° hill slope some 490 m, before destroying several buildings including a junior school, and continuing a further 100 m. In the area where the slide finally came to rest it had a thickness of some 9 m (Bishop *et al.*, 1969).

Two technical factors hampered the rescue operation. Firstly, following the rotational slip in the tip face, which had involved an element of foundation failure in the glacial drift covering the valley sides, and which stripped off the cover to the underlying sandstones, water was released. In conjunction with flow from a broken 31-inch water main, this turned some of the slide debris

Figure 10.2 Aerial view, Aberfan flow slide Tip No 7. Photo: National Museum of Wales.

into a mudflow. However, this had little to do with the original rapid movements and was a feature not present in the earlier slides, both at Aberfan and at Cilfinnydd Common. Furthermore, the clear evidence that the development of the mudflow was a secondary event came from the relatively dry material outside the mudslide area (Figures 10.2, 10.3). The second technical factor was that the flowing material consolidated to a relatively dense state when it came to rest. Excavations during the rescue were therefore made with difficulty.

The flowslide involved only about $110\,000\,m^3$ of material, of which only some $40\,000\,m^3$ crossed the railway (Figure 10.4) and entered the village. Some 140 deaths, including 112 children in the school, were recorded. The remedial measures in the aftermath of the disaster comprised the complete removal of

Figure 10.3 Aberfan tip flowslide debris.

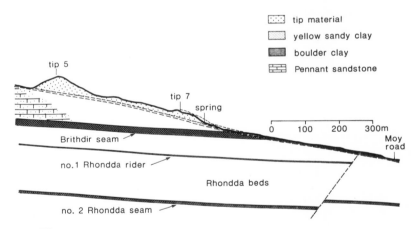

Figure 10.4 Cross-section of Aberfan Tip No 7 complex, and slide details.

the still-threatening tip complex as well as the debris from the town, and the installation of drainage in the hillside on the former site of the tips.

Nor should it be imagined that flow slides are confined to tips of colliery discard: Bishop (1973) reviews a number of examples in a variety of mine wastes. These included limestone waste from a quarry in Derbyshire; mica sand from china-clay workings in Cornwall; and fly ash from a tip at Jupille in Belgium. They have also occurred in 'engineered' slopes, and the examples quoted are the failure of the hydraulic fill at the Fort Peck Dam with a loss of 80 lives in 1938, and the failure of the tailings dam across Middle Fork, Buffalo Creek, West Virginia where a series of earth dams were breached in a cascade failure costing 118 lives and 50 million dollars' worth of damage.

10.2 Construction period failures of earthfill slopes: Carsington Dam and Acu Dam

The failure of the Carsington Dam in Derbyshire in June 1984 is a classic example of an end-of-construction failure in an earth-fill slope. The dam itself was designed to provide a storage reservoir for water pumped over 9 km from the river Derwent, and had to be capable of being drawn down over a few months if the impounded water was needed—in a drought for instance.

This scheme is designed to increase the water resources of the Severn–Trent Water Authority by providing an additional 225 million litres per day to nearly 3 million people in the East Midlands. The reservoir, of some 35 million cubic metres capacity, is to be impounded behind a 1225 m dam, which has a maximum height of 35 m. Side slopes were 1:3 upstream and 1:2.5 downstream.

The design cross-section had three main fill zones, each made from materials excavated from the reservoir area. Highly weathered Carboniferous shales were used for the core, with less weathered shales used in the transition zones, and unweathered shale used for the shoulders. Finger-type drains were used in both shoulders to control porewater pressures; these and the wave protection on the upstream slope were made from crushed Carboniferous limestone excavated from a nearby quarry.

The dam was extensively instrumented with piezometers in the core, foundation, and downstream zones (180 in all, primarily on four main cross-sections), but was under-provided with instruments for displacement monitoring, although there were a small number of USBR cross-arm settlement gauges and some downstream extensometers. Construction took place in the three good weather seasons of 1982, 1983 and 1984, and was almost complete when a slide occurred in the upstream face of the dam (Figures 10.5, 10.6, 10.7) over a length of over 400 m of the crest.

Displacements of 10–15 m took place in a matter of days, but with the placement of berm at the toe of the slide (placed as much to protect the drawoff tower and forebay area as to stop the slide) the slide was brought to rest. The chronology of the slide is interesting, as it shows that even though there was a fleet of muck-shifting plant on site, and fill was placed at the toe of the developing slide with scarcely any delay, it proved impossible to arrest the movements before the dam was completely ruined.

Placing fill had restarted in the 1984 season, and was almost complete by

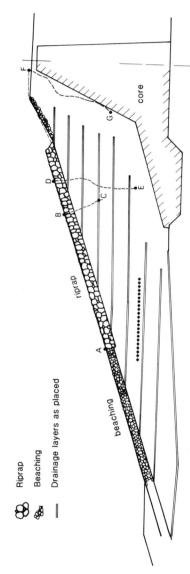

Figure 10.5 Cross-section, Carsington Dam, showing mode of failure: (1), unfailed section (blocks ABC, EDFG from section after failure transposed to probable initial locations). The slip surface followed the upstream core extension and also made use of the weak path provided by foundation clays. Elsewhere, slip surfaces passed through the base of the shoulder fill. See also Figure 10.6.

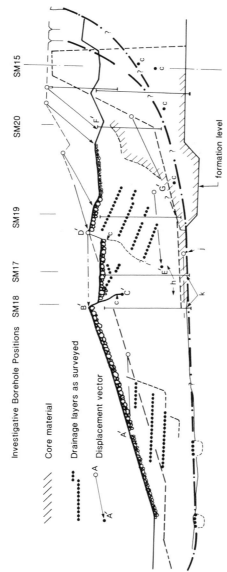

Figure 10.6 Cross-section, Carsington Dam, showing mode of failure: (2), section after failure. Information from site investigation (preliminary). ——————, original profile of dam; ———·———, shear surface; ———·——— ?, shear surface (inferred).

Figure 10.7 Carsington: rear scarp of slide showing slip surface.

June. Work stopped, partly because of heavy rainfall (40 mm), and partly because of a Sunday, and was to restart on Monday 4 June. At 7.30 a.m., a routine inspection showed a tension crack, 20–50 mm wide and 220 m long, parallel to, but downstream of, the dam centre line. Within two hours this had widened, and heaving at the toe of the slope was observed.

Instrument readings taken during the day showed marked increases in porewater pressures, both in the upstream parts of the core and in the foundation; taken in conjunction with surface deformation measurements, this showed that a massive slide was developing. Work started that afternoon on the construction of an additional toe berm, going on into the evening, at which time the tension crack had widened to some 130 mm, but without vertical displacement.

By the following morning, the crack had widened to 350 mm, and had dropped 300 mm on the upstream side, and it was followed by a second crack 2 m downstream of the original one. Despite the continuation of berm construction at the toe, the crack widening accelerated, up to 150 mm per hour in the late afternoon. At the end of the day, the main crack was 3 m wide and the sliding mass had settled by 3 m, and in addition, the slip was increasing in length as shown by the extension of the crack, and changes in the instrument readings, including additional blockages in several USBR settlement gauge access tubes.

At this point the main berm extension was abandoned, and work concentrated on filling around the forebay area and drawoff tower, as these were becoming threatened by a southward extension of the slip. When that area was stabilized, the main slide had come to rest, having moved up to 13 m along the slip surface. In places the dam crest had settled 10 m, and elsewhere toe heaves of 2.5 m were recorded.

The slide was not entirely unexpected, since the previous December a report commissioned by the contractors had thrown serious doubts on the stability of the dam, mainly because of the shape of the core. This was designed with an upstream extension which seemed to invite the formation of the eventual slide. Furthermore, even though the dam was built with some of the more weathered shale left in the foundation, the role of this in possibly destabilizing the dam, although covered in the stability report, was ignored by the dam's designers. Construction continued and failure, in retrospect, was inevitable.

When preliminary design assumptions for the dam were first published, they showed that extremely conservative porewater-pressure assumptions had been made: in combination with higher than normal shear strengths, the quoted factors of safety did not look unreasonable on cursory inspection. However, growing doubts as to the stability of the dam arose during the second of the planned three seasons' construction, partly because of some strains in the upstream face of the dam, and partly due to an almost intuitive distrust of the core shape by the contractor's geotechnical consultant.

In British practice, the contractor has little control over design decisions, and repeated requests for serious technical discussion of the overall stability issue were denied by the dam's designers. Unfortunately, relations between designer and contractor were already strained: there were claims issues, a number of tragic deaths on site due in part to a failure of the designers to consider the ventilation of some inspection chambers adequately, and problems with the construction of the main supply tunnel. When the designers produced some computer output to justify their assessment of the dam's stability, and the premise on which these were based was found to be very shaky indeed, the contractors commissioned their own stability investigation.

The writer was fortunate enough to be requested to make the necessary

Figure 10.8 Carsington dam failure.

computations. At first an analysis was made using the combination of shear strengths and porewater pressures assumed in the original design. This not only failed to produce the quoted factor of safety, it clearly demonstrated substantial calculation errors! Although the porewater pressures measured in the dam were significantly less than those assumed in the original design, the shear strengths proved also to be less. A report on stability bringing all of these factors into the open was totally ignored, the contractor was persuaded to continue with the third season's fill placement, and the scene was set for the inevitable collapse.

Following the failure, a detailed investigation into its causes was undertaken. When complete, the borings, trial pits and testing cost far more than it would have done to complete the dam to a modified version of the original plans. During the investigation it also came to light that settlements of the dam crest in the off season had been measured, and tentatively attributed to the onset of a large landslide, but these and other premonitory signs were dismissed.

The nature of the materials on site was also investigated at this stage, it being found that oxidation of the pyritic Carboniferous shales was taking place at a significant rate, and that the resulting acidic conditions were attacking the limestones used in drainage blankets. Carbon dioxide from this reaction had in fact caused the fatalities in inspection chambers below the downstream toe of the dam the previous year. These reactions had not affected the operational strengths of any of the dam fills, however. It also came to light that the weak foundation materials had been emplaced on some of the valley slopes under the dam by periglacial solifluction, and were sheared to a greater or lesser extent (Skempton, 1985). Early recognition of such a feature should have led the designers to have this material removed, since it was likely to contribute to instability, although, once identified as a structural weakness in terms of stability, the mode of origin of such material becomes academic.

No doubt the failure of this dam will have repercussions which will affect UK embankment-dam practice for years to come. Two desirable outcomes will be that the problems of treatment of dam foundations will receive more attention, and the regrettable trend of recent years towards weak upstream blankets and core extensions will be reversed. These two factors contribute significantly to embankment-dam stability problems. Additional lessons to be learnt from this failure are the role of brittleness in the performance of clay fills (all too often fills are seen as non-brittle), and the importance of not only installing adequate instrumentation, but also of reading it and paying attention to the results as soon as they are available.

An equally disastrous case of an earth dam which failed during the construction period is given by de Mello (1982), and summarized by Penman (1985). This dam, the Acu Dam in Brazil, was to have been 40 m high, but it failed in December 1981 while 5.2 m below crest level, during construction. In contrast to Carsington, failure was rapid (taking about 1/2 hour) with crest settlements of the order of 15 m, and lateral movements of 25 m. To facilitate speedy construction, the cutoff through a sand foundation was to be made concurrently with the embankment raising, and this cutoff was constructed in a dewatered open cut some 22 m deep upstream of the dam centre line. (A feature broadly similar to this was used in the Derwent Dam, and may be seen in the section illustrated in Figure 7.19. Such a cutoff needs to be connected to the core of the dam, and this connection, or upstream blanket, is an easy path for slip surfaces to follow. In the Derwent Dam, designed with flat side slopes because the weakness of its foundation had been recognized, no additional problems are introduced by this.) This cut had 1 : 2 sideslopes, and a black silty clay was placed against the downstream slope as backfilling proceeded. The location of the foundation cutoff, and the remainder of the original dam details, are shown in Figure 10.8. At this stage it was decided that the foundation cutoff and the core of the dam were not sufficiently linked, and a design modification, introducing a 7 m thick horizontal blanket of the same black silty clay underneath the upstream shoulder of the dam, between foundation cutoff and core, was made.

Two slips occurred in temporary slopes, each about 150 m long: major construction failures in their own right, but shortly to be eclipsed by an even more disastrous event. These slips involved about 160 000 m³ of the black clay fill.

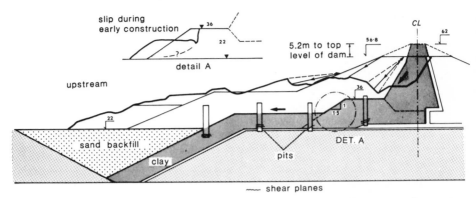

Figure 10.9 Cross-section, Acu Dam, showing mode of failure. As at Carsington, core shape and upstream blankets played a role in this failure.

Figure 10.10 Acu Dam. Photo: World Water/Jose Pessoa.

The whole of the upstream slope failed in December 1981, almost exactly a year after the cofferdam failures. As at Carsington, tension cracks formed in the construction surface downstream of the core, with the core and downstream shoulder sinking and pushing the side slopes and toe of the dam upstream over a length of some 600 m. Photographs (Figure 10.10) taken at the time show a situation similar to that existing after the failure at Carsington, the size and shape of the failed mass being broadly similar. The difference was that the failure of the Acu Dam only took about 30 minutes.

In both cases, the failures were predictable in the sense that conventional stability analyses carried out systematically and properly, using realistic shear strength data readily available on site had it only been interpreted correctly, should have given rise to serious doubts in the design engineer's mind about stability. Similarly, in both cases, the failures could have been avoided by comparatively small, and relatively inexpensive, remedial measures taken at a late stage in construction, had the stability problems been foreseen, and the requisite actions taken early enough. Naturally, it would have been possible to evolve satisfactory design from scratch as an alternative. In the event, the additional costs and delays incurred as a result of the failures have in both cases been considerable.

10.3 Failure of a reservoir slope leading to overtopping: the Vaiont Dam

In contrast to Carsington, the Vaiont Dam survived loadings vastly more than those anticipated in design. Constructed in a steep river gorge in northern Italy, the problem there was the instability of the reservoir sides. There had been several small slides from the valley sides as the reservoir was filled, but these served only to alarm local inhabitants (they caused seiches of such a scale that local fishermen became afraid to fish from the lake's edge lest they be caught in them), and to destabilize a massive rock block which slid into the reservoir at high speed during the evening of 9 October 1963.

The dam, a doubly curved arch dam, was constructed between 1956 and 1960. At a height of some 265.5 m it was the world's highest thin arch. Running 190 m from abutment to abutment at the crest was a road, partly cantilevered from the arch dam structure. Some 150 million cubic metres of water was impounded behind the dam, which at its thinnest was only some 3.4 m thick, although at the valley floor the dam reached a thickness of 22.7 m.

The immediate cause of the disaster was a landslide from the flanks of Monte Toc, a peak which rose to one side of the reservoir. The sliding mass caused a wave some 260 m high above reservoir level to surge up the opposite side of the lake. Subsequently, there were other waves which overtopped the dam by 100 m, destroying the dam crest road and the downstream powerhouse (Vaiont was one of five dams forming the northwestern Piave hydroelectric scheme). Water then surged down the Vaiont valley where the flood wave was still over 70 m high at the confluence with the main Piave valley which it

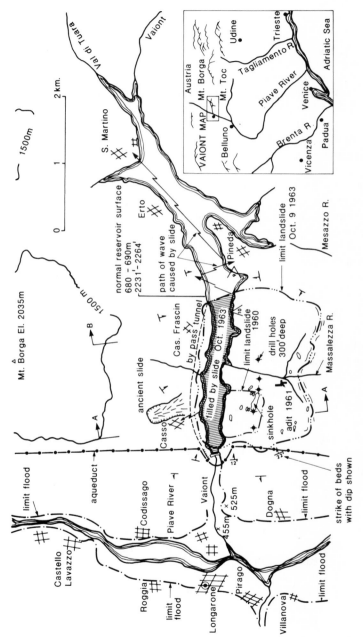

Figure 10.11 Vaiont dam: plan showing affecting area. (After Kiersch, 1964.)

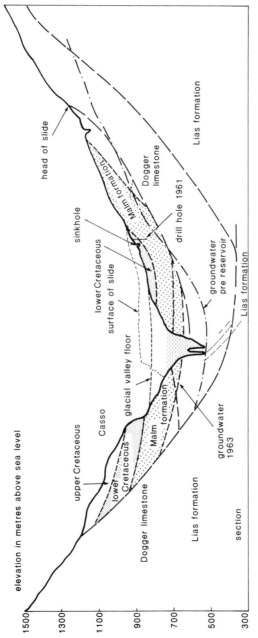

Figure 10.12 Vaiont dam: section through slide.

crossed, entirely obliterating the small town of Longarone in the main valley. Subsequently, the villages of Pirago, Rivalta, Villanova and Fae in the path of the flood were demolished. Kiersch (1964) records a death toll of 3000 from this disaster; the monument at Vaiont itself records 2018 fatalities.

Large-scale landslides in the slopes of the Vaiont valley were not unexpected: there was evidence of prehistoric slides, and some premonitory evidence for the later disastrous event. The dips of the strata were also unfavourable for stability (see Figure 10.11 for a plan, and Figure 10.12 for sketch sections of the slide). In 1960, there was a slide of about one million cubic metres near the dam, accompanied by ground deformations over a much larger area. These were monitored by geodetic survey pillars which showed that initial movements of 250 to 300 mm per week gradually slowed to about 10 mm per week displacement. A diversion tunnel was constructed in the opposite bank lest a landslide prevent the water from reaching the dam's outlet works, and the reservoir was kept low. An attempt was made to investigate the slide with an adit, and with a 90 m-deep drill hole (Figure 10.12). In retrospect, these were too shallow to encounter the main slide surface, and were inevitably inconclusive.

Deformations increased again about 18 September, 1963 to about 10 mm per day, and then, following a period of heavy rainfall, continued to increase, reaching 800 mm per day on 9 October, whereupon the monitoring team realized that all their geodetic stations were moving *en masse*, indeed, that the developing slide was about five times bigger in extent than they had expected. In this respect, animals grazing on the slopes of Monte Toc were better judges than the engineers: they had left the slide area on 1 October!

It was decided to lower the reservoir in anticipation of a slide, but heavy inflows, and the reduction in reservoir volume with the slide movement so frustrated this, that, despite discharges of about $200 \, m^3/s$, the reservoir level actually increased slightly. The main slide took place in less than 30 seconds, within which time a large part of the $135 \, M \, m^3$ of the reservoir contents were pushed aside to form upstream and downstream seiches. The intensity of the latter suggests that the upstream part of the slide moved first, with a hingeing action in plan. This is concordant with the geology.

In the sketch cross-sections it will be seen that the dam is set within a steep post-glacially eroded gorge cut through the bottom of a U-shaped glacial valley. These valleys had picked out a fault zone, and the folding of the beds allowed slides to follow bedding surfaces throughout their length. No doubt sliding was encouraged by tectonic shears in the more argillaceous members of the formations involved, but attention has also been drawn to the presence of stress relief fractures, both as 'old' sets associated with the U-shaped valley, and 'new' sets resulting from the rapid downcutting of the later gorge. However, the rapidity of the slide suggests strongly that it contained a large element of 'first-time failure', i.e. was highly brittle. Existing discontinuities can only play a small part in this, with their low brittleness (Skempton, 1966).

Also noteworthy is that this slide occurred under conditions of reservoir filling. Ordinarily, an increase in stability is computed when the reservoir load is applied (section 6.3) unless significant porewater-pressure changes are experienced. It is probable that this effect occurred in the Vaiont landslide.

10.4 Construction problems in natural slopes: Sevenoaks road construction and the Lower Greensand escarpment in Kent

Sevenoaks road construction

The Weald–Boulonnais anticlinal dome structure in south-east England and north-east France is eroded deeply in its centre, exposing successively older rocks as its centre is reached. The most prominent feature is the ring of Chalk hills which have gentle dip slopes trending away from the Weald, and much sharper scarps facing inwards. There is, however, a second, less well-developed inner ring of hills, which mark the outcrop of the Lower Greensand.

Two sites, discussed later in this chapter, are found where these two rings of hills are breached by the English Channel. These sites are in one case a present coastal slope, and in the other, a former coastal slope, in the Chalk and Lower Greensand respectively. To set these comparatively recently developed slopes into proper context, consider the inland slopes of the Lower Greensand near Sevenoaks Weald.

Figure 10.13 Hythe Beds escarpment at Lympne showing the remains of the landslipped Roman Fort. The topography is similar to the upper parts of the escarpment at Sevenoaks, but the solifluction lobes are absent.

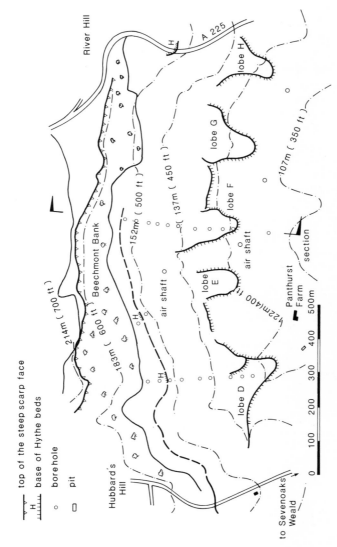

Figure 10.14 Sevenoaks: plan of the escarpment.

In this neighbourhood, there is the strong, dissected *cuesta* of the Hythe Beds capping slopes of Atherfield and Weald Clays. This was to be crossed by the A21 Sevenoaks bypass, a dual-carriageway road under construction in the mid 1960s. The landslipped nature of the scarp slopes was not properly appreciated at that time, and some instabilities were triggered, which led to the abandonment of a number of engineering structures, and the realignment of a length of the road which was originally designed to sweep obliquely down the hill slopes (as the finished road does today, albeit on a different alignment).

A number of streams are thrown out of the Hythe Beds, and these have dissected the lower slopes of the escarpment forming ridges and isolated hills which in many cases stand 20–30 m above stream bed level. Investigation of these revealed that they are capped with Head deposits consisting of angular Chert fragments, and other stones derived from the Hythe Beds, set in a clay matrix. Further detailed work showed them to be remnants of what were originally extensive sheets of periglacially soliflucted debris, probably of Wolstonian age, extending for distances of at least 2 km from the escarpment, and inclined at extremely low slope angles of as little as 1.5°. These sheets were then dissected, and the hill top and ridge top deposits are all that remain.

The Hythe Beds are cambered, and large-scale structural disturbances occurred in the escarpment. These features are preserved as spurs in the Sevenoaks area. (Elsewhere along the escarpment, the cambered blocks of Hythe Beds were quarried for building stone. In many of the now abandoned quarry workings, the sites are still criss-crossed by ridges of soil which formed the infilling to wide *gulls* and which the quarrymen did not see fit to remove.) At the toe of the slope, valley-bulge related features were formed. The large-scale structures are probably contemporary with the solifluction described above.

Following the events described above, a significant erosion of the surface took place. This was accompanied by retreat of the escarpment, principally in embayments formed between the spurs. Further, and later, solifluction debris, this time comprising largely clays of the Atherfield and Weald clay units in the

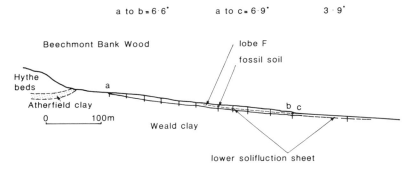

Figure 10.15 Sevenoaks: section through the escarpment.

slope, but with some embedded angular chert fragment, moved up to 1 km from the scarp slope. This deposit, which is about 2 m in thickness, is inferred to have been formed in the Devensian. It overlies brecciated Weald Clay within which shear surfaces have developed, and is contemporary with the shallow slides occurring in the embayments of the escarpment.

Lobate forms which are relics of a third phase of solifluction are still preserved on the slopes. These have in places buried a soil dated as 12000 radiocarbon years BP developed on top of the second solifluction sheet. These lobes have a much lesser 'run out' from the escarpment of around 300 m, and a steeper overall angle of approximately 7° (Skempton and Weeks, 1976).

Road construction stimulated movement of some of the soliflucted soils. It was found impractical to stabilize the slopes with the original design alignment, and a new alignment for the road, incorporating split carriageways, was constructed downslope of the original line. This required a number of already constructed overbridges to be abandoned and demolished, increased costs significantly, and delayed completion of the road.

This situation contrasts strongly in its subsurface details to that revealed by investigations at Lympne on the edge of the outcrop of the Hythe Beds where the Weald–Boulonnais dome is cut by the English Channel.

Lympne abandoned cliff

The slope at Lympne is part of an extensive abandoned cliff, running from Aldington in the west to Hythe in the east, now protected from marine erosion by Romney Marsh and its associated shingle spits. Lympne is just east of centre in this stretch, which is formed by a continuation of the Lower Greensand escarpment in the eastern part of the Weald from Sevenoaks along the coast of southern England. There is a slight tendency for the average inclination and signs of instability to increase from west to east, which suggests that marine erosion ceased more recently in the east, but which may alternatively reflect lateral changes in the geology.

At Lympne, the slope is almost exactly 100 m high, and consists of two main elements: a 15 m high scarp in the Hythe Beds, inclined at 35°, below which a 550 m long, slightly irregular slope extends down to the marsh at an average inclination of nearly 9°. The upper part of this slope is formed in the Atherfield Clay, the lower parts in the Weald Clay.

In contrast to the geologically analogous escarpment at Sevenoaks, the absence of periglacial solifluction features at Lympne indicates that marine erosion at its foot ceased after the Younger Dryas period of the Late Glacial (8850–8050 BC). Indeed, radiocarbon dates on wood recovered from just above a now buried former sea cliff, suggest a much later abandonment, c. 2500 BC. On close inspection, it proves possible to identify an upper, degradation, zone and a lower, accumulation, zone occupying respectively 60% and 40% of

Figure 10.16 Lympne: relics of landslipped Roman Fort.

the downslope length of the cliff. The slope appears to have developed to its long-term stable angle by virtue of the fact that the slope angles in the degradation and accumulation zones are so nearly equal. However, the presence of numerous slip scarps, soil waves and undulations in the upper slopes, in contrast to the smoother profile of the lower slopes is a reminder of their different mechanics.

Particular interest in the slopes at Lympne arises from the presence there of a Roman fort, now much disrupted by landsliding, but which formerly was one of a sequence of 'Saxon shore forts'. The fort *Lemanis*, or to give it its more modern name, Stutfall Castle, was probably constructed *c.* 275–280 AD and abandoned unusually early in *c.* 340–350 AD. It is tempting to see in this early abandonment signs of landsliding, or even in the scant evidence of a possible earlier fort at the site (in the form of a dedicated altar reused in the later construction, and stamped markings in tiles) the possibility of earlier stability problems, but this is not supported by firm evidence.

It is believed that at the time of the construction of the fort, the process of burial of the former sea cliff under debris slipping down from the degradation zone was relatively far advanced, and that the degradation zone must have been fairly similar to, although slightly fresher (perhaps a degree or two steeper) than that of the present. It would therefore have been somewhat less stable.

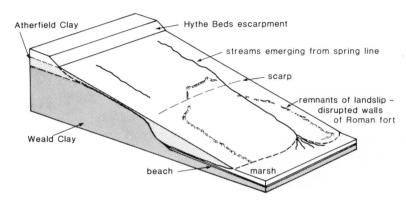

Figure 10.17 Lympne, block diagram.

A combined geotechnical and archaeological investigation of the slopes was undertaken (Hutchinson *et al.* 1985) which showed the subsurface details of Figure 10.17. It was expected, by analogy with the abandoned cliff at Hadleigh (Hutchinson and Gostelow, 1976), and with the slopes to the east where such features are readily recognizable, that the upper parts of the slope immediately below the Hythe Beds outcrop would be occupied by the remains of rotational slips. Instead, it was found that such features were absent, probably because of quarrying of the slipped and *in-situ* Hythe Beds, both in Roman times, and later, when Lympne Castle, which is perched on the escarpment, was built.

With both boreholes and trial pits it was possible to delineate the base of slipping in the degradation zone, the shape of the buried former sea cliff and shore platform, and the extent to which the latter has been overridden by the movement of the accumulation zone. Landslide debris has advanced approximately 130 m seaward of the toe of the former sea cliff, burying the former beach and parts of the marsh to a depth of up to 20 m. Evidence in the samples recovered from depth, and in the dislocated positions of the wall fragments, suggested that there had been at least 35 m of movement since Roman times.

In the more archaeological elements of the investigation, it proved possible to locate sections of oak piling on which the fort walls and bastions had originally been constructed, and from which they had been moved by the effects of shallow landsliding. Fortuitously, the piles had been driven into *in-situ* Weald Clay, and, before being sheared off them, the northern walls of the fort had been tilted so parting company with the uppermost piles before sliding downslope.

Detailed appraisal of the geomorphological evidence made it possible to identify a number of different phases of slipping, some of which had moved parts of the fort walls and bastions (or towers) between 5 and 32 m. It is by no means clear that these slides were entirely local to the fort, but also there is no evidence to suggest a more widespread slipping in the escarpment. From evidence of the reactivation of accumulation zone slides at Wierton, Ohio (D'Appolonia *et al.*, 1967) and Hadleigh (Hutchinson and Gostelow, 1976), where movement followed as a result of tampering with the toe of the slope, it is tempting to ascribe at least the initiation of the disruptive slope movements to the activities of man, and to blame the Romans themselves for it, possibly through diggings to build quays, or dredging of a silting harbour.

10.5 Construction failures in cut slopes with high brittleness: the Panama Canal

Excavation of the Panama Canal began with a French enterprise, stopped, restarted, and was eventually completed under U.S. supervision. Among problems faced by the early workers were disease and other privations, but perhaps more serious and disheartening were the landslides triggered by the canal excavations. Some of these are active at present, and represent a significant maintenance cost, as well as from time to time interfering with traffic flow through the canal.

Reports on the activity of these slides are extensive, and include the accounts of Binger (1948), various U.S. Army Corps of Engineers' Waterways Experimental Station reports, and the series of papers culminating with Lutton *et al.* (1979). Attention has been concentrated in the Gaillard Cut area, wherein lie the deepest cuttings of the scheme. Work started there with the first French company in about 1884, but within the next two years, as work began to assume a reasonable degree of efficiency, the slide problems started to occur. The problems of this, combined with the interference with trunk rail lines, caused the first French company to fail, and no further works were attempted for a six-year period ending in 1895, when work was taken over by the second French company. This continued until the U.S. takeover in 1903, initially by civilian engineers, but from about 1908, until the opening of the canal in 1914, by U.S. Army Engineers. The canal was closed repeatedly by slides in the first two years of operation, but subsequently these tailed off. In the 1950s and 1960s, the canal was more than doubled in width in recognition of increasing vessel size, and many of the troublesome areas were removed.

Initially, excavation was 'in the dry', latterly using steam shovels under the U.S.-controlled phase of the work. These removed about 75 million m^3, about 25% of which was slide debris, up until 1913 when the canal was flooded, and excavation was completed with dredging. A total of about 45 million m^3 of extra muck-shifting has been required because of slide movements up until the present day.

Reasons for slide activity in the Tertiary volcanic sediments of the Canal Zone are varied, but include oversteep design slopes, inflow of water from river and stream diversions, and the placement of spoil heaps too close to the edge of cuttings. Many of the slides are controlled in whole or part by the faulting, with

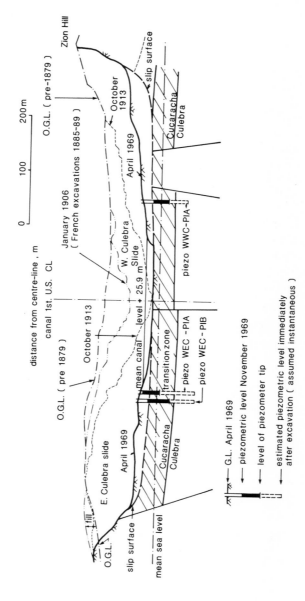

Figure 10.18 Panama Canal cut.

which the canal area abounds, but the bedding plane dips, although generally low, are in many cases towards the canal, and also have a contributory role.

The high proportions of montmorillonites (up to about 55%) in the shales lead to excessive brittleness in the stress–strain behaviour of these materials, and to a need to flatten the slopes substantially to stabilize the slides once they have occurred. Residual strengths as low as $c'_r = 0$, $\phi'_r = 4°$ have been measured, and also back-analysed from the slides themselves. These are probably the lowest residual strengths ever measured, and the resulting engineering difficulties can be imagined. No wonder that the engineers responsible for the construction, although at first dismayed by the landslide problem, eventually adopted an attitude of complete fatalism and gave up attempts to stabilize the slides in place, resorting to their complete removal where at all possible.

The slide classification system adopted entails the estimation of the volume of the slide in million yd^3, and assigning a class number as the next lowest integer number to the log to base 10 of this volume. Hence a slide of between 1 and 10 million yd^3 would be of class 6. The preponderant landslides in the Gaillard cut were of class 5, although three reached class 7, and the number of serious slides (i.e. exceeding class 4) experienced in this reach of the canal since construction commenced was 55.

A typical section through the canal, showing both the design cut, and that achieved following slides, is shown for the Culebra Cut in Figure 10.18.

10.6 Active coastal landslide complexes: Folkestone Warren, Ventnor and the Undercliff

The geological sequence including the Chalk and Upper Greensand, overlying the Gault and Lower Greensand, occurs at a number of locations on the southern coast of England and on the northern coast of France. Where this sequence is combined with steep slopes, and shallow coastwise dips in the strata, so that the sequence occurs along an extensive stretch of coast, landslides occur on a large scale. The principal localities where this happens are east of Folkestone in Kent, and in the vicinity of Ventnor on the Isle of Wight (the 'Undercliff'). Near to Folkstone, the dips are a couple of degrees to the east, because the beds are on the eastern side of the Weald–Boulonnais anticlinal dome, and the outcrop of Gault overlain by Chalk is the site of massive landslides, forming a coastal feature known as Folkestone Warren. In the 1840s, a railway was constructed to connect the two towns of Dover and Folkestone, and, although built partly in tunnel, the route crossed the Warren largely in cut.

The town of Folkestone is updrift of the Warren with respect to the prevailing direction of littoral drift. It is therefore not surprising to note that a series of progressive extensions to the harbour breakwater throughout the

L

Figure 10.19 Folkestone Warren: aerial view. The scar of the big chalk fall of 1915 is clearly seen in the foreground.

Figure 10.20 Folkestone Warren: chalk fall. One of the Chalk Falls associated with the 1915 landslide not only overran the railway, but crossed the whole Warren, and ran out to sea about 400 m. (Contemporary photograph.)

nineteenth century led to the entrapment of shingle against the breakwater, with a consequent denudation of the beach fronting the Warren. This in turn led to a series of movements, reactivating ancient landslides, some of which affected the railway. An account of the close connection of beach denudation

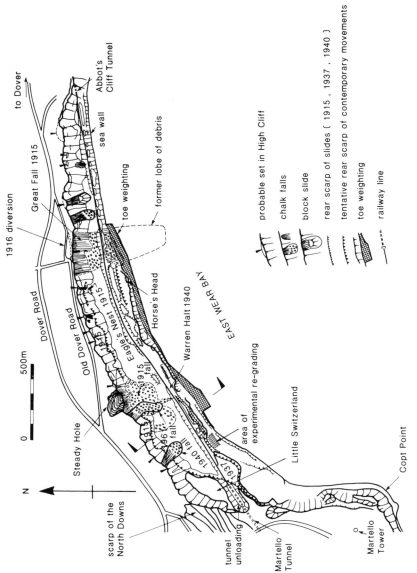

Figure 10.21 Folkstone Warren: geomorphological map.

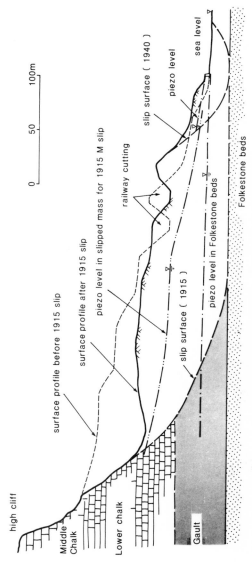

Figure 10.22 Folkstone Warren: section through the slides.

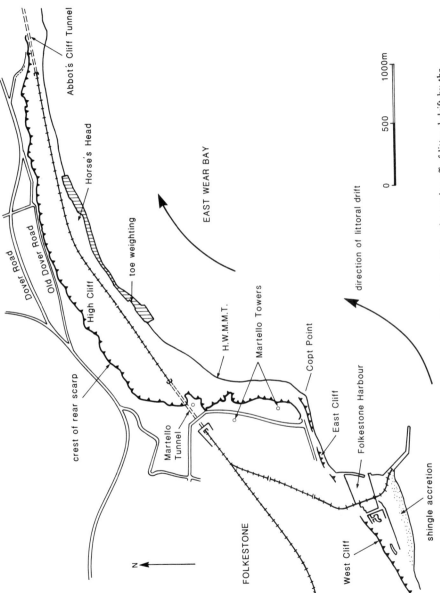

Figure 10.23 Folkestone Warren, and its relationship to the cutting-off of littoral drift by the Folkestone Harbour Pier.

in the Warren area with the extension of harbour piers updrift at Folkestone is given by Hutchinson *et al.* (1980).

As well as rotational landslides, the Warren is subject to falls of chalk from the High Cliff (rear scarp). Although strong *in-situ*, the chalk breaks up readily in rockfalls, becoming very mobile, and forming tongues and lobes of debris with a very large run-out, similar to mountain avalanche-flow-slides (sturzstroms—see below).

In his paper setting the study of the Warren landslides in a modern context, Hutchinson (1969) classified the various mass movements which had been observed into three main types:

M-type (massive multiple rotational) slides

R-type (rotational) landslides in the sea cliff

F-movements or falls of chalk from the high cliff or rear scarp

In addition to noting the effect of beach denudation, Hutchinson *et al.* (1980) showed that the chalk falls had affected the Warren much more than was previously recognized, and not only provided hazards immediately beneath the High Cliff, but as a result of the enormous runout of the larger falls, could pose a danger anywhere on the Warren as well as close inshore. Following the construction of the railway in 1846, there was a very sensitive indicator of movement present, and records of landslide reactivation which were sketchy until that date, become better thereafter. Movements of R-type landslides, which do not immediately affect the railway, were most common, but a series of M-type landslides in the late nineteenth century, culminating in the major slide of 1915, took place.

The railway enters and leaves the Warren through tunnels (the Martello tunnel in the west, and Abbotscliff tunnel in the east), and following an M-type landslide in 1896 (there had been an earlier one in 1877) which cracked the Martello tunnel near its eastern portal, a programme of stabilization works was carried out. This included the construction of some drainage headings into the landslide mass, and the opening out of the eastern end of the Martello tunnel to prevent the possibility of a train becoming trapped there. Some sea-wall construction was also commenced.

Unfortunately, the works undertaken were insufficient to prevent the large M-type landslide of the evening of the 19 December 1915. This appears to have commenced as an ordinary M-type slide, which as a renewal of movement on a pre-existing slip surface should have suffered only small relative displacement. However, the series of M-type slides (1877, 1896 and now 1915) undermined the High Cliff which forms the rear scarp of the Warren landslide complex, and this triggered off a number of falls of chalk. The largest of these, from the vicinity of a depression in the cliff top known locally as Steady Hole, added an undrained load to the head of the main slide and precipitated the large movements of the Warren that took place that night—from 10 to 20 m over much of the Warren, but 50 m in the vicinity of the Steady Hole Chalk fall.

Another large fall of chalk, to the east of Steady Hole, of a mass of chalk which had moved very slightly in 1877, was transformed into a flowslide which buried the railway over a width of 250 m or more and ran out to sea some 400 m. At the time, this was referred to as 'the Great Fall'. Other, smaller, chalk falls took place, one of which carried down a house built near to the cliff edge, together with its occupants.

The damage to the railway and its installations took nearly five years to rectify, and movements in the High Cliff led to the diversion of the Folkestone–Dover road. It is surprising, therefore, in a landslide disaster of this magnitude, to record that there was no loss of life. This was entirely fortuitous, for on the Warren were railwaymen and a detachment of soldiers, as well as the occupants of the house which slid down the High Cliff. The soldiers, in particular, were fortunate not to be overwhelmed by the Great Fall which swept close by the signal box in which they were stationed.

Equally fortunate were the passengers in a train (Figure 10.20) which had luckily left Folkestone a little late, and which was stopped by signalmen at the Martello tunnel portal. Subsequently, continuing movements of the slide derailed the train.

In addition to rectifying the immediate damage done to the Warren railway and installations, the railways undertook a programme of research investigations into the landslides, involving Professor Terzaghi at one stage, and

Figure 10.24 Foreshore at Folkestone Warren after the 1915 landslide, showing beach heaves. (Contemporary photograph.)

Figure 10.25 St Catherines, Undercliff, Isle of Wight.

culminating (after an interruption in the period 1939–45) in an extensive coastal defence, toe weighting and drainage scheme (Wood, 1955; Viner-Brady, 1955). Even this did not completely correct the problem, and at the time of writing, a further programme of investigation had just been completed. This latest investigation was addressed to the problems of chalk falls as well as to further movements in the Warren landslides.

At Ventnor in the Isle of Wight, and along the coastal landslide complex locally termed the 'Undercliff', the geological succession is broadly similar to that at Folkestone. The lowermost beds of the Chalk are present over much of the rear scarp, but the Upper Greensand (comprising weak sandstones together with limestones and Cherts) is much better developed on the Isle of Wight than at the Warren where it is represented only by a thin stratum of the Glauconitic Marl. The Gault too, is different, being much sandier on the Isle of Wight.

The beds present dip south-south-east in a gently plunging syncline. Thus, at each end of the Undercliff, the various strata are higher relative to sea level than in the remainder. At the ends of the Undercliff too, there has been a markedly higher incidence of mass movements than in the centre of the stretch of coast, partly because the supply of landslide debris resistant to marine erosion forms a smaller proportion of the whole at these locations, and partly because the lower rocks are themselves more erodible. The mass movements

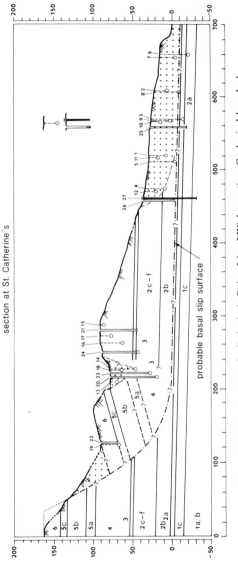

Figure 10.26 Landslides at St Catherine's Point, Isle of Wight: section. Geological key: 1*a*, *b*, sandy strata in ferruginous sands; 2*a*, *c–f*, predominantly sandy strata in sandrock; 2*b*, clayey stratum in sandrock; 3, carstone; 4, Gault; 5*a*, passage beds (Upper Greensand); 5*b*, 'Malm Rock' (Upper Greensand); 5*c*, chert beds (Upper Greensand); 6, Lower Chalk; stipple, landslipped debris.

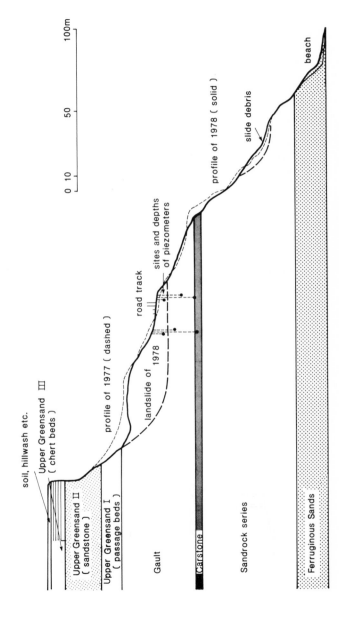

Figure 10.27 Gore Cliff, Isle of Wight: section.

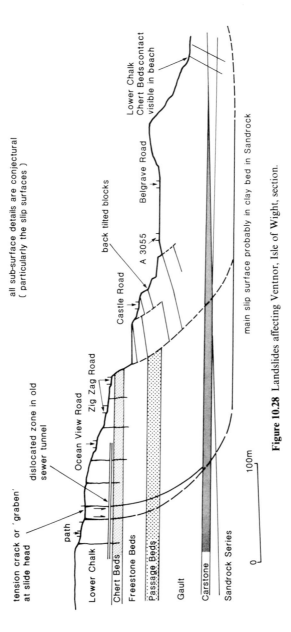

Figure 10.28 Landslides affecting Ventnor, Isle of Wight, section.

which do occur at these ends of the Undercliff are multiple rotational landslides with a flat sole seated in the lower half of the Gault. Unlike many other similar slides in interbedded clays and other rocks, the slip surface does not penetrate to the base of the argillaceous stratum, but this seems to be because the lower part of the Gault is much stronger than the remainder. The junction between the upper, weaker, Gault and the stronger, lower, part, seems to encourage the formation of slip surfaces in the same way that more obvious changes in lithology do elsewhere.

Evidence from historical records is that the movements that have occurred have been renewals of movement on pre-existing slip surfaces, either in response to toe erosion, or in conjunction with cliff falls from the Upper Greensand and Chalk which forms the rear scarp of the slide complex. In no recorded instance has there been a slide involving previously unslipped material, i.e. there has been no significant retrogression of the landslides except for the incorporation of rockfall debris from the rear scarp.

Along the main Undercliff, which has a length of about 10 km, and a width from rear scarp to toe of often in excess of 500 m, there are mainly signs of shallow movement. These include damage to roads, services and to a certain extent to housing, but on a minor scale, and at such a rate that it is repaired as and when it happens.

In contrast to the movement patterns which are shown over much of the

Figure 10.29 Chale Cliffs, Isle of Wight. This area is perhaps the most rapidly eroding part of the coastline of the Isle of Wight, and retreat is under the effects of seepage erosion, falls, and slides.

Figure 10.30. Ventnor graben feature. Formerly the site of a row of cottages; movements in 1960 led to their demolition. The playing field which has replaced them is repeatedly broken by slide scars. Strictly, this is a tension crack.

Undercliff, there is one particular zone of movement (in Upper Ventnor) which is of a different character, and which is intrinsically more hazardous. This is a fracture zone, traceable with certainty for several hundred metres, but possibly nearly a kilometre in total length, approximately 200 m landward of the previously recognized position of the rear scarp of the Undercliff slide complex. Without doubt, it is the tension crack at the rear of a retrogressive movement in that area, and it would seem from a strictly limited amount of deformation monitoring to be accelerating in both widening and settlement.

Damage to a road which crosses the fracture zone is extensive, needing attention almost on an annual basis. A sewer, which formerly ran beneath the road in tunnel, deflected so much that the costs of at first relining and later piping on adjustable jacking trestles became prohibitive, and the sewer had to be rerouted. Numerous houses in the vicinity of the feature were irreparably damaged following a particularly severe period of movement in 1960, and were pulled down. The land was reused as a playing field (Figures 10.28, 10.30).

10.7 Mountain avalanches: Andean sturzstroms and landslide dams

Plafker and Eriksen (1978) give an account of a number of catastrophic avalanches emanating from Nevados Huascaran, the highest peak of the

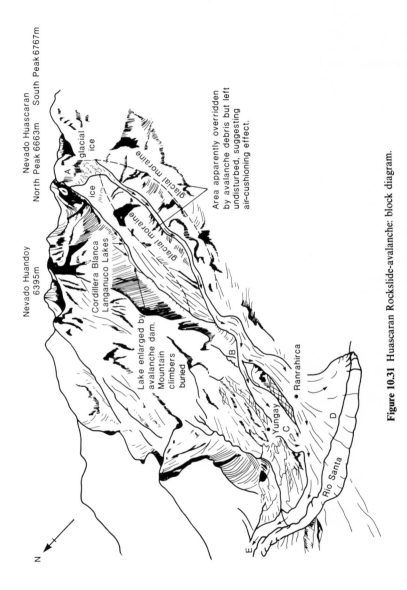

Figure 10.31 Huascaran Rockslide-avalanche: block diagram.

Peruvian Andes. The evidence is that two recent avalanche–rockslides in 1962 and 1970 took place in an area covered by deposits from a huge pre-Columbian event, although there had been no recorded avalanches prior to the 1962 event since the arrival of the Spaniards in the early sixteenth century. The peak of the mountain remains in a dangerously unstable state.

In the 1962 avalanche the death toll was 4000, mainly in the city of Ranrahirca. The later event killed about 18 000 in Yungay: evidence of the larger scale of the later, earthquake-triggered, movement at some 50–100 million cubic metres, as against 13 million in the earlier one. The pre-Columbian event was of significantly greater size.

The debris originated from between 5400 and 6500 m elevation on the west face of Nevados Huascaran, and travelled 16 km to the Rio Santa which there has an elevation of approximately 2400 m. Velocities must have averaged 280 km/h (1970), 170 km/h (1962) and possibly well in excess of 300 km/h in the pre-Columbian event. These velocities have been computed from such data as missile mass, crater size, etc.

An even larger total mass of material involved in the Mayunmarca rockslide and debris flow (Kojan and Hutchinson, 1978) involved about 1000 million cubic metres of material and dammed the Rio Mantaro in the Peruvian Andes in April 1974. This landslide dam was 3.8 km long, and impounded a 38 km long lake to a maximum depth of 150 m. When breached, 44 days after the slide, the peak discharge was about 10 000 m^3/s. The flood wave travelled 800 km downstream in about three days, and flood waves were 20 to 35 m high 15 to 100 km away. The initial rockslide took place in about 3 minutes, and had a runout of 8 km. The death toll was 451.

An example of a happier ending to the formation of landslide dam in a mountainous region is given by Byce (1984). This particular landslide dam, over 65 m high, 270 m wide and 460 m thick, was formed in April 1983 from the accumulated debris of a landslide on the western slope of Spanish Fork Canyon in north central Utah. By the time it blocked the river, the debris had travelled approximately 2.4 km from its source. The town of Thistle was flooded by the rising water levels impounded by the dam, and two important highways and an essential rail link were destroyed. There was also the hazard of flooding downstream when eventually the landslide dam was breached and overtopped.

Not only was it essential to reroute the railway, but also as a first priority it was essential to divert the rapidly rising water in the landslide-dam-impounded lake. The 145 m long diversion tunnel, of 3 m by 2.4 m section, was excavated in a week through the right abutment of the slide, just below the crest of the landslide dam, and connected to a culvert 165 m long to carry its discharge just three days before rising water levels in the lake had risen to the level of the diversion inlet, and the flow of water (about 60 m^3/s) through the tunnel commenced. In order to ensure that the slide was not overtopped and breached by a rapid rise in the lake level as heavy snows in the surrounding mountains melted, barge-mounted pumps working close to the upstream face lifted water up and over the slide dam.

The completion of the first railway diversion tunnel was ahead of schedule, and saved trains a time-consuming and costly 1600 km detour. A second, parallel, tunnel was then excavated to accommodate the other rail track.

Permanent diversion works were undertaken to drain the lake. These consisted of a low-level tunnel with a raise-bored shaft 5 m in diameter and 50 m high. Once the tunnel and shaft were complete, a trench from the lake to the shaft was excavated, and lowered progressively as the lake level fell. By early January 1984 the lake had been drained, and although retained as a permanent stream diversion tunnel so that the landslide dam did not have to be removed, the diversion tunnel had done its work.

11 Design recommendations for man-made slopes

11.1 Embankments and cut slopes

Embankment details

Embankments constructed of freely draining granular materials need only small factors of safety against shallow slope movements. They usually operate quite effectively without internal drainage layers, but may benefit from the inclusion of soil reinforcement to prevent lateral strains on soft foundations.

Sands and silts when used in any proportion in embankments need careful protection against erosion due to rainwater runoff, at least until a vegetation cover has become established. Care should be taken in the discharge of collected rainwater (from a road pavement, for instance).

Embankments formed of clays, or of clay shales, need careful treatment. Compaction dry of optimum will minimize porewater-pressure generation, causing the formation of initially helpful suctions in most cases indeed, but subsequently the dissipation of these suctions can lead to long-term, shallow failures of the embankment side slopes if this phenomenon is not appreciated. Compaction wet of optimum will cause loss of shear strength and increased consolidation settlement, due to the generation and subsequent dissipation of positive porewater pressures. Compaction in this manner is of most use when forming the impermeable barrier (core) in an earth or rockfill dam.

Porewater pressures in the surface layers of a clay embankment can rise to quite high values, say equivalent to $r_u = 0.3$ in the top metre or so (Crabb and West, 1985). This is made worse when shrinkage cracks have formed and then are flooded in an intense rainfall. The surface layers of such a slope are therefore correspondingly vulnerable to shallow slides unless extremely flat side slopes are adopted. Since these flat side slopes are often uneconomic if merely the cost of additional land acquisition and construction are taken into account, the engineer may be prepared to accept some failures, and repair these as and when they happen. Alternatively, the selection of flattened side slopes may be justified on aesthetic grounds, or because they mean that some land can be returned to agriculture, rather than be sterilized as an embankment side slope with a fence at the bottom.

Although large berms are inevitably the response to indications of critical deep-seated modes of failure, a series of small berms up the side slopes of a large embankment may be used where the critical stability problem is failure of the embankment material on a large scale, and where the overall slope angle must be reduced. For construction reasons, or because small-scale local failures of the individual berms are less likely, these will stand with the increased slope angle. Widely used on the steep downstream face of rockfill dams, these berms are not generally liked, on aesthetic grounds mainly, because they introduce obviously artificial features on a large and visually intrusive scale. On the other hand, there is no necessity for the flat tops to the berms to be horizontal, or even parallel, and they can form useful access to gauge houses or surface survey monuments on the slope.

Foundation treatments

Foundation failure in an embankment may come about due to a number of causes. Simplest of these is inadequate shear strength under undrained conditions, in which case the appropriate response is either to build slowly, allowing time for porewater-pressure dissipation, or to install some deep drainage measures to dissipate the pore pressures more rapidly. There is a possibility that the installation of deep drains like sand drains or wicks may disturb the soil and lead to increased settlements. Soft alluvial deposits often have a desiccated crust which will play an important role in stability, and it may be expedient to omit topsoil stripping, or to replace the crust with a layer of reinforced soil. In either case, strains in the subsoil will be reduced.

Deep modes of failure must always be checked where the foundation properties are weaker than the fill, no matter how thin the weak foundation is (a layer of weathered rock, for instance). Displacements in this should be measured, and any signs of accelerating movement should call for a reassessment of stability.

A second source of foundation modes of failure is the presence of slip surfaces in the subsoil where the embankment is constructed on sidelong ground. The initial site investigation should have detected these, and they should be considered in the design of the embankment. The range of techniques for stabilizing slides has been covered earlier; the only one which enters the category of a 'foundation treatment' is to dig out and recompact the sheared material. This will only be effective where the remoulded strength is appreciably higher than the residual strength: if it is not, replacement of excavated material with imported fill may be appropriate. Care must be taken not to trigger instability with the excavations, and construction of individual 'keys', or digging out and backfilling small sections at any time, should be considered.

Cut slopes

Due regard in the design of cut slopes should be paid to their tendency to destabilize with the passage of time, and with the equilibration of their stress-relieved and depressed porewater pressures to a generally higher level. This means that simple and cheap undrained shear strength measurements are of no use in design except possibly in the very short term.

Cuts in sidelong ground should be undertaken with caution, and only after appropriate investigations to check on the presence of sheared superficials. Even on initially flat sites, there is the possibility that tectonic shears may exist, if the geological sequence and history has been favourable to their development. Generally speaking, cuts are more susceptible to the effects of the presence of these shears than are embankments because the latter partly offset the increased levels of shear stress which they impose by raising the normal effective stresses on potential sliding surfaces.

The stability of rock slopes is even more closely controlled by the discontinuities which may be present in the rock mass. These include bedding, faulting, jointing, tectonic shearing and solution or weathering. Many of these features cannot be found economically before excavation commences, and it is better to log the exposed faces and put remedial measures in hand on a 'design as you go' basis. It is important, nevertheless, not to worsen stability unnecessarily during construction, and the use of minimum blasting charges, or pre-split blasting, assists in this. Even in the most competent rocks, but particularly in strongly jointed or shattered rock faces, there is the problem of individual boulder movement. Attention must be given to the control of such details.

11.2 Control of water

Water-retaining embankments

Upstream cores are beneficial for the stability of the downstream slopes of an earth or rockfill dam, since the water pressures impose a net downwards load and increase the effective stresses. The concept is useful where steep side slopes are possible and there is a possibility of the whole dam sliding downstream under the lateral thrust of the water. As an extreme, the impermeable membrane may be placed on the upstream face of the dam: such a membrane might be asphalt—it most certainly would not be a clay fill—and would only be used where settlements of the rest of the dam would be small.

Upstream clay cores have an adverse effect on upstream slope stability. Together with upstream clay blankets, which are a potentially economic method of increasing the seepage path length under the dam, they represent a comparatively easy path for slip surfaces to follow, and may need toe-weighting berms to offset this effect.

Rapid drawdown is an important 'loading' case for many dams. Appropriate drainage measures in the upstream shoulder of the dam can be installed to allow the escape of reservoir water, and these can be combined with construction drains. It is, however, essential to accommodate the discharges from such drains whether used for temporary conditions, or to improve stability during drawdown.

Lagoons used to store fines (or 'tailings') from washing operations at a mine or quarry usually have to operate as storage for the water which is then recycled, as, in most developed countries, contaminated water cannot be discharged into watercourses. This generally rules out the possibility of omitting the impermeable barrier in the retaining embankment, and designing it with the appropriate filter-protected drains to be 'permeable'. If these lagoons have just grown at a mine or quarry, they tend to lack adequate drawoff or spillway arrangements, and are constructed with a minimum of freeboard. Designed structures will probably be more adequate in this respect.

Where hydraulic gradients are high, there is the possibility of seepage erosion. This is normally confined to fine sands, and to silts, where the particles are small enough to be easily carried, but which have no cohesion to hold them in place against the plucking action of the water. Some cohesive soils may crack (e.g. stiff dam cores) or rocks may have open joints, and the passage of water through these can enlarge them by erosion or solution. Appropriate filters can control this activity.

11.3 Tips and spoil heaps

The key factor to remember in the design of spoil heaps is the compaction requirement. Bishop (1973) suggests that lack of adequate compaction at Aberfan, and elsewhere, was a primary cause of the development of flow-slide mechanisms. Compaction has other advantages too: it reduces air voids, which makes the spontaneous combustion of carbonaceous material less likely, and enables a greater mass of waste to be placed in a tip of a given size. Furthermore, the compaction requirement brings with it the need to raise a tip in small lifts just like a controlled fill, and in most cases this reduces plant costs because not only may it reduce the haul distance, but dump-truck drivers feel more confident in backing to the small slopes of a single lift and there is not therefore the need to keep a dozer continually on hand to double-handle spoil dumped short of the edge.

The second factor to keep clearly in mind is the foundation condition at a tip site. Where a tip is formed on a soft alluvial foundation, this can give rise to serious stability problems not in character with the types of spoil used. Raising the tip in lifts can utilize consolidation of the foundation with its ensuing gain in strength to enable higher and/or steeper tips to be formed.

When tips are constructed on sidelong ground they can interfere with the natural drainage pattern. This can be the result of damming surface water channels, or it may be due to the disruption of subsurface flow paths. In either case, it can be deleterious to the stability of both the tip itself, and of adjacent slopes otherwise unaffected by tipping.

Finally, there is always the problem of trying to incorporate slurry from small settling lagoons in tips. This has to be carefully watched, or overall stability can be affected seriously. Such materials can be dealt with safely, provided that some forethought is exercised. They need to be placed in bounded enclosures at the centre of a tip, and filled over before the deposits become too thick. Thereafter, it is essential to control the rate of placement of additional fill on top so that undrained pore-pressure changes can dissipate.

Where large quantities of fines are introduced into tips, it is desirable to monitor the generation and dissipation of porewater pressures, and to use this to control the rate of other fill placement on that part of the tip.

11.4 Control of construction with instrumentation

Part of the rationale behind the adoption of effective stress methods for the calculation of stability during construction is that the effect of either a rise of porewater pressure, or its dissipation, can be immediately included in any assessment of stability. Thus, the incorporation in any large fill (for example in an embankment dam) of a number of piezometers to yield this data is an essential first step in any instrumentation scheme. In addition to the porewater-pressure generation and dissipation in a fill, the foundation conditions can be equally important, particularly in the case of embankments on soft foundations.

Piezometers also give information on the amount of porewater pressure remaining to be dissipated, and thus the likely further amount of consolidation settlement which can be expected.

When making a stability assessment on the basis of measured porewater pressures, it should be remembered that such an estimation is made *a posteriori* and that in addition to using the computed factor of safety, the supervising engineer must also consider the rate of generation of porewater pressure related to the rate of fill placement immediately preceding the measurement of those porewater pressures.

An alternative (or addition) to the use of porewater-pressure based control of construction is to measure displacements of the soil structure. Most crucial measurements involve the lateral deformations in the soil; settlements or heaves are often much less telling indicators of the onset of failure, although they do act equally effectively in demonstrating the gross magnitude of movements which are taking place. In the sections on slope instrumentation the use of inclinometers is mentioned. These are particularly useful in that they not only show movements at isolated positions, as do surface monitoring stations, but give an idea of the profile of movement along the length of the access tubing. They are therefore good indicators of the onset of movements in concentrated zones in a fill, such as accompany the first stages of a failure.

Provided that the deformation measurements are taken regularly enough, and are interpreted correctly and without delay, they may be used in construction control in many cases. Here, however, a note of caution must be sounded. The use of deformation measurements to control construction must be limited entirely to non-brittle soils. In soils which have drained or undrained brittleness, the detection of excessive rates of deformation may take place too late to take effective remedial action, especially in the case where large muck-shifts would be required.

Effective interpretation of deformation measurements includes the consideration of changes in the rate of deformation, taken both on its own and in combination with the rate of fill placement, and in terms of its distribution throughout the slope section. This requires that plots of deformation be kept up to date, sometimes more frequently even than on a daily basis in crucial situations. Furthermore, the method can only be used where one can be sure that the cessation of filling will be followed almost immediately by the arresting of movement.

11.5 General considerations on the choice of factor of safety

What are adequate factors of safety for earth slopes? This is a source of practical difficulty for all geotechnical engineers and engineering geologists involved in this work on a daily basis, and poses a philosophical problem of considerable magnitude. Examine the reasons for adopting factors of safety in engineering in general:

the materials may not have the strengths assumed in design

loads may be more than those assumed

construction may impose severe conditions locally or for short periods

the analytical tools used may have shortcomings which the factor of safety helps to accommodate

limitation of deflections.

In many branches of engineering, some of these factors may be far more readily controlled than others, and this leads to certain philosophical approaches being adopted. In structural engineering, for instance, the engineer has the opportunity to specify the use of materials whose performance under particular stress conditions he knows to a relatively high precision. He can additionally design a structure which is amenable to simple analysis, and can control construction carefully to prevent overstress during this phase of activity.

Where the structural engineer finds difficulty is in assessing the *loads* which his structure will have to carry in service. This is overcome in the case of commonplace structures by the use of tabulated loadings (often far in excess of probable loadings, but which bear the approval of the majority of engineers); and in the case of large, complex or important structures, by detailed experimental work. This uncertainty in loading allows the engineer to define his factor of safety in terms of loading: i.e. the factor of safety is the ratio of the actual load borne by the structure to the load which causes the occurrence of an undesirable condition (collapse?).

In contrast, the geotechnical engineer has much more confidence in the loading, which most of the time includes very little more than the self-weight of the soil, but the shear strength is much more in doubt. He therefore chooses a factor of safety based on the ratio of available shear strength to that which is required to prevent the undesirable condition. This factor of safety may be identical to the structural engineer's *load factor*, but may not always be so.

An example of this was the case of a bridge abutment resting on a bank seat at the top of a cut slope, alluded to in section 5.14. Along much of the length of the cutting the factor of safety against sliding on the worst slip circle was 1.30, but in the vicinity of the bank seat the factor of safety calculated on this basis was 1.25. In contrast, the 'load factor' against failure under the additional applied load was of the order of 6.

A partial safety factor approach is sometimes used in the assessment of an overall factor of safety which should be achieved in a particular case. Each individual design parameter is assigned its own safety factor, and these are then summed or multiplied together. With so many factors in a soil slope, even relatively small partial factors produce an end result that is genuinely overconservative. Furthermore, since the unknowns must be more numerous in the case of a small insignificant slope, compared with such an important structure as an earth-fill dam, thoroughly investigated and instrumented, the paradoxical result is that the former must be given a higher overall factor of safety than the latter!

Another feature of the selection of absolute values of safety factors is that quite simply, large increases in safety factor may not be achievable either economically or in absolute terms in the case of large slopes—natural landslides for example. Is it merely pragmatic to settle for lower factors of safety in this case?

My view is that calculated safety factors are meaningless (except for those that imply failure). They only exist as a relative index of stability when considering remedial measures, e.g. 'if *this* is done, rather than *that*, which gives the better (more reliable/cost-effective/permanent) result?' What counts is the ability of the slope to resist a certain destabilizing event (or events) singly, serially, or in combination. Loading, except for seismic loading, is rarely a significant factor. The important destabilizing factors in order of importance in most cases are as follows:

increase in porewater pressures

removal of support at the toe, either naturally by erosion, or by excavation

changes in the soil or rock shear strength. Predominant in this is the loss of strength from peak to residual with strain, but it could include chemical effects, or change in soil fabric or structure

loading at the head of the slope by other mass movement forms principally, but by fill placement or 'live loading' in exceptional cases

seismic loading.

Thus, if one were to design all slopes with the highest conceivable porewater pressures, assume excavation at the toe, and some additional loading at the top of the slope, use residual strength and consider some reasonable seismic accelerations, a factor of safety of 1, if achievable, might be perfectly acceptable. This is because factors of safety are already built into the analysis.

11.6 Choice of a factor of safety for stabilizing existing landslides

An alternative approach is illustrated by example in the following. A coastal landslide is subject to intermittent movements. The factor of safety is therefore of the order of 1, so a back analysis to evaluate a field-mobilized residual strength is undertaken: this involves appropriate subsurface investigations. Recession of the slope crest threatens a highway. It is found that there are a number of possible schemes:

a sea wall close to the shore line, with a little fill behind it which acts as a toe load. This brings about a marginal improvement in safety factor to 1.05

drainage alone. This increases the normal factor of safety to 1.40 without a sea wall, although erosion will gradually decrease this as support from the toe of the slope is removed

a cut and fill solution involving remodelling the slope section and toe loading. This also requires a sea wall, but increases the factor of safety to 1.15. Some surface water drains are included

relocate the road: the factor of safety of the slope is then immaterial.

The schemes are ranked in order of cost: which to choose? The first scheme is cheapest, and it is thought that movements of the slide will stop. The toe is protected, so that erosion will not take place. Detailed investigation shows that the wall will fail with beach scour equivalent to the 100-year storm, or the slide will move and destroy the wall if groundwater levels rise to those likely with a wet winter of 50-year return period. The highway authority may then have to budget for repair of this section at almost total replacement cost.

The second scheme has a definite life. What is of concern is the retreat of the cliff crest, not its toe, and this is the result of the landslide movements. Drainage will protect the slope from the 200-year groundwater conditions, but toe erosion will continue, and at the end of 50 years the road is left in an indefensible position.

In the third scheme, which is more than twice as expensive as the first, but not so expensive as combining the first two, all of the design objectives are met; it is the technically preferred scheme. Road realignment inland is even more expensive, and subject to protest from local interest groups.

It will be obvious that the technically preferred scheme was never in the running for serious consideration! A gambling approach would be to do the first and cheapest scheme, relying on it to hold erosion so that options for the future were kept open. If the deep drainage has to be done later, this will come out of a separate year's funding, and thus appear 'cheaper'. Adoption of the second scheme would not keep those options available for more than a couple of years at most, and requires a definite decision on policy. Of course the most satisfactory approach from the point of view of the highway authority is to do nothing. That saves money in the short term. In the long term, it removes many of the technical alternatives so that ultimately, regardless of protest, cost, or any other consideration, the road must be moved.

The factor of safety is of little account in this procedure. In the event of works being undertaken, the technically weakest remedial scheme is likely to be chosen, and the one with the lowest factor of safety!

It should not be thought that a factor of safety of 1.05 is totally inadequate: where the residual strength is all that is mobilized, and the design takes into account relatively adverse porewater-pressure conditions, this might be a highly acceptable figure. In the case of the Folkestone Warren M-type landslides (Chapter 10) that is all that is provided by the massive toe weighting, but that construction would survive the worst conceivable storm, or a rise in the groundwater level of many metres. The required factor of safety is therefore dependent on the *scale* of the affected slope.

One final point is this: the factor of safety to be sought to provide *overall* stability does not necessarily guarantee the absence of subsequent ground strains in the future. This might be acceptable in the case of the stabilization of a slope to prevent cliff retreat, to prevent encroachment on a roadway or garden at its toe, but most certainly will not be adequate if buildings are to be placed on the slope after 'stabilization'. Dwellings are very brittle; their occupants will register deformations in the fabric of the building with extreme sensitivity. Much higher factors of safety are needed in such a case, often combined with positive measures to control structural movement such as deep piled foundations, retaining walls and so on.

11.7 Choice of a factor of safety for earthworks

Dealing with slopes containing slip surfaces is relatively easy in respect of the choice of factor of safety: any value discernibly in excess of unity should prevent significant further movement. The choice is not so clear with earthworks which could possibly be involved in progressive failure. Bishop showed in 1951 on the basis of a relatively crude elastic analysis of a slope that overstress might be expected where the factor of safety established by limit equilibrium techniques was less than about 1.8. The significance of the overstress—it being possible to have a zone of overstress which was stable in its extent, and not growing—is related to the brittleness index of the soil. One needs larger factors of safety where the possibility of progressive failure is a serious risk.

I would disagree with the often stated opinion that the residual strength is irrelevant to the stability of compacted earth fills which have not been pre-sheared. Its impact is on the selection of an initial factor of safety. This can be treated in one of several ways: firstly, by a partial factor of safety approach. In this case a basic factor of safety is selected, and this is then multiplied by a 'brittleness index allowance'. In extreme cases this additional factor might be $1 + I_b$ or $1/I_b$.

Secondly, one might assess the results of failure in terms of likely deformations, and the cost of remedial measures, and select the factor of safety accordingly. Since the result of failure is a function of the brittleness index, the resulting choice of factor of safety will reflect that parameter, albeit in an *ad hoc* fashion. Either of these two approaches will confirm the view that, with non-brittle materials, relatively low overall factors of safety can be acceptable.

The third approach is to neglect limit equilibrium design and its factors of safety altogether, and to rely more heavily on continuum solutions. If the stress–strain and porewater-pressure responses have been modelled to a suitable degree of accuracy, then a design which shows no undesirable deformations will be as acceptable as one analysed via slip-circle techniques and showing an adequate factor of safety. Advances in computer technology will make this possible in a short space of time, although I suspect that engineers' prejudices will be far longer lived, and the relative cheapness of limit equilibrium techniques will ensure their survival long into the future.

11.8 Checking and validation, especially of computer analyses

Checking is an integral part of the engineering analysis and design process. In slope engineering, the checks take a number of forms. First of all, the interpretation of the data from a site investigation needs to be checked. This is as much the case in other branches of geotechnical engineering as in slope stability, and I will have to assume that all competent engineers and geologists will have some procedure in their practice to prevent things being missed at this stage. It might be forgiveable, for instance, not to find slip surfaces in cores or samples: they may well have been missed between runs. They could equally well be detected using instruments such as inclinometers if they were moving, or there could be some indication of their presence in missing sections of the core. What would be unforgiveable would be not to even look for them where their presence might have an impact on an engineering scheme. That demands a certain level of competence to be exhibited throughout this stage of a project.

It is perhaps in the more detailed work that errors can creep in and not be found in the routine day-to-day interaction of technical personnel working on the same project. Failure to select appropriate samples for testing, or the use of unsuitable testing techniques, are both sources of error which will bring calamitous consequences to the eventual slope. An example of this would be to measure only the undrained strength when a cutting was proposed, or to use multistage tests in a soil of high brittleness. This is an error made in basic *philosophy*. The higher the position of the decision-maker, the less likely his decisions are to be questioned, especially in an organization where such questioning is discouraged. Checking such facets of slope engineering is more a matter of keeping an open mind, and being abreast of modern developments in the theory, than in having a set institutional procedure for checking.

Such institutional checking procedures are most effective when applied to calculations than where applied to concepts; indeed, rigorous proscription of methods stifles creative input from engineers, halts the development of new techniques, and leads to such general intellectual boredom that numerical errors increase rather than the reverse. Slope stability analyses *per se* are intrinsically easy to have checked through a system, because the procedures are so well documented. This should pose little difficulty, but in practice it does. A reason for this is the use of computers. Checking in this case involves the two steps: checking that the input data fairly represents the analysis desired, and validation of the internal workings of the computer program.

My personal preference for checking the validity of input data is to have this replotted from the computer input for comparison with the working drawings from which the data has been extracted. Long experience of this suggests that where the checking is done by hand, it is no good at all going from the original data sheets: this does not check the input to the program which may differ in respect of keying errors. The program must print out the data given to it, and this must be used for plotting. I have found some engineers who detach this data from the printout and discard it; this is a reckless step, since it makes the subsequent error-checking process much longer and more difficult.

Again, the replotting has to be carried out by a separate engineer, and on clean paper—not back on to the original section. This then minimizes the effects of plotting what the analyst thought were his intentions, compared to what he actually did. Such a bias is unconscious (almost always), but is ever present. The only collaboration should be in specifying the scale so that the two sections can be overlaid. Then the differences show immediately. Tracing paper can be used for one of the plots, or both can be overlaid on a light table, or just stuck up on a convenient window.

The process is speeded up by using a pen plotter to draw out the section. Figure 11.1 shows my own microcomputer workstation complete with an A3 size pen plotter (left). This takes less than two minutes to draw a check plot of any data set for a stability analysis, almost regardless of its complexity. Errors

Figure 11.1 Microcomputer workstation for slope stability computations. The essential features are an 'Apricot' 16-bit microcomputer, an A3 plotter tablet and a dot matrix printer.

in data preparation can then be found quite simply and quickly. A typical plot from this is shown in Figure 11.2. For the impecunious, fairly high-quality graphics is available now on dot matrix printers, if at the cost of precise control over scale (Figure 11.3).

Data errors are easy to make regardless of the program used, but they become easier when the program has certain features. There are two main ones: if the program input is not in the form of coordinates, and where the coordinate system does not measure levels increasing upwards. The former case is usually found in extremely simple programs developed in-house at short notice and to meet a specific need, but the latter case is found in some proprietary software. Where a 'downwards increasing' y-coordinate system is utilized, the program will almost certainly be difficult and frustrating to use. Errors that do arise will tend to occur in the translation from the site coordinate system (where levels increase upwards) to the program coordinate system. Usually, they will cause the program to crash, but they may lead to extremely misleading results in some circumstances.

The whole process of checking is facilitated where the original data sets are preserved. This used to be easy, but inconvenient, when punched paper media were used. It can be costly where on-line data storage is used with modern computer systems, but now that floppy-disk-based systems (both as terminals and as stand-alone computers) are becoming more widely available, the data

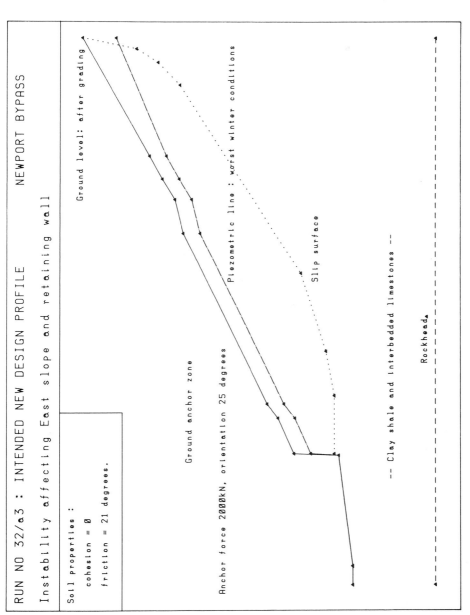

Figure 11.2 Pen plotter picture of a slope as a data check prior to analysis.

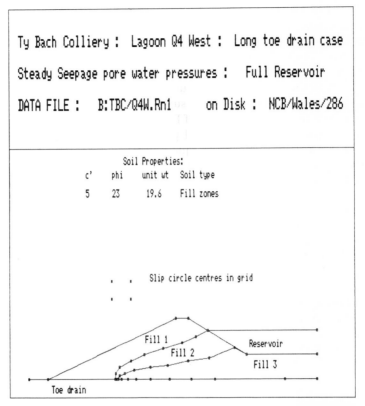

Figure 11.3 Print-out on a dot matrix printer for data checking. It is not usually possible to obtain precisely-scaled output in this format, but a good impression of the data can be gained. Gross errors in data preparation are usually visible.

files can be kept on one or more disks, taking up very little filing space. This is easy and cheap, but does require discipline. It is also facilitated where programs expect a *data file* and do not utilize interactive, 'question-and-answer' type input.

Validation of the program code itself is a more vexed question. The first step must be a comparison with hand calculations (where possible) and with other programs. However, such simple matters as the subdivision of the slip mass into slices can influence the final result, and some manipulation, particularly of the porewater-pressure data, may be required to suit the general format of a number of programs from different sources. Some authorities may operate a validation scheme. There are some drawbacks to this. The first is that the test problems fall into two categories: trite problems attemptable by any program, however simple; and more complex problems, culled from the literature. It is easy to be lulled into a false sense of security by the former class of problem, since they test so few of the facilities of the program. Getting the 'right' answer for a single slip surface in a one-layer problem with a dry soil is not a very difficult test. Some caution is needed when using published data, unless it is published specifically for the purpose of validating other programs. The diagrams are usually redrafted specifically for publication, and then reproduced at a small scale. Coordinate accuracy is therefore poor. More serious than that, however, is the inadvertent supply of incorrect results.

I have several times seen the slope of Figure 5.4 analysed. Regrettably, the section is accompanied in the original paper by a factor of safety. Careful reading of the text reveals that the quoted factor of safety is not for the illustrated slip circle at all. This accounts for the differences found by Bell (1968) and Sarma (1973) when they have tried to analyse the same section. Other cases may be worse, and caution must be exercised when comparing test cases to published data.

The Geotechnical Control Office of Hong Kong (GCO, 1983) recently ran a validation exercise on slope stability analysis programs in use in Hong Kong. The results were disturbing. Many of the submitted answers quite clearly had input data which were incorrect, despite the sections being simple. Furthermore, many of the analyses contained the same fundamental defect. They had used Janbu's method (section 5.7), but instead of applying the correction factor at the end of the procedure, had incorporated it in every iterative step. This came about because the incorrect procedure is advocated in a textbook with a wide circulation in Hong Kong. Some further work by the same group showed up the extremely simplistic basis for Janbu's correction factor chart itself (Figure 5.7), which should therefore be used with extreme caution.

11.9 Conclusion

It is quite possible to engineer economic and safe slopes using quite elementary design concepts and techniques, provided that some fundamental principles are borne in mind. The application of these principles must be supplemented with acute observation, an open mind and a measure of humility.

Research into the stability of slopes is not, however, a sterile field. If its only result lay in quantifying the fundamental parameters for a range of soil and rock tapes, such work would be useful. However, that is a limited objective, and in exploring those intriguing problems where the theory seems not to apply, extensions to the general theory are produced. Finally, it is in the research into natural slopes that we discover more about the Quaternary and recent history of our surroundings, the processes that acted then, and how, even today, in a country as 'stable' as the United Kingdom, these may have an impact on our lives and livelihoods.

The worst calamities to befall us are those which are avoidable, or are those which are our own fault. I hope that this book has gone some way towards helping the geotechnical profession to avoid the avoidable in the engineering of slopes, and to reduce both the number of failures which occur, and their impact on the public. Whether or not this expectation will be met, it is certain that the steady march of technology and knowledge in this field will lead to some of the concepts becoming obsolete, and new ones taking their place. This is to be welcomed, and I look forward to the incorporation of these new ideas in future editions of this book.

Appendix A

Equations in the Morgenstern and Price method

In equation (5.25), where

$$E'_{i+1} = \frac{1}{(L+Kb)}\left(E'_i L + Pb + \frac{1}{2}Nb^2\right) \tag{A.1}$$

the coefficients K, L, N and P are

$$K = \lambda k\left(\frac{\tan\phi'}{F} + A\right) \tag{A.2a}$$

$$L = 1 + A\frac{\tan\phi'}{F} + \lambda m\left(\frac{\tan\phi'}{F} + A\right) \tag{A.2b}$$

$$N = \frac{\tan\phi'}{F}\{p - r(1+A^2) + 2At\} - \{2t - Ap\} \tag{A.2c}$$

$$P = \frac{1}{F}\{(c - s\tan\phi')(1+A^2) + q\tan\phi'$$

$$+ Au\tan\phi'\} + \{Aq - u\} \tag{A.2d}$$

(Note that in their 1965 paper, Morgenstern and Price express N and P in terms of r_u values. The above formulation is thought to be more general.) Hence, starting with assumed F and λ values, the complete range of E' forces from the toe to the head of the slide may be evaluated. This includes the resultant out-of-balance force at the head of the slide.

In the moment equation (5.21)

$$X = \frac{d}{dx}(E'y'_t) - y\frac{dE'}{dx} + \frac{d}{dx}(P_w h) - y\frac{dP_w}{dx} \tag{A.3}$$

This can be rewritten (after some manipulation) as

$$X = \frac{d}{dx}E'(y'_t - y) + E'\frac{dy}{dx} + \frac{d}{dx}P_w(h - y) + P_w\frac{dy}{dx} \tag{A.4}$$

Now, expressing $E'(y'_t - y)$ as R, the moment of the E' force about the level of the slip surface, it is found that

$$R_{i+1} = R_i + \int_0^b (X - AE')\,dx - \int_0^b AP_w\,dx$$

$$- [P_w(h - y)]_0^b \tag{A.5}$$

$$= R_i + \int_0^b (X - AE')\,dx + Q \tag{A.6}$$

In this, Q is effectively a moment of water-pressure forces on the sides of the slice and their point of action. Where it is possible to write for the variation of P_w force across the slice

$$P_w = tx^2 + ux + v \tag{A.7}$$

and hP_w is known at both sides of the slice, then

$$Q = \int_0^b tx^2 + ux + v\,dx + [P_w(h - y)]_0^b \tag{A.8}$$

In principle, therefore, it should prove possible to integrate across each of the slices in turn, finally arriving at an out-of-balance moment at the head of the slide to match the out-of-balance force there. New F and λ values could then be selected to make a systematic elimination of the force and moment errors. This would best be done using two-variable Newton formulae.

Derivation of expressions for ΔE and ΔR in terms of partial derivatives

$$\Delta E = \frac{\partial E}{\partial \lambda}\Delta\lambda + \frac{\partial E}{\partial F}\Delta F$$

$$\Delta R = \frac{\partial R}{\partial \lambda}\Delta\lambda + \frac{\partial R}{\partial F}\Delta F \tag{A.9}$$

Dividing by $\partial E/\partial\lambda$ and $\partial R/\partial\lambda$ respectively

$$\frac{\Delta E}{\partial E/\partial\lambda} = \Delta\lambda + \frac{\partial E}{\partial F}\bigg/\frac{\partial E}{\partial\lambda}\Delta F$$

$$\frac{\Delta R}{\partial R/\partial\lambda} = \Delta\lambda + \frac{\partial R}{\partial F}\bigg/\frac{\partial R}{\partial\lambda}\Delta F \tag{A.10}$$

Eliminating $\Delta\lambda$

$$\frac{\Delta E}{\partial E/\partial\lambda} - \left(\frac{\partial E}{\partial F}\bigg/\frac{\partial E}{\partial\lambda}\right)\Delta F = \frac{\Delta R}{\partial R/\partial\lambda} - \left(\frac{\partial R}{\partial F}\bigg/\frac{\partial R}{\partial\lambda}\right)\Delta F$$

$$\frac{\Delta E}{\partial E/\partial\lambda} - \frac{\Delta R}{\partial R/\partial\lambda} = \left[\frac{\partial E}{\partial F}\bigg/\frac{\partial E}{\partial\lambda} - \frac{\partial R}{\partial F}\bigg/\frac{\partial R}{\partial\lambda}\right]\Delta F \tag{A.11}$$

Multiply throughout by $(\partial E/\partial \lambda)\cdot(\partial R/\partial \lambda)$

$$\Delta E\frac{\partial R}{\partial \lambda} - \Delta R\frac{\partial E}{\partial \lambda} = \left[\frac{\partial E}{\partial F}\cdot\frac{\partial R}{\partial \lambda} - \frac{\partial R}{\partial F}\cdot\frac{\partial E}{\partial \lambda}\right]\Delta F$$

$$\Delta F = \frac{\Delta E\dfrac{\partial R}{\partial \lambda} - \Delta R\dfrac{\partial E}{\partial \lambda}}{\dfrac{\partial E}{\partial F}\cdot\dfrac{\partial R}{\partial \lambda} - \dfrac{\partial R}{\partial F}\cdot\dfrac{\partial E}{\partial \lambda}} \qquad (A.12)$$

Similarly,

$$\Delta\lambda = \frac{\Delta E\dfrac{\partial R}{\partial F} - \Delta R\dfrac{\partial E}{\partial F}}{\dfrac{\partial E}{\partial F}\cdot\dfrac{\partial R}{\partial \lambda} - \dfrac{\partial R}{\partial F}\cdot\dfrac{\partial E}{\partial \lambda}} \qquad (A.13)$$

Derivatives of both R and E with respect to F and λ are required. Differentiating equation (A.1) with respect to F and λ gives

$$\frac{\partial E'_{i+1}}{\partial F} = \frac{1}{(L + Kb)}\left\{L\frac{\partial E'_i}{\partial F} - \left(\frac{\partial L}{\partial F} + b\frac{\partial K}{\partial F}\right)E'_i\right.$$

$$\left. + E'_i\frac{\partial L}{\partial F} + b\frac{\partial P}{\partial F} + \frac{1}{2}b^2\frac{\partial N}{\partial F}\right\} \qquad (A.14)$$

together with

$$\frac{\partial E'_{i+1}}{\partial \lambda} = \frac{1}{(L + Kb)}\left\{L\frac{\partial E'_i}{\partial \lambda} - \left(\frac{\partial L}{\partial \lambda} + b\frac{\partial K}{\partial \lambda}\right)E'_i\right.$$

$$\left. + L\frac{\partial E'_i}{\partial \lambda} + E'_i\frac{\partial L}{\partial \lambda}\right\} \qquad (A.15)$$

The derivatives of K, L, N and P may be obtained from equations (A.2). The integral part of the moment equation, expanded thus

$$\int_0^b \frac{(\lambda(kx + m) - A)\{E'_iL + Px + \frac{1}{2}Nx^2\}}{(L + Kx)}\,dx \qquad (A.16)$$

may be expressed as

$$\int_0^1 \frac{1}{(1 + Hx)}\left\{\sum_{j=0}^{3} T_j x^j\right\}dx \qquad (A.17)$$

in which

$$H = \frac{kb}{L}$$

$$T_0 = b(\lambda m - A)E'_{i-1}$$

$$T_1 = \frac{b^2}{L}((\lambda m - A)P + \lambda k E'_{i-1}L)$$

$$T_2 = \frac{b^3}{L}\left(\frac{1}{2}(\lambda m - A)N + \lambda k P\right)$$

$$T_3 = \frac{b^4}{L} \cdot \frac{1}{2}\lambda k N \tag{A.18}$$

Derivatives with respect to F and λ may also be obtained, e.g.

$$\frac{\partial R_{i+1}}{\partial F} = \frac{\partial R_i}{\partial F} + \int_0^1 \frac{1}{(1 + Hx)^2}\left\{\sum_{j=0}^4 U_j x^j\right\} dx \tag{A.19}$$

and

$$\frac{\partial R_{i+1}}{\partial \lambda} = \frac{\partial R_i}{\partial \lambda} + \int_0^1 \frac{1}{(1 + Hx)^2}\left\{\sum_{j=0}^4 V_j x^j\right\} dx \tag{A.20}$$

where the coefficients $U_0 \cdots U_4$ and $V_0 \cdots V_4$ are given as follows:

$$U_0 = b(\lambda m - A)\frac{\partial E'_{i-1}}{\partial F}$$

$$U_1 = \frac{b^2}{L^2}\left\{(\lambda m - A)\left[K\left(\frac{\partial E'_{i-1}}{\partial F}L + E'_{i-1}\frac{\partial L}{\partial F}\right) + L\frac{\partial P}{\partial F}\right.\right.$$
$$\left.\left. - E'_{i-1}L\frac{\partial K}{\partial F} - P\frac{\partial L}{\partial F}\right] + \lambda k L^2 \frac{\partial E'_{i-1}}{\partial F}\right\}$$

$$U_2 = \frac{b^3}{L^2}\left\{\lambda k\left[K\left(\frac{\partial E'_{i-1}}{\partial F}L + E'_{i-1}\frac{\partial L}{\partial F}\right) + L\frac{\partial P}{\partial F}\right.\right.$$
$$\left. - E'_{i-1}L\frac{\partial K}{\partial F} - P\frac{\partial L}{\partial F}\right]$$
$$\left. + (\lambda m - A)\left[\frac{1}{2}L\frac{\partial N}{\partial F} + K\frac{\partial P}{\partial F} - P\frac{\partial K}{\partial F} - \frac{1}{2}N\frac{\partial L}{\partial F}\right]\right\}$$

$$U_3 = \frac{b^4}{L^2}\left\{\lambda k\left[\frac{1}{2}L\frac{\partial N}{\partial F} + K\frac{\partial P}{\partial F} - P\frac{\partial K}{\partial F} - \frac{1}{2}N\frac{\partial L}{\partial F}\right]\right.$$

$$+ (\lambda m - A)\left[\frac{1}{2}K\frac{\partial N}{\partial F} - \frac{1}{2}N\frac{\partial K}{\partial F}\right]$$

$$U_4 = \frac{b^5}{L^2}\lambda k\frac{1}{2}\left[K\frac{\partial N}{\partial F} - N\frac{\partial K}{\partial F}\right] \tag{A.21}$$

$$V_0 = \frac{b}{L^2}\left\{mE'_{i-1}L^2 + (\lambda m - A)L^2\frac{\partial E'_{i-1}}{\partial \lambda}\right\}$$

$$V_1 = \frac{b^2}{L^2}\left\{kL^2E'_{i-1} + m(LP + E'_{i-1}LK)\right.$$

$$+ \lambda k\frac{\partial E'_{i-1}}{\partial \lambda}L^2 + (\lambda m - A)\left(KL\frac{\partial E'_{i-1}}{\partial \lambda}\right.$$

$$\left.\left. + KE'_{i-1}\frac{\partial L}{\partial \lambda} - E'_{i-1}\frac{\partial K}{\partial \lambda}L - P\frac{\partial L}{\partial \lambda}\right)\right\}$$

$$V_2 = \frac{b^3}{L^2}\left\{m\left\{\frac{1}{2}LN + KP\right\} + k(LP + E'_{i-1}LK)\right.$$

$$+ \lambda k\left(KL\frac{\partial E'_{i-1}}{\partial \lambda} + KE'_{i-1}\frac{\partial L}{\partial \lambda} - E'_{i-1}L\frac{\partial K}{\partial \lambda}\right.$$

$$\left.\left. - P\frac{\partial L}{\partial \lambda}\right) - (\lambda m - A)\left(P\frac{\partial K}{\partial \lambda} + \frac{1}{2}N\frac{\partial L}{\partial \lambda}\right)\right\}$$

$$V_3 = \frac{b^4}{L^2}\left\{\frac{1}{2}mKN + k\left(\frac{1}{2}LN + KP\right)\right.$$

$$\left. - \lambda k\left(P\frac{\partial K}{\partial \lambda} - \frac{1}{2}N\frac{\partial L}{\partial \lambda}\right) - \frac{1}{2}(\lambda m - A)N\frac{\partial K}{\partial \lambda}\right\} \tag{A.22}$$

$$V_4 = \frac{1}{2}\frac{b^5}{L^2}kN\left(K - \lambda\frac{\partial K}{\partial \lambda}\right)$$

If H lies outside the range $-0.4 < H < 0.5$, then use may be made of the following formulae to perform the integration.

(a) Moments:

$$\int_0^1 \frac{1}{(1 + Hx)}\sum_{j=0}^3 T_j x^j$$

Let $$B_j = \int_0^1 \frac{x^j}{(1 + Hx)}$$

then $$B_0\frac{1}{H}\log_e(1 + H)$$

$$B_{j+1} = \frac{1}{H}\left(\frac{1}{j+1} - B_j\right)$$

Hence the integral is $\sum\limits_{j=0}^{3} B_j T_j$ (A.23)

(b) Derivatives:

Let
$$C_j = \int_0^1 \frac{x^j}{(1 + Hx)^2}$$

then
$$C_0 = \frac{1}{1 + H}$$

$$C_{j+1} = \frac{1}{H}(B_j - C_j)$$

and the integrals are $\sum\limits_{j=0}^{4} C_j U_j (\text{or } V_j)$ (A.24)

Where H lies within this range, numerical integration must be used. A five-point Gauss rule is suggested, so that the integrals become as follows.

(c) Moments:

$$\sum_{n=1}^{5} W_n \cdot \frac{1}{(1 + Hx_n)} \sum_{j=0}^{3} T_j x_n^j$$ (A.25)

(d) Derivatives:

$$\sum_{n=1}^{5} W_n \cdot \frac{1}{(1 + Hx_n)^2} \sum_{j=0}^{4} U_j (\text{or } V_j) x_n^j$$ (A.26)

in which the abscissae x_n and weights W_n are:

n	x_n	W_n
1	0.046 910 077	0.118 463 042
2	0.230 765 345	0.239 314 335
3	0.5	0.284 444 444
4	0.769 234 855	as W_2
5	0.953 089 923	as W_1

Since the evaluation of the moment integrals in the foregoing is a complex procedure, in the early iterations, considerable savings in computer time can be obtained by using approximate integration formulae:

$$I_m = b(\lambda m - A)E'_{i+1} + b(\lambda(m + bk) - A)E'_i$$

$$I_F = b(\lambda m - A)\frac{\partial E'_{i+1}}{\partial F} + b(\lambda(m + bk) - A)\frac{\partial E'_i}{\partial F}$$

$$I_\lambda = b(\lambda m - A)\frac{\partial E'_{i+1}}{\partial \lambda} + b(\lambda(m + bk) - A)\frac{\partial E'_i}{\partial \lambda}$$ (A.27)

The procedure to be followed is this.

(1) Evaluate the terms K, L, N and P, so that E' for each slice boundary and its derivatives can be found.
(2) Using the approximate integration formulae (A.27) at first, evaluate R and its derivatives for each slice boundary
(3) Compute the required changes to F and λ using equations (A.12) and (A.13).

Repeat steps (2) and (3) until suitably converged, and then switch to the use of the exact integration formulae. Final convergence is obtained either when the errors in E and R are suitably small, or when the computed changes to F and λ are below the threshold of importance.

A starting value for F of about 1.3 to 1.4, and for λ in the range $-0.3/f_{\max}$ to $-0.4/f_{\max}$ (where $f_{\max}$ is the largest absolute $f(x)$ value for the slide) will often give a speedy solution. However, either this initial combination or some combination that arises during the solution process can cause numerical problems to occur. These manifest themselves by a change in sign from positive to negative for L or $L + Kb$ in one or more slices, which implies the occurrence of a zero or (less commonly) infinite E' force within a slice or some similar peculiarity in the moments. Apart from the first iteration, the solution is to step back halfway from the current F and λ values to those which were satisfactory previously. Further stepping back, of half the remaining differences, may be required. The problem is symptomatic of having overshot the 'correct' F and λ values.

In the first iteration, the problem reflects a poor initial choice of F and λ. A so-called 'feasible region' may be defined within the range $F_{\min}$ to $F_{\max}$ where $F_{\min}$ is the maximum of

$$\frac{(A - \lambda f(x))\tan\phi}{1 + \lambda f(x)A}$$

subject to $1 + \lambda f(x)A < 0$, and $F_{\max}$ is the minimum of the function subject to $1 + \lambda f(x)A > 0$.

The choice of modified factor of safety with which to continue the solution, F, depends on the results of $F_{\max}$ and $F_{\min}$.

$$F_{\max} > F_{\min} + 0.1 \qquad\qquad F = F_{\min} + 0.1$$

$$F_{\min} + 0.1 > F_{\max} > E_{\min} \qquad F = \frac{F_{\max} + F_{\min}}{2}$$

$$F_{\max} \leqslant F_{\min} \qquad\qquad F = \text{maximum of } A\tan\phi + 0.1$$
$$\text{or } F_{\max} \text{ not found} \qquad\qquad\qquad\qquad\qquad\qquad (A.28)$$

During the iterative process, overshooting is a problem. Morgenstern and Price suggest two controls on this. Firstly, the maximum iterative step size is limited to 0.5; secondly, the weighted sum of the squares of the out-of-balance force (E'_n) and moments (R_n) at the head of the slide, $E'^2_n + D \cdot R^2_n$, must reduce if the process is genuinely converging. (D is a weighting factor to reduce the significance of moment terms which tend to be of much larger magnitude than the forces.) A suitable form for D is

$$D = \frac{\left(\dfrac{\partial E'_n}{\partial \lambda}\right)^2 + \left(\dfrac{\partial E'_n}{\partial F}\right)^2}{\left(\dfrac{\partial R_n}{\partial \lambda}\right)^2 + \left(\dfrac{\partial R_n}{\partial F}\right)^2} \tag{A.29}$$

This coefficient is re-evaluated only when the sum of squares has reduced, and in an iteration where it increases, the working F and λ values of the last two iterations are averaged for the next try. It will be found that the change from approximate to exact integration of the moment functions will sometimes lead to a slight increase in the residuals, and the check should not be carried out.

In particularly difficult cases, it may be necessary to hold λ constant, and to iterate with F only, computing ΔF from

$$\Delta F = - R/(\partial R/\partial F) \tag{A.30}$$

This brings F back to a value where the two-variable Newton procedure can rapidly converge.

Appendix B

B.1 Derivation of force and moment equilibrium equations in Maksumovic's method

Take first the three equilibrium conditions:

$$\sum H = 0$$

$$E'_i - E'_{i+1} - N'_i \cdot \sin \alpha_i + S_i \cdot \cos \alpha_i - H_i = 0 \tag{B.1}$$

$$\sum V = 0$$

$$X_i - X_{i+1} + N'_i \cdot \cos \alpha_i + S_i \cdot \sin \alpha_i - V_i = 0 \tag{B.2}$$

$$\sum R = 0$$

$$E'_{i+1}\left(r_i + \frac{b_i}{2}\cdot \tan \alpha_i\right) - E'_i\left(r_{i+1} - \frac{b_i}{2}\cdot \tan \alpha_i\right) - \frac{b_i}{2}(X_i + X_{i+1}) + M_i = 0 \tag{B.3}$$

together with the Mohr failure criterion

$$S_i = \frac{N'_i \cdot \tan \phi'}{F} + \frac{b_i c_i}{\cos \alpha_i \cdot F}. \tag{B.4}$$

Substituting for the shear force in each of the above:

$$E'_i - E'_{i+1} - N'_i \cdot \sin \alpha_i + \left[\frac{N'_i \cdot \tan \phi'}{F} + \frac{b_i c_i}{\cos \alpha_i \cdot F}\right]\cos \alpha_i - H_i = 0 \tag{B.5}$$

$$E'_i - E'_{i+1} - N'_i \cdot \sin \alpha_i + \frac{N'_i \cdot \tan \phi'}{F}\cdot \cos \alpha_i + \frac{b_i c_i}{F} - H_i = 0 \tag{B.6}$$

$$N'_i \cdot \cos \alpha_i\left(\frac{\tan \phi'}{F} - \tan \alpha_i\right) = H_i + (E'_{i+1} - E'_i) - \frac{b_i c_i}{F} \tag{B.7}$$

$$X_i - X_{i+1} + N'_i \cdot \cos \alpha_i + \left[\frac{N'_i \cdot \tan \phi'}{F} + \frac{b_i c_i}{\cos \alpha_i \cdot F}\right]\sin \alpha_i - V_i = 0 \tag{B.8}$$

$$N'_i\left[\cos \alpha_i + \frac{\tan \phi' \cdot \sin \alpha_i}{F}\right] = V_i + (X_{i+1} - X_i) - \frac{b_i c_i \cdot \tan \alpha_i}{F} \tag{B.9}$$

Making the Morgenstern and Price assumption for the interslice forces,

$$X_i = \lambda f_i \cdot E'_i \tag{B.10}$$

and substituting:

$$N_i' \cos \alpha_i \left[1 + \frac{\tan \phi' \cdot \tan \alpha_i}{F} \right] = V_i + \lambda(f_{i+1} \cdot E_{i+1}' - f_i \cdot E_i')$$
$$- \frac{b_i c_i \cdot \tan \alpha_i}{F} \qquad \text{(B.11)}$$

This then gives the opportunity to eliminate $N_i \cos \alpha_i$ between equations (B.11) and (B.7):

$$\frac{\dfrac{\tan \phi'}{F} - \tan \alpha_i}{1 + \dfrac{\tan \phi' \cdot \tan \alpha_i}{F}} = \frac{H_i + (E_{i+1}' - E_i') + \dfrac{b_i c_i}{F}}{V_i + \lambda(f_{i+1} \cdot E_{i+1}' - f_i \cdot E_i') - \dfrac{b_i c_i \tan \alpha_i}{F}} \qquad \text{(B.12)}$$

Let

$$a_i = \frac{\tan \phi'}{F} - \tan \alpha_i \bigg/ 1 + \frac{\tan \phi' \cdot \tan \alpha_i}{F} \qquad \text{(B.13)}$$

so that

$$a_i = \frac{H_i + (E_{i+1}' - E_i') - \dfrac{b_i c_i}{F}}{V_i + \lambda(f_{i+1} \cdot E_{i+1}' - f_i \cdot E_i') - \dfrac{b_i c_i \tan \alpha_i}{F}} \qquad \text{(B.14)}$$

$$V_i \cdot a_i + \lambda \cdot a_i(f_{i+1} \cdot E_{i+1}' - f_i \cdot E_i') - a_i \cdot \frac{b_i c_i \cdot \tan \alpha_i}{F}$$
$$= H_i + (E_{i+1} - E_i)\frac{b_i c_i}{F} \qquad \text{(B.15)}$$

$$E_{i+1}'(1 - \lambda \cdot f_{i+1} \cdot a_i) = E_i'(1 - \lambda \cdot f_i \cdot a_i) + \frac{b_i c_i}{F}$$
$$\times (1 - \tan \alpha_i \cdot a_i) + a_i V_i - H_i \qquad \text{(B.16)}$$

This is the force equilibrium equation.
The moment equilibrium equation is developed from equation (B.3):

$$E_{i+1}' \left(r_{i+1} + \frac{b_i \cdot \tan \alpha_i}{F} \right) = E_i' \left(r_i - \frac{b_i}{2} \tan \alpha_i \right)$$
$$+ \lambda \frac{b_i}{2}(f_i \cdot E_i' + f_{i+1} \cdot E_{i+1}') - M_i \qquad \text{(B.17)}$$

Rearranging terms and putting $R = Er$

$$R_{i+1} = R_i - E_{i+1}' \cdot \frac{b_i \cdot \tan \alpha_i}{2} + \lambda \frac{b_i}{2}(f_{i+1} \cdot E_{i+1}' + f_{i+1} \cdot E_i')$$
$$- M_i - E_i' \frac{b_i \cdot \tan \alpha_i}{2} \qquad \text{(B.18)}$$

$$R_{i+1} = R_i - \frac{b_i \cdot \tan \alpha_i}{2}(E'_{i+1} + E'_i) + \lambda \frac{b_i}{2}(f_i \cdot E'_i + f_{i+1} \cdot E'_{i+1}) - M_i \quad \text{(B.19)}$$

This is the moment equilibrium equation.

B.2 Differentiation of force and moment equilibrium equations

$$E'_{i+1}(1 - \lambda \cdot f_{i+1} \cdot a_i) = E'_i(1 - \lambda \cdot f_i \cdot a_i) + \frac{b_i c_i}{F}$$

$$\times (1 - \tan \alpha_i \cdot a_i) + a_i V_i - H_i$$

Differentiate with respect to λ:

$$\frac{\partial E'_{i+1}}{\partial \lambda}(1 - \lambda \cdot f_{i+1} \cdot a_i) - f_{i+1} \cdot a_i \cdot E'_{i+1} = \frac{\partial E'_i}{\partial \lambda}(1 - \lambda \cdot f_i \cdot a_i) - f_i \cdot a_i \cdot E'_i$$

Differentiate with respect to F:

where
$$a_i = \frac{\dfrac{\tan \phi'}{F} - \tan \alpha_i}{1 + \dfrac{\tan \alpha_i \tan \phi'}{F}} \quad \text{(B.20)}$$

$$\text{LHS} = \frac{\partial E'_{i+1}}{\partial F}\left(1 + \frac{\tan \alpha_i \tan \phi'}{F} - \frac{\lambda \cdot f_{i+1} \cdot \tan \phi'}{F} + \lambda \cdot f_{i+1} \cdot \tan \alpha_i\right)$$

$$+ E'_{i+1}\left(\frac{\tan \alpha_i \tan \phi'}{F^2} + \frac{\lambda f_{i+1} \cdot \tan \phi'}{F^2}\right) \quad \text{(B.21}a)$$

$$\text{RHS} = \frac{\partial E'_i}{\partial F}\left(1 + \frac{\tan \alpha_i \tan \phi'}{F} - \lambda \cdot f_i \cdot \frac{\tan \phi'}{F} + \lambda \cdot f_i \cdot \tan \alpha_i\right)$$

$$+ E'_i\left(\frac{\tan \alpha_i \cdot \tan \phi'}{F^2} + \frac{\lambda \cdot f_i \cdot \tan \phi'}{F^2}\right)$$

$$- \frac{b_i c_i}{F^2}(1 + \tan^2 \alpha_i) - \frac{\tan \phi'}{F^2}V_i + \frac{\tan \alpha_i \cdot \tan \phi'}{F^2}H_i \quad \text{(B.21}b)$$

$$\frac{\partial E'_{i+1}}{\partial F} = \frac{1}{1 + \dfrac{\tan \alpha_i \cdot \tan \phi'}{F} - \lambda \cdot f_{i+1}\left(\dfrac{\tan \phi'}{F} - \tan \alpha_i\right)}$$

$$\times \left\{\frac{\partial E'_i}{\partial F}\left[1 + \frac{\tan \alpha_i \cdot \tan \phi'}{F} - \lambda \cdot f_i\left(\frac{\tan \phi'}{F} - \tan \alpha_i\right)\right]\right.$$

$$\left. + E'_i\left(-\frac{\tan \alpha_i \cdot \tan \phi'}{F} + \frac{\lambda f_i \cdot \tan \phi'}{F^2}\right)\right.$$

$$- E'_{i+1}\left(-\frac{\tan\alpha_i \cdot \tan\phi'}{F} + \frac{\lambda f_{i+1} \cdot \tan\phi'}{F^2}\right)$$

$$\left. -\frac{b_i c_i}{F^2}(1+\tan^2\alpha_i) - \frac{\tan\phi'}{F^2}V_i + \frac{\tan\alpha_i \cdot \tan\phi'}{F^2}H_i\right\} \quad \text{(B.22)}$$

$$R_{i+1} = R_i - \frac{b_i \cdot \tan\alpha_i}{2}(E'_{i+1} + E'_i) + \frac{\lambda b_i}{2}(f_i \cdot E'_i + f_{i+1} \cdot E'_{i+1}) - M_i$$

Differentiate with respect to λ:

$$\frac{\partial R_{i+1}}{\partial\lambda} = \frac{\partial R_i}{\partial\lambda} - \frac{b_i \cdot \tan\alpha_i}{2}\left(\frac{\partial E'_{i+1}}{\partial\lambda} + \frac{\partial E'_i}{\partial\lambda}\right)$$

$$+ \frac{\lambda b_i}{2}\left(f_i\frac{\partial E'_i}{\partial\lambda} + f_{i+1}\frac{\partial E'_{i+1}}{\partial\lambda}\right) + \frac{b_i}{2}(f_i \cdot E'_i + f_{i+1} \cdot E'_{i+1}) \quad \text{(B.23)}$$

Differentiate with respect to F:

$$\frac{\partial R_{i+1}}{\partial F} = \frac{\partial R_i}{\partial F} - \frac{b_i \cdot \tan\alpha_i}{2}\left(\frac{\partial E'_{i+1}}{\partial F} + \frac{\partial E'_i}{\partial F}\right)$$

$$+ \frac{\lambda b_i}{2}\left(f_i\frac{\partial E'_i}{\partial F} + f_{i+1}\frac{\partial E'_{i+1}}{\partial F}\right) \quad \text{(B.24)}$$

Appendix C

Displacement of a slide under a single acceleration pulse

The acceleration is dependent on the net force acting on the slide, hence

$$\frac{d^2 s}{dt^2} = g \cdot \frac{\cos(\phi' - \beta)}{\cos \phi'}(k - k_{crit}) \tag{B.1}$$

Since k is a function of time, this is integrable, as follows.

Case (i) Rectangular pulse of duration $T/2$ and amplitude $k_{max}g$

$$S_{max} = \frac{1}{8}\left(\frac{k_{max}}{k_{crit}} - 1\right)\frac{\cos(\phi' - \beta)}{\cos \phi'}k_{max}gT^2 \tag{B.2}$$

or

$$\left(\frac{k_{max}}{k_{crit}} - 1\right)\frac{Q}{8}$$

Case (ii) Triangular pulse of duration $T/2$ and amplitude $k_{max}g$

$$S_{max} = \left(\frac{k_{max}}{k_{crit}}\right)\left\{4\left(1 - \frac{k_{crit}}{k_{max}}\right)\left(1 - \lambda\frac{k_{crit}}{k_{max}}\right)\right.$$
$$\left. - \left(1 - \lambda\left(\frac{k_{crit}}{k_{max}}\right)^2\right)\right\}\frac{Q}{96} \tag{B.3}$$

for
$$0 \leqslant \frac{k_{crit}}{k_{max}} \leqslant \{1 - \sqrt{1 - \lambda}\}/\lambda$$

and

$$S_{max} = \left\{\left(1 - \frac{k_{crit}}{k_{max}}\right)^3(2 - 2\sqrt{1 - \lambda} - \lambda)\right\}\frac{Q}{24}$$

for
$$\{1 - \sqrt{1 - \lambda}\}/\lambda \leqslant \frac{k_{crit}}{k_{max}} \leqslant 1 \tag{B.4}$$

Case (iii) Half-sine pulse of duration $T/2$ and amplitude $k_{max}g$

$$S_{max} = \left\{ \frac{k_{crit}}{k_{max}} - \pi + \cos^2(\alpha/2)\cot(\alpha/2) \right\} \frac{Q}{4} \qquad \text{(B.5)}$$

for
$$0 \leqslant \frac{k_{crit}}{k_{max}} \leqslant 0.725$$

and

$$S_{max} = \left(\frac{k_{crit}}{k_{max}} - \sin q \right)^2 \cdot \frac{Q}{\left(8\pi^2 \dfrac{k_{crit}}{k_{max}} \right)} \qquad \text{(B.6)}$$

for
$$0.725 \leqslant \frac{k_{crit}}{k_{max}} \leqslant 1$$

in which

$$\alpha = \sin^{-1}\left(\frac{k_{crit}}{k_{max}} \right)$$

$$q = \alpha + \frac{k_{crit}}{k_{max}}(\cos \alpha - \cos q)$$

The author is indebted to Dr. S. K. Sarma for the original derivation of these formulae.

Appendix D

Conferences and symposia in slope stability and related disciplines

A.1 International Conferences in Soil Mechanics and Foundation Engineering

No	Date	Location	No	Date	Location
1	1936	Cambridge, Mass.	7	1969	Mexico
2	1948	Rotterdam	8	1973	Moscow
3	1953	Zurich	9	1977	Tokyo
4	1957	London	10	1981	Stockholm
5	1961	Paris	11	1985	San Francisco
6	1965	Montreal			

A.2 European Conferences in Soil Mechanics and Foundation Engineering

No	Date	Location	No	Date	Location
1	1954	Stockholm	5	1972	Madrid
2	1958	Brussels	6	1976	Vienna
3	1963	Wiesbaden	7	1979	Brighton
4	1967	Oslo	8	1982	

A.3 Regional Meetings of the Engineering Group of the Geological Society of London

No	Date	Location	Subject
1	1965	Sheffield	Rock mechanics
2	1966	Newcastle	Coal measures: mining and construction
3	1967	Cardiff	Glacial deposits
4	1968	Cambourne	Engineering geology in SW England
5	1969	Strathclyde	Engineering geology with respect to regional planning
6	1970	Dublin	Engineering geology
7	1971	Leeds	Route planning, design construction
8	1972	Bristol	Engineering geology of slopes
9	1973	Durham	Engineering geology of reclamation redevelopment
10	1974	Liverpool	Engineering geology of tunnels
11	1975	Norwich	Foundations on Quaternary deposits
12	1976	Exeter	Methods of investigation in engineering geology
13	1977	Cardiff	Engineering geology of soluble rocks
14	1978	Southampton	Coastal engineering geology
15	1979	Newcastle	Engineering geological mapping
16	1980	Kingston	Engineering geology applied to construction

17	1981	Bangor	Engineering geology of construction materials
18	1982	Birmingham	Engineering implications of earth surface processes: engineering geomorphology
19	1983	Hull	Engineering geology of tidal rivers
20	1984	Guildford	Site investigation practice
21	1985	Sheffield	Engineering geology and groundwater

A.4 International Symposium on Landslides

No	Date	Location
1	1972	Japan
2	1977	Japan
3	1980	New Delhi, India
4	1984	Toronto, Canada

A.5 Other conferences

International Congress on Large Dams

A.6 Journals

Bulletin of the International Association of Engineering Geology
Canadian Geotechnical Journal
Géotechnique
Ground Engineering
Proceedings, Institution of Civil Engineers
Proceedings, American Society of Civil Engineers
Quarterly Journal of Engineering Geology

References

Al-Dhahir, Z.A., Kennard, M.F. Morgenstern, N.R. (1970) Observations on porewater pressures beneath the ash lagoon embankments at Fiddler's Ferry power station, *Proc. Conf. on In-situ Testing of Soils and Rocks*, Institution of Civil Engineers, London, 265–276.

Allan, A.R. (1968) Coast protection and stabilisation at Herne Bay. *Civil Engineering & Public Works Review* **63**, 860–861.

Arber, M.A. (1941) The coastal landslips of West Dorset. *Proc. Geol. Assoc.* **52**, 273–283.

Ambraseys, N. Sarma, S.K. (1967) The response of earth dams to strong earthquakes. *Géotechnique* **17**, 181–213.

Atkinson, J.H. (1981) Foundations and Slopes. McGraw-Hill.

Ayres, D.J. (1961) The treatment of unstable slopes and railway track formations. *J. Trans. Soc. Engrs.* **52**, 111–138.

Ayres, D.J. (1985) Treatment of shallow underground fires. Technical Note 23, *Proc. Int. Symp. on Failures in Earthworks*, Institution of Civil Engineers, London.

Ayres, D.J. (1985) Stabilisation of slips in cohesive soil by grouting. Technical Note 9, *Proc. Int. Symp. on Failures in Earthworks*, Institution of Civil Engineers, London.

Bagnold (1954, 1956) Quoted by Hsu (1978) (q.v.)

Barron, R.A. (1948) Consolidation of fine-grained soil by drain wells. *Trans. ASCE* **113**, 718–754.

Barratt, M.J. (1985) Isle of Wight—Shoreline erosion and protection. *Proc. Conf. on Problems associated with the Coastline*, Isle of Wight County Council.

Barton, M.E. (1973) The degradation of the Barton Clay cliffs of Hampshire. *Q. J. Eng. Geol.* **6**, 423–440.

Barton, M.E. (1977) Landsliding along bedding planes. *Bull. Int. Assoc. Eng. Geol.* **16**, 5–7.

Barton, M.E. Coles, B.J. (1984) The characteristics and rates of the various slope degradation processes in the Barton Clay cliffs of Hampshire. *Q. J. Eng. Geol.* **17**, 117–136.

Bayley, M.J. (1971) Cliff stability at Herne Bay. *Civil Engineering and Public Works Review* **67**, 788–792.

Beles, A.A. Stanculescu, I.I (1958) Thermal treatment as a means of improving the stability of earth masses. *Géotechnique* **8**, 158–165.

Bell, J.M. (1968) General slope stability analysis. *Proc. ASCE J. Soil Mechanics & Foundations Div.* **94**, 1253–1270.

Berkeley-Thorne, R. Roberts, A. (1981) *Coastal Defence and Marine Works*. Thomas Telford, London.

Biczysko, S (1981) Relic landslip in West Northamptonshire. *Q. J. Eng. Geol.* **14**, 169–174.

Binger, W.V. (1948) Analytical studies of the Panama Canal slides. *Proc. 2nd Int. Conf. on soil Mechanics & Foundation Engineering*, Rotterdam, **2**, 54–60.

Binnie, G.M., Gerrard, R.T. Eldridge, J.G., Kirmani, S.S., Davis, C.V., Dickinson, J.C., Gwyther, J.R., Thomas, A.R., Little, A.L., Clark, J.F.F. Sneddon, B.T. (1967) Mangla: Part 1, Engineering of Mangla. *Proc. Inst. Civ. Engrs* **38**, 343–544, but see especially §§157–161, pp. 398–400.

Bishop, A.W. (1955) The use of the slip circle in the stability analysis of earth slopes. *Géotechnique* **5**, 7–17.

Bishop, A.W. (1967) Progressive failure—with special reference to the mechanism causing it. Panel discussion. *Proc. Geotech. Conf. Oslo* **2**, 142.

Bishop, A.W., Green, G.E., Garga, V.K., Andresen, A. Brown, J.D. (1971) A new ring shear apparatus and its application to the measurement of residual strength. *Géotechnique* **21**, 273–328.

Bishop, A.W. (1971) The influence of progressive failure on the choice of the method of stability analysis. Technical Note, *Géotechnique* **21**, 168–172.

Bishop, A.W., Hutchinson, J.N., Penman, A.D.M. Evans, H.E. (1969) *Geotechnical Investigations into the Causes and Circumstances of the Disaster of 21st October, 1966.* A selection of the technical reports submitted to the Aberfan Tribunal, Welsh Office. HMSO London.

Bishop, A.W. Morgenstern, N. (1960) Stability coefficients for earth slopes. *Géotechnique* **10**, 129–150.

Bishop, A.W. (1973) The stability of tips and spoil heaps. *Q. J. Eng. Geol.* **6**, 335–376.

Bishop, A.W. Bjerrum, L. (1960) The relevance of the triaxial test to the solution of stability problems. *Proc. Res. Conf. on the shear strength of cohesive soils, Denver, Colorado* (Special Publ. American Society of Civil Engineers) 439–501.

Bishop, A.W. (1973) The influence of an undrained change in stress on the pore pressure in porous media of low compressibility. Technical Note, *Géotechnique* **23**, 435–442.

Bishop, A.W. Wesley, L.D. (1974) A hydraulic triaxial apparatus for controlled stress path testing. *Géotechnique* **24**, 657–670.

Bishop, A.W. Henkel, D.J. (1957, 1971) *The Measurement of Soil Properties in the Triaxial Test* (edns. 1 & 2). Edward Arnold, London.

Bjerrum, L. (1954) Stability of natural slopes in quick clay. *Proc. Eur. Conf. on Stability of Earth Slopes*, Stockholm, **2**, 16–40; and *Géotechnique* **5**, 101–119.

Bjerrum, L. (1971) *Subaqueous Slope Failures in Norweigian Fjords.* NGI Publication 88.

Bromhead, E.N. (1972) A large coastal landslide at Herne Bay, Kent. *The Arup Journal* (Ove Arup & Partners, London) **7**, 12–14.

Bromhead, E.N. (1978) Large landslides in London Clay at Herne Bay, Kent. *Q. J. Eng. Geol.* **11**, 291–304

Bromhead, E.N. (1979) Factors affecting the transition between the various types of mass movement in coastal cliffs consisting largely of overconsolidated clay, with special reference to Southern England. *Q. J. Eng. Geol.* **12**, 291–300.

Bromhead, E.N. (1979) A simple ring shear apparatus. *Ground Engineering* **12**, 5, 40–44.

Bromhead, E.N. Vaughan, P.R. (1980) Solutions for seepage in soils with an effective stress dependent permeability. *Proc. 1st Int. Conf. on Numerical Methods for Non-linear Problems*, Swansea. Pineridge Press, 567–578.

Bromhead, E.N. Curtis, R.D. (1983) A study of alternative methods for measuring the residual strength of London Clay. *Ground Engineering* **16**, 39–41.

Bromhead, E.N. (1984) Stability of slopes and embankments. In Chapter 3 in *Ground Movements Their Effects on Structures*, R.K. Taylor, P.B. Attewell, eds., Surrey University Press.

Bromhead, E.N. Dixon, N. (1984) Pore water pressure observations in the coastal clay cliffs at the Isle of Sheppey, England. *Proc. 4th Int. Symp. on Landslides*, Toronto **1**, 385–390.

Bromhead, E.N. (1984) An analytical solution to the problem of seepage into counterfort drains. *Can. Geotech. J.* **21**, 657–662.

Bromhead, E.N., Chandler, M.P. Hutchinson, J.N. (1985) Coast erosion and landsliding at Gore Cliff, Isle of Wight. (in prep).

Bromhead, E.N., Havouzari, A.L., Brice, M. Rofe, B.R. (1985) Groundwater modelling by microcomputer—applications to dam and reservoir slope stability. *Conf. on Groundwater in Engineering Geology, Sheffield.*

Brunsden, D. Jones, D.K.C. (1976) The evolution of landslide slopes in Dorset. *Phil. Trans. Roy. Soc. London* **A283**, 605–631.

Brunsden, D. Jones, D.K.C. (1972). The morphology of degraded landslide slopes in South West Dorset. *Q. J. Eng. Geol.* **5**, 205–222.

Butterfield, R., Harkness, R.M., Andrawes, K.Z. (1970). A stereo-photogrammetric method for measuring displacement fields. *Géotechnique* **20**, 308–314.

Byce, J. (1984) Rapid tunnelling solves Utah's landslide crisis. *Tunnels & Tunnelling*, September 1984, 15–17.

Cedergren, H.R. (1967) *Seepage, Drainage and Flow Nets.* John Wiley, New York.

Chandler, M.P. Hutchinson, J.N. (1984) Assessment of relative slide hazard within a large, pre-existing coastal landslide at Ventnor, Isle of Wight. *Proc. 4th Int. Symp. on Landslides, Toronto*, **2**, 517–522.

Chandler, M.P., Parker, D.C. Selby, M.J. (1981) An open-sided field direct shear box. British Geomorphological Research Group Technical Bulletin 27.

Chandler, R.J. (1966) The measurement of residual strength in triaxial compression. *Géotechnique* **16**, 181–186.

Chandler, R.J. (1970) The degradation of Lias Clay slopes in an area of the East Midlands. *Q. J. Eng. Geol.* **2**, 161–181.

Chandler, R.J. (1970) Solifluction on low-angled slopes in Northamptonshire. *Q. J. Eng. Geol.* **3**, 65–69.

Chandler, R.J. (1970) A shallow slab slide in the Lias clay near Uppingham, Rutland. *Géotechnique* **20**, 253–260.

Chandler, R.J. (1972) Periglacial mudslides in Vestspitzbergen and their bearing on the origin of fossil 'solifluction' shears in low angled clay slopes. *Q. J. Eng. Geol.* **5**, 223–241.

Chandler, R.J. (1974) Lias Clay: the long-term stability of cutting slopes. *Géotechnique* **24**, 21-38.

Chandler, R.J., Pachakis, M.J., Mercer, J. Wrightman, J. (1974) Four long-term failures of embankments founded on areas of landslip. *Q.J. Eng. Geol.* **6**, 405–422.

Chandler, R.J. Skempton, A.W. (1974) The design of permanent cutting slopes in stiff fissured clays. *Géotechnique* **24**, 457–466.

Chandler, R.J. (1977) Back analysis techniques for slope stabilisation works: a case record. *Géotechnique* **27**, 479–495.

Chandler, R.J. (1979) Stability of a structure constructed in a landslide. *Proc. 7th Eur. Conf. on Soil Mechanics & Foundation Engineering, Brighton*, **3**, 175–182.

Chandler, R.J. (1984) Delayed failure and observed strengths of first time slides in stiff clay: a review. *Proc. 4th Int. Symp. on Landslides, Toronto*, **2**, 19–25.

Chandler, R.J. (1984) Recent European experience of landslides in overconsolidated clays and soft rocks. *Proc. 4th Int. Symp. on Landslides, Toronto*, **1**, 61–81.

Chen, Z.Y. Morgenstern, N.R. (1983) Extensions to the generalized method of slices for slope stability analysis. *Can. Geotech. J.* **20**, 104–119.

Clayton, C.R.I. Matthews, M. (1984) The use of oblique aerial photography to investigate the extent and sequence of landslipping at Stag Hill, Guildford, Surrey. In *Proc. Conf. Site Investigation Practice: Assessing BS 5930*. University of Surrey, 319–330.

Clayton, C.R., Simons, N.E. Matthews, M. (1982) *Site Investigation*. Granada.

Collin, A. (1846) Glissements spontanes des terrains argileux. Paris, Carilian-Goeury et Dulmont. (English translation by Schriever, University of Toronto Press, 1956).

Crabb, G.I. West, G. (1985) Monitoring pore water pressures in an embankment slope. Technical Note 4, *Proc. Int. Symp. on Failures in Earthworks*, Institution of Civil Engineers, London.

Darcy, (1856) *Les fontaines publiques de la Ville de Dijon*. V. Dalmont, Paris.

D'Appolonia, E., Alperstein, R. and D'Appolonia, D.J. (1967) Behaviour of a colluvial slope. *Proc. ASCE, J. Soil Mechanics & Foundations Div.* **93**, 447–473.

Dearman, W.R., (1974) Presentation of information on engineering geological maps and plans. *Q. J. Eng. Geol.* **7**, 317–320.

Dearman, W.R. Fookes, P.G. (1974) Engineering geological mapping for civil engineering practice in the United Kingdom. *Q. J. Eng. Geol.* **7**, 223–256.

De Freitas, M. and Watters, R. (1973) Some field examples of toppling failure. *Géotechnique* **23**, 495–513.

Dumbleton, M.J. West, G. (1970) Air photograph interpretation for road engineers in Britain. Road Research Laboratory Report LR369, Dept. of the Environment, London.

Dumbleton, M.J. West, G. (1971) Preliminary sources of information for site investigations in Britain. Road Research Laboratory Report LR401, Dept. of the Environment, London.

Dumbleton, M.J. West, G. (1972). Preliminary sources of site information for roads in Britain. *Q. J. Eng. Geol.* **5**, 7–14.

Dunbaven, M. (1983) Discussion on R.S. Evans (1981) *Q. J. Eng. Geol.* **16**, 243–245.

Duvivier, J. (1940) Cliff stabilisation works in London Clay. *J. Inst. Civ. Engrs* **14**, paper 5233.

Early, K.R. Skempton, A.W. (1972) Investigations of the landslide at Walton's Wood. *Q. J. Eng. Geol.* **5**, 19–41.

Eckel, E.B. (1958) (ed.) Landslides in engineering practice. Highway Research Board, Special Publ. 29.

Ellis I.W. (1985) The use of reticulated pali radice structures to solve slope stability problems. Technical Note 11, *Proc. Int. Symp. on Failures in Earthworks*, Institution of Civil Engineers, London.

Eigenbrod, K.D. (1975) Analysis of the pore pressure changes following the excavation of a slope. *Can. Geotech. J.* **12**, 429–440.

Evans, R.S. (1981) An analysis of secondary toppling rock failures—the stress redistribution method. *Q. J. Eng. Goel.* **14**, 77–86.

Fellenius, W. (1936) Calculation of the stability of earth dams. *Trans. 2nd Congr. on Large Dams, Washington*, **4**, 445–459.

Fookes, P.G. Sweeney, M. (1976) Stabilisation and control of local rockfalls and degrading rock slopes. *Q. J. Eng. Geol.* **9**, 37–56.

Forster, A. Northmore, K.J. (1985) Landslide distribution in the South Wales Coalfield. *Proc. Symp. Landslides in the South Wales Coalfield*, Polytechnic of Wales.

Franklin, J.A. Denton, P.E. (1973) Monitoring for rock slopes. *Q. J. Eng. Geol.* **6**, 259–286.

Freeze, R.A. and Witherspoon, P,A. (1966) Theoretical analysis of regional groundwater flow: 1— analytical and numerical solutions to the mathematical model. *Water Resources Research* **2**, 641–656.

Freeze, R.A. and Witherspoon, P.A. (1967) Theoretical analysis of regional groundwater flow: 2— effect of water table configuration and subsurface permeability variations. *Water Resources Research* **3**, 623–634.

Freeze, R.A. and Witherspoon, P.A. (1968) Theoretical analysis of regional groundwater flow: 3— Quantitative interpretations. *Water Resources Research* **4**, 581–590.

Geotechnical Control Office (GCO), Hong Kong (1982) Final report on comparison and checking of computer/calculator programs for slope stability analysis. Geotechnical Control Office, Engineering Development Department, Hong Kong.

Gibson, R.E. Morgenstern, N.R. (1962) A note on the stability of cuttings in normally consolidated clays. *Geotechnique* **12**, 212–216.

Gibson, R.E. and Shefford, G.C. (1968) The efficiency of horizontal drainage layers for accelerating consolidation of clay embankments. *Géotechnique* **18**, 327–335.

Gibson, R.E. Henkel, D.J. (1954) Influence of duration of tests at constant rate of strain on measured 'drained' strength. *Géotechnique* **4**, 6–15.

Goodman, R.E. Bray, J.W. (1976) Toppling of rock slopes. *Proc. Specialty Conf. on Rock Engineering for Foundations and Slopes, Boulder, Colorado*, American Society of Civil Engineers, **2**.

Gregory, C.H. (1844) On railway cuttings and embankments; with an account of some slips in the London Clay, on the line of the London and Croydon Railway. *Minutes, Proc. Inst. Civ. Engrs* **3**, 135–145.

Haefeli, R. (1948) The stability of slopes acted upon by parallel seepage. *Proc. 2nd Int. Conf. on Soil Mechanics, Rotterdam*, **1**, 57–62.

Hanna, T.H. (1973) *Foundation Instrumentation*. Trans Tech Publications, Cleveland.

Harr, M.E. (1962) *Groundwater and Seepage*. McGraw-Hill, New York.

Hawkins, A.B. (1973) The geology and slopes of the Bristol Region. *Q. J. Eng. Geol.* **6**, 185–206.

Hawkins, A.B. (1977) The Hedgemead landslip, Bath. In *Large Ground Movements and Structures*, J.D. Geddes, ed. Pentech Press.

Hawkins, A.B. Privett, K. (1978) A building site on cambered ground at Radstock, Avon. *Q. J. Eng. Geology* **14**, 151–168.

Head, K.H. (1980, 1981, 1985) *Manual of Soil Laboratory Testing*. Vols. 1, 2 & 3, Pentech Press.

Heim, A. (1881, 1932) Cited by Hsu, (1978) (q.v.).

Henkel, D.J. (1956) Discussion on Watson (1956) *Proc. Inst. Civ. Engrs* **5**, 320–323.

Henkel, D.J. (1957) Investigations of two long term failures in London Clay slopes at Wood Green and Northolt. *Proc. 4th Int. Conf. on Soil Mechanics & Foundation Engineering, London*, **2**, 315–320.

Henkel, D.J. (1960) The shear strength of saturated remoulded clays. *Proc. Res. Conf. on the shear strength of cohesive soils, Boulder, Colorado*.

Henkel, D.J. (1967) Local geology and the stability of natural slopes. *Proc. ASCE, J. Soil Mechanics Div.*, **4**, 437–446.

Henkel, D.J. Skempton, A.W. (1954) A landslide at Jackfield, Shropshire, in an overconsolidated clay. *Proc. Eur. on Stability of Earth Slopes, Stockholm*, **1**, 90–101, and *Géotechnique*, **5**, 131–137.

Higginbottom, I.E. Fookes, P.G. (1970) Engineering aspects of periglacial features in Britain. *Q. J. Eng. Geol.* **3**, 85–118.

Hodge, R.A.L. Freeze, R.A. (1977) Groundwater flow systems and slope stability. *Can. Geotech. J.* **14**, 455–476.

Hoek, E. (1973) Methods for the rapid assessment of the stability of three dimensional rock slopes. *Q. J. Eng. Geol.* **6**, 243–255.

Hoek, E. Bray, J.M. (1974) *Rock Slope Engineering*. Inst. of Mining Metallurgy, London.

Hoek, E. Brown, E. (1974) *Large Underground Openings in Rock*. Inst. of Mining Metallurgy, London.

Hoek, E. Franklin, J.A. (1968) A simple triaxial cell for field and laboratory testing of rock. *Trans. Inst. Min. Metall. London*, **A77**, 22–26.

Hollingsworth, S.E., Taylor, J.H. Kellaway, G.A. (1944) Large scale structures in the Northamptonshire Ironstone Field. *Q. J. Geol. Soc. London* **100**, 1–44.

Holmes, A. (1978) *Principles of Physical Geology*. 3rd edn., Thomas Nelson.

Horswill, P. Horton, A. (1976) Cambering and valley bulging in the Gwash valley at Empingham. *Phil. Trans. Royal. Soc. London* **A283**, 427–462.

Hsu, K.J. (1978) Albert Heim: Observations on landslides and relevance to modern interpretations. In *Rockslides and Avalanches*, Vol. 1, B. Voight, ed. 70–93.

Hutchinson, J.N. (1961) A landslide on a thin layer of quick clay at Furre, Central Norway. *Géotechnique* **11**, 69–94.

Hutchinson, J.N. (1967) The free degradation of London Clay cliffs. *Proc. Geotech. Conf. on Shear Strength of Natural Soils and Rocks*. NGI, Oslo, **1**, 113–118.

Hutchinson, J.N. (1967) Mass movement. In *Encyclopaedia of Geomorphology*, R.W. Fairbridge, ed. Reinhold, New York, 668–696.

Hutchinson, J.N. (1969) A reconsideration of the coastal landslides at Folkestone Warren, Kent. *Géotechnique* **19**, 6–38.

Hutchinson, J.N. (1970) A coastal mudflow on the London Clay cliffs at Beltinge, North Kent. *Géotechnique* **20**, 412–438.

Hutchinson, J.N. (1973) The response of London Clay cliffs to differing rates of toe erosion. *Geologia Applicata e Idrogeologia* **8**, 221–239.

Hutchinson, J.N. Bhandari, R.K. (1971) Undrained loading, a fundamental mechanism of mudflows and other mass movements. *Géotechnique* **21**, 353–358.

Hutchinson, J.N., Somerville, S., Pettley, D.J. (1973) A landslide in periglacially disturbed Etruria Marl at Bury Hill, Staffordshire. *Q. J. Eng. Geol.* **6**, 377–404.

Hutchinson, J.N., Prior, D.B. Stevens, N. (1974) Potentially dangerous surges in an Antrim mudslide. *Q. J. Eng. Geol.* **7**, 363–376.

Hutchinson, J.N. Gostelow, T.P. (1976) The development of an abandoned cliff in London Clay at Hadleigh, Essex. *Phil. Trans. Roy. Soc. London* **A 283**, 557–604.

Hutchinson, J.N. (1977) Assessment of the effectiveness of corrective measures in relation to geological conditions and types of slope movement. *Bull. Int. Assoc. Eng. Geol.* **16**, 131–155.

Hutchinson, J.N., Bromhead, E.N. Lupini, J.F. (1980) Additional observations on the Folkestone Warren landslides. *Q. J. Eng. Geol.* **13**, 1–31.

Hutchinson, J.N., Chandler, M.P., Bromhead, E.N. (1981) Cliff recession on the Isle of Wight S-W coast. *Proc. 10th Int. Conf. on Soil Mechanics & Foundation Engineering, Stockholm* 3/12, 429–434.

Hutchinson, J.N. (1983) Methods of locating slip surfaces in landslides. *Bull. Assoc. Eng. Geol.* **XX**, 3, 235–252.

Hutchinson, J.N. (1984) Landslides in Britain and their countermeasures. *J. Japan Landslide Soc.* **21**, 1–25.

Hutchinson, J.N. Chandler, M.P., Bromhead, E.N. (1985) A review of current research on the coastal landslides forming the Undercliff of the Isle of Wight, with some practical implications. *Proc. Conf. on Problems associated with the Coastline*, Isle of Wight County Council.

Hutchinson, J.N., Bromhead, E.N., Chandler, M.P. (1985) Investigation of the coastal landslides at St Catherine's Point, Isle of Wight (In prep.)

Hutchinson, J.N., Poole, C., Lambert, N. Bromhead, E.N. (1985) Combined archaeological and geotechnical investigations of the Roman Fort at Lympne. Brittannia, in press.

Jakobson, B, (1952) The landslide at Surte on the Gota River. *Roy. Swedish Geotech. Inst. Proc.* **5**, 1–121.

Janbu, N. (1957) Stability analysis of slopes with dimensionless parameters. Harvard University Soil Mechanics Series, No. 46.

Janbu, N. (1973) Slope stability computations. In *Embankment Dam Engineering, Casagrande Memorial Volume*, eds. Hirschfield & Poulos. John Wiley, New York, 47–86.

Janbu, N., Bjerrum, L. Kjaernsli, B. (1956) Soil mechanics applied to some engineering problems. Norwegian Geotechnique Institute Publication 16.

Kellaway, G.A. Taylor, J.H. (1968) The influence of landslipping on the development of the city of Bath, England. *Proc. 23rd Int. Geol. Conf.* **12**, 65–76.

Kalaugher, P.G. (1984) Stereoscopic fusion of a photographic image with a directly observed view: applications to geomorphology. *Z. Geomorph. N.F.*, Suppl.-Bd. **51**, 123–140.

Kealy, C.D. Busch, R.A. (1971) Determining seepage characteristics of mill-tailings dams by the finite element method, US Bureau of Mines Report R17477.

Kenney, T.C. (1956) An examination of the methods of calculating the stability of slopes. MSc Thesis, Imperial College.

Kenney, T.C. (1964) Pore pressures and the bearing capacity of layered clays. *Proc. ASCE, J. Soil Mechanics & Foundations Div.* **90**, 27–55.

Kiersch, G.A. (1964) Vaiont reservoir disaster. *Civil Engineering*, ASCE, March 1964, 32–39.

Knox, G. (1939) Landslides in the South Wales Valleys. *Proc. South Wales Inst. Engrs* **43**, 161–233.

Kojan, E. Hutchinson, J.N. (1978) Mayunmarca rockslide and debris flow. *Rockslides and Avalanches.* Vol. 1, B. Voight, ed., 315–364.

Lafleur, J. Lefebvre, G. (1980). Groundwater regime associated with slope stability in Champian Clay deposits. *Can. Geotech. J.* **17**, 44–53.

Lapworth, H. (1911) The geology of Dam Trenches. *Trans. Inst. Water Engrs*, 1–26.

Leadbeater, A.D. (1985) A57 Snake Pass, remedial work to slip near Alport Bridge. *Proc. Int. Symp. on Failures in Earthworks*, Institution of Civil Engineers, London, Paper 3, 29–38.

Little, A.L. Price, V.E. (1958) The use of an electronic computer for slope stability computation. *Géotechnique* **8**, 113–120.

Littlejohn, G.S. (1970) Soil Anchors. *Proc. Conf. on Ground Engineering.* Institution of Civil Engineers, London, 33–44.

Littlejohn, G.S. (1980) Design estimation of the ultimate carrying capacity of ground anchors. *Ground Engineering* **13**, 25–39.

Lowe, J. Karafiath, L. (1960) Stability of earth dams upon drawdown. In *Proc. 1st Pan-Am. Conf. Soil. Mech. Foundation Engng (Mexico)*, **2**, 537–560.

Lupini, J.F., Skinner, A.E. Vaughan, P.R. (1981) The drained residual strength of cohesive soils. *Géotechnique* **31**, 181–214.

Lutton, R.J., Banks, D.C. Strohm, W.E. Jr. (1979) Slides in Gaillard Cut, Panama Canal Zone. In *Rockslides and Avalanches*, Vol. 2, B. Voight, ed., Elsevier, Amsterdam, Chapter 4.

McKecknie Thompson, G. Rodin, S. (1972) Colliery spoil tips—after Aberfan. *Proc. Inst. Civil Engrs*, paper 7522.

Maksimovic, M. (1970) A new method of slope stability analysis. Private communication.

de Mello, V.F.B. (1977) Reflections on design decisions of practical significance to embankment dams. *Géotechnique* **23**, 281–355.

de Mello, V.F.B. (1982) *A Case History of a Major Construction Period Dam Failure.* Amici et Alumni Em. Prof E de Beer, 1040 Bruxelles, Belgie, 63–78

Moran, Proctor, Mueser & Routlege, Consulting Engineers (1955). Study of deep soil stabilization by vertical sand drains. Report to the Bureau of Yards & Docks, Department of the Navy, Washington, D.C.

Morgan, C.S. (1985) Field trip notes. *Proc. Symp. Landslides in the South Wales Coalfield*, Polytechnic of Wales.

Morgenstern, N.R. (1963) Stability charts for earth slopes during rapid drawdown. *Géotechnique* **13**, 121–131.

Morgenstern, N.R. Price, V.E. (1965) The analysis of the stability of general slip surfaces. *Géotechnique* **15**, 79–93.

Morgenstern, N.R. Price, V.E. (1967) A numerical method for solving the equations of stability of general slip surfaces. *Computer Journal* **9**, 388–393.

Morgenstern, N.R. Guther, H. (1972) Seepage into excavations in a medium possessing stress dependent permeability. *Proc. Symp. on Percolation through Fissured Rock*, International Association Rock Mechanics & International Association for Engineering, Geology, Stuttgart, T2-C, 1–15.

Murray, B., Malin, M.C. Greeley, R. (1981) *Earthlike Planets: Surfaces of Mercury, Venus, Earth, Moon, Mars.* Freeman, San Francisco.

Naylor, D.J. (1974) Stresses in nearly incompressible material by finite elements with application to the calculation of excess porewater pressure. *Int. J. For Numerical Methods in Engineering* **8**, 443–460.

Neuman, S.P. Witherspoon, P.A. (1971) Analysis of non-steady flow with a free surface using the finite element method. *J. Water Resources* **7**, 611–623.

Newbery, J. Baker, D.A. (1981) Stability of cuts on the M4 north of Cardiff. *Q. J. Eng. Geol.* **14**, 195–206.

Newmark, N.M. (1965) Effects of earthquakes on dams and embankments. *Géotechnique* **15**, 139–159.

O'Connor, M.J. Mitchell, R.J. (1977) An extension to the Bishop and Morgenstern slope stability charts. *Can. Geotech. J.* **14**, 144–151.

Penman, A.D.M. (1985) The failure of Acu Dam. Technical Note 6. *Proc. Int. Symp. on Failures in Earthworks*, Institution of Civil Engineers, London.

Penman, A.D.M. (1985) Notes to introduce an informal discussion on tailings dams. Institution of Civil Engineers, London.

Penman, A.D.M. Charles, J.A. (1974) Measuring movements of embankment dams. In *Field Instrumentation in Geotechnical Engineering*, British Geotechnical Society, 341–358.

Petterson, K.E. (1955) The early history of circular sliding surfaces. *Géotechnique* **5**, 275–296.

Pilot, G. Moreau, M. (1973) *La stabilité des remblais sur sols mous*. Editions Eyrolles, Paris.

Pitts, J. (1979) Discussion. *Q. J. Eng. Geol.* **12**, 277–279.

Pitts, J. (1979) Morphological mapping in the Axmouth-Lyme Regis Undercliffs. *Q. J. Eng. Geol.* **12**, 205–217.

Plafka, G. Eriksen, G.E. (1978) Nevados Huascaran avalanches, Peru. *Rockslides and Avalanches*, Vol. 1, B. Voight, ed., Chapter 9.

Polubarinova-Kochina, P.Y. (1962) *Theory of Groundwater Movement*. Translated by R. De Wiest, Princeton University Press.

Pugh, R.S. Bromhead, E.N. (1985) Design of seepage control measures for an embankment dam using the finite element method. *Proc. 15th Int. Congr. on Large Dams, Lausanne*, 1167–1183.

Richardson, L.F. (1908) The lines of flow of water in saturated soils. *Proc. roy. Dublin Soc.* **11**, 295–316.

Rowe, P.W. (1968) The influence of geological features of clay deposits on the design and performance of sand drains. *Proc. Inst. Civil engrs*, Supplement 70585.

Rowe, P.W. (1970) Derwent Dam—embankment stability and displacements. *Proc. Inst. civil Engrs* **45**, 423–453.

Rowe, P.W. (1972) Embankments on soft alluvial ground. *Q. J. Eng. Geol* **5**, 127–142.

Sarma, S.K. (1973) Stability analysis of embankments and slopes. *Géotechnique* **23**, 423–433.

Sarma, S.K. (1976) Response and stability of earth dams during strong earthquakes. Report prepared for US Army Procurement Agency (Europe).

Sarma, S.K. Bhave, M.V. (1974) Critical acceleration versus static factor of safety in stability analysis of earth dams and embankments. *Geotechnique* **24**, 661–665.

Savigear, R.A.G. (1971) Report accompanying exhibition of geomorphological maps and mapping techniques. University of Reading, British Geomorphological Research Group.

Schuster, R.L. (1978) Introduction. In Landslides, Analysis and Control. HRB Special Report 176, US National Academy of Sciences, Washington, D.C.

Scott, R.F. (1965) *Principles of Soil Mechanics*. Addison-Wesley.

Seed, H.B. Sultan, H.A. (1967) Stability of sloping core earth dams. *P. Proc. ASCE* **93**, 45–68.

Sharpe, C.F.S. (1938) *Landslides and Related Phenomena*. Columbia University Press, New York.

Shreve, R.L. (1966) Sherman landslide, Alaska. *Science* **154**, 1639–1643.

Shreve, R.L. (1968) *The Blackhawk Landslide*. Geol. Soc. America, Special Paper 108.

Siddle, H.J. (1984) South Wales spoil stabilisation tunnels in sandstone. *Tunnels & Tunnelling*, December 1984, 17–19.

Skempton, A.W. (1948) The $\phi = 0$ analysis of stability and its theoretical basis. *Proc. 2nd Int. Conf. on Soil Mechanics and Foundation Engineering, Rotterdam*, **1**, 72–78.

Skempton, A.W. Golder, H.Q. (1978) Practical examples of the $\phi = 0$ analysis of stability of clays. *Proc. 2nd Int. Conf. on Soil Mechanics and Foundation Engineering, Rotterdam*, **2**, 63–70.

Skempton, A.W. (1954) The pore pressure coefficients A and B in saturated soils. *Géotechnique* **4**, 143–147.

Skempton, A.W. Delory, F.A. (1957) Stability of natural slopes in London Clay. *Proc. 4th Int. Con. on Soil Mechanics Foundation Engineering, London*, **2**, 378–381.

Skempton, A.W. (1964) Long term stability of clay slopes. *Géotechnique* **14**, 77–101.

Skempton, A.W. (1966) Bedding plane slip, residual strength and the Vaiont landslide. *Géotechnique* **16**, 82–84.

Skempton, A.W. and Petley, D.J. (1967) The strength along structural discontinuities in stiff clays. *Proc. Geotechnical Conference, Oslo*, **2**, 29–46.

Skempton, A.W. Hutchinson, J.N. (1969) Stability of natural slopes embankment foundations.

Proc. 7th Int. Conf. on Soil Mechanics & Foundation Engineering, State of the art volume, 291–340.

Skempton, A.W. (1970) First time slides in overconsolidated clays. *Géotechnique* **20**, 320–324.

Skempton, A.W. Chandler, R.J. (1975) The design of permanent cutting slopes in stiff fissured clays. *Géotechnique* **25**, 425–427.

Skempton, A.W. Weeks, A.G. (1976) The Quaternary history of the Lower Greensand Escarpment and Weald Clay Vale near Sevenoaks, Kent. *Phil. Trans. Royal Soc. London* **A 283**, 493–526.

Skempton, A.W. (1977) Slope stability of cuttings in brown London Clay. *Proc. 9th Int. Conf. on Soil Mechanics & Foundation Engineering, Tokyo*, **3**, 261–270.

Skempton, A.W. (1979) Landmarks in early soil mechanics. In *Proc. 7th Eur. Conf. Soil Mechanics, Brighton*, **5**, 1–26.

Skempton, A.W. Coats, D.J. (1985) Carsington Dam failure. *Proc. Int. Symp. on Failures in Earthworks*, Institution of Civil Engineers, London.

Skempton, A.W. (1985) Residual strength of clays in landslides, folded strata and the laboratory. *Géotechnique* **35**, 3–18.

Skinner, A.E. (1969) Note on the influence of interparticle friction on the shearing strength of a random assembly of spherical particles. *Géotechnique* **19**, 150–157.

Snedker, E.A. (1985) The stabilisation of a landslipped area to incorporate a highway by the use of a system of bored piles. Technical Note 12, *Proc. Int. Symp. on Failures in Earthworks*, Institution of Civil Engineers, London.

Spencer, E.E. (1967) A method of the analysis of the stability of embankments assuming parallel inter-slice forces. *Géotechnique* **17**, 11–26.

Spencer, E.E. (1973) The thrust line criterion in embankment stability analysis. *Géotechnique* **23**, 85–100.

Squire, J.M., Money, M.S. Sayer, P.R. (1978, in press). Building on coastal landslips: two case histories. *Q. J. Eng. Geol.*

Strucom (Structures & Computers Ltd.) (1984) *User Manual for Flonet: Finite element analysis of seepage*. Structures & Computers Ltd., London.

Taylor, D.W. (1937). Stability of earth slopes. *J. Boston Soc. Civil Engrs* **24**, 197–246.

Taylor, R.L. Brown, C.B. (1967) Darcy flow solutions with a free surface. *Proc. ASCE, J. Hydraulics Div.* **93**, 25–33.

Tomlin, G.R. (1970) Seepage analysis of zoned anisotropic soils by computer. *Géotechnique* **16**, 220–230.

Toms, A.H. (1953) Recent research into the coastal landslides of Folkestone Warren, Kent, England. *Proc. 3rd Int. Conf. on Soil Mechanics Foundation Engineering, Zurich*, **2**, 288–293.

Varnes, D.J. (1958) Landslide types and processes. In *Landslides in Engineering Practice*, E.B. Eckel, ed., US National Academy of Sciences, Highway Research Board, Special Report **29**, 20–47.

Varnes, D.J. and the Commission on Landslides and other Mass Movements. *Landslides Hazard Zonation: a review of principles & practice*. International Association of Engineering Geology & UNESCO (in press)

Vaughan, P.R. Soares, H.F. (1982) Design of filters for clay cores of dams. *Proc. ASCE, J. Geotechnical Div.*, 17–31.

Vaughan, P.R. Walbancke, H.J. (1973) Pore pressure changes and delayed failure of cutting slopes in overconsolidated clay. *Géotechnique* **23**, 531–539.

Viner-Brady, N.E.V. (1955) Folkestone Warren landslips: remedial measures 1948–1950. *Proc. Inst. Civil Engrs*, Railway paper No 57, 429–441.

Voight, B. (ed). (1978) *Rockslides and Avalanches*. 2 vols., Elsevier, Amsterdam.

Voight, B. Janda, R.J., Glicken, H. Douglass, P.M. (1983) Nature and Mechanics of Mount St Helens rockslide–avalanche of 18 May 1980. *Géotechnique* **33**, 243–273.

Warren, C.D. (1985) Private communication.

Watson, J.D. (1956) Earth movements affecting LTE railway in deep cutting east of Uxbridge. *Proc. Inst. Civil Engrs* **11**, 320–323.

Weeks, A.G. (1969) The stability of slopes in south-east England as affected by periglacial activity. *Q. J. Eng. Geol.* **2**, 49–61.

Whitman, R.V. Bailey, W.A. (1967) Use of computers for slope stability analysis. *Proc. ASCE, J. Soil Mechanics Div.*, **93**, 475–498.

Wilson, R.L. Smith, A.K.C. (1983) The construction of a trial embankment on the foreshore at Llandulas. In *Shoreline Protection*, Thomas Telford, London, 223–233.

Wilson, S.D. Mikkelsen, P.E. (1978) Field instrumentation. In *Landslides: Analysis Control,* Transportation Research Board Special Publ. 176 US National Academy of Sciences Chapter 5.

Wood A.M.M. (1955) Folkestone Warren Landslips: investigations, 1948–1950. *Proc. Inst. Civil Engrs,* Railway paper No. 56, 410–428.

Zaruba, Q. Mencl, V. (1969) *Landslides and their control.* Elsevier, Amsterdam and Academia, Prague.

Zienkiewicz, O.C. (1972) *The Finite Element Method in Engineering Mechanics.* 2nd edn., Wiley International.

Zienkiewicz, O.C. Cheung, Y.K. (1965) Finite elements in the solution of field problems. *The Engineer* **220**, 507–520.

Zienkiewicz, O.C., Mayer, P. Cheung, Y.K. (1966) Solution of anisotropic seepage by finite elements. *Proc. ASCE, J. Engineering Mechanics Div.* **92**, 111–142.

Zienkiewicz, O.C. Parekh, C.J. (1970) Transient field problems: two dimensional and three dimensional analysis by isoparametric finite elements. *Int. J. Numerical Methods in Engineering* **2**, 61–71.

Subject index

Index to place names

Index of geological units and names